THE WORLD OF QUANTUM CHEMISTRY

The Duke de Broglie

ACADÉMIE INTERNATIONALE
DES SCIENCES MOLÉCULAIRES QUANTIQUES

INTERNATIONAL ACADEMY
OF QUANTUM MOLECULAR SCIENCE

THE WORLD OF QUANTUM CHEMISTRY

PROCEEDINGS OF THE FIRST INTERNATIONAL CONGRESS
OF QUANTUM CHEMISTRY HELD AT MENTON, FRANCE,
JULY 4–10, 1973

Edited by

RAYMOND DAUDEL

Centre de Mécanique Ondulatoire Appliquée

and

BERNARD PULLMAN

Institut de Biologie Physico-Chimique (Fondation Edmond de Rothschild)

D. REIDEL PUBLISHING COMPANY

DORDRECHT-HOLLAND / BOSTON-U.S.A.

Library of Congress Catalog Card Number 73–91429

ISBN–13:978–94–010–2158–6   e–ISBN–13:978–94–010–2156–2
DOI: 10.1007/978–94–010–2156–2

Published by D. Reidel Publishing Company,
P. O. Box 17, Dordrecht, Holland

Sold and distributed in the U.S.A., Canada, and Mexico
by D. Reidel Publishing Company, Inc.
306 Dartmouth Street, Boston,
Mass. 02116, U.S.A.

TABLE OF CONTENTS

PART V / EFFECTS OF ENVIRONMENT ON THE BEHAVIOUR OF MOLECULES

(Chairman: Alberte Pullman)

ADRESSE DU DUC LOUIS DE BROGLIE (PRIX NOBEL)

Membre de l'Académie Française et de l'Académie Internationale des Sciences Moléculaires Quantiques

En 1948 se tenait à Paris, selon l'expression du Professeur Robert Mulliken, le Premier Colloque de Chimie Quantique d'après guerre. Depuis cette date de nombreuses réunions du même type furent organisées dans de nombreux pays.

Aujourd'hui pour la première fois, s'ouvre une réunion qui par son ampleur et par sa structure, comportant cinq symposiums et huit sections functionnant en parallèle, constitue un véritable Congrès rassemblant des spécialistes de tous les domaines de la Chimie Quantique.

En organisant ce Congrès mes Confrères de l'Académie Internationale des Sciences Moléculaires Quantiques ont voulu atteindre plusieurs objectifs. Ils ont désiré rendre hommage à mes premiers travaux sur la Mécanique Ondulatoire publiés en 1923, il y a exactement cinquante ans. Je tiens à leur dire combien je suis sensible à cet hommage. Ils ont aussi voulu montrer que la chimie quantique constitue maintenant une Science adulte aux applications multiples.

Permettez-moi d'évoquer brièvement le bilan de ce demi-siècle depuis les premiers résultats signalés dans mon mémoire de thèse d'où il ressort que l'hypothèse d'une onde associée à l'électron de l'hydrogène suffit à démontrer que pour un état stable de l'atome l'action par tour doit être nécessairement un multiple entier de la constante de Planck.

En 1927, Burrau traitait selon les méthodes de la mécanique ondulatoire la molécule-ion hydrogène. La même année Heitler er London effectuaient leurs travaux classiques sur la molécule d'hydrogène. La méthode des orbitales moléculaires due principalement aux travaux de Hund et de Mulliken permit d'obtenir rapidement des résultats importants notamment dans le domaine des spectres des petites molécules. Peu de temps après, Pauling, Slater et Hückel jetèrent les bases d'une théorie quantique de la chimie organique.

Mais c'est par la conjonction de la naissance de théories élaborées et de l'avènement des ordinateurs que la chimie quantique est entrée dans sa phase décisive.

Avec l'aide de ces puissants moyens de calcul, permettant de déterminer et de manipuler plusieurs milliards d'intégrales moléculaires en quelques heures, il est devenu de nos jours possible de prévoir théoriquement la conformation nucléaire et la structure électronique de molécules aussi complexes et aussi importantes que les principaux constituants des acides nucléiques qui portent le message génétique et jouent un rôle fondamental dans les sciences de la vie : la biochimie quantique est née.

On a pu aussi montrer l'existence de groupes d'atomes, les pharmacophores, qui porteraient l'essentiel des propriétés pharmacodynamiques de nombreux médicaments. Un aspect quantique de la pharmacologie se développe et il a été récemment possible d'établir un lien entre la structure électronique de certaines molécules et leur activité antibiotique.

Le recours aux ordinateurs permet aussi d'étudier, en tenant compte de tous les électrons, des molécules contenant des atomes très divers y compris des atomes lourds. C'est pourquoi les méthodes de la chimie quantique deviennent maintenant utiles en chimie minérale.

Le problème des relations entre la structure des molécules et leur réactivité chimique reste au centre des préoccupations des quanto-chimistes. La théorie des collisions permet la détermination des trajectoires réactionnelles, donc des sections efficaces de réaction. La belle méthode des jets moléculaires croisés qui permet de réaliser en physicien des réactions chimiques produit de nouvelles données expérimentales très précises qui ne manqueront pas de contribuer au développement de la théorie quantique de la réactivité chimique.

Les methodes de la chimie quantique sont, en effet, parvenues à un tel degré d'élaboration que le recoupement de leurs résultats à l'aide des données provenant des techniques expérimentales les plus délicates devient fructueux.

On commence à pouvoir étudier théoriquement l'effet de l'environnement sur le comportement des molécules et des perspectives nouvelles s'ouvrent sur les frontières entre la chimie quantique, la physique des liquides et celle des solides.

A beaucoup de ces travaux les quanto-chimistes français animés par les plus anciens d'entre eux, actuellement membres de notre Compagnie, ont apporté d'importantes contributions et ce fut sans doute une raison de plus du choix d'une ville française pour réunir le Premier Congrès International de Chimie Quantique qui commence aujourd'hui.

ALLOCUTION DU PROFESSEUR RAYMOND DAUDEL

Président de l'Académie Internationale des Sciences Moléculaires Quantiques

MONSIEUR LE SÉNATEUR MAIRE, MESDAMES, MES CHERS CONFRÈRES, MES CHERS COLLÈGUES,

En 1963, grâce à l'intérêt de M. Francis Palméro, pour notre activité scientifique s'ouvrait en ce site exceptionnel la première école d'été de chimie quantique organisée sous les auspices de l'O.T.A.N.

L'accueil de la municipalité fut tel qu'en 1965, puis en 1968, nous organisâmes, Bernard Pullman et moi, deux autres écoles d'été de chimie quantique et nombreux parmi vous sont ceux qui participèrent à ces réunions.

En 1970, c'est encore Menton que Madame Alberte Pullman et moi-même choisissions pour organiser sous les auspices du C.N.R.S. un colloque international pour célébrer le 25ème anniversaire de la fondation du Centre de Chimie Théorique de France. Si Menton est ainsi devenu (comme Upsal, Sanibel et Jérusalem) un des centres de gravité de la Chimie Quantique Internationale, c'est que cette belle cité est toujours et n'est jamais identique à elle-même.

Toujours pareille par la chaleur de son hospitalité.

Toujours différente par sa structure d'accueil.

En 1963, pour recevoir 100 personnes, j'avais dû faire venir de Paris les chaises munies d'écritoire de la salle de conférences de notre laboratoire.

Aujourd'hui vous êtes reçus dans un théâtre comparable à celui de la Vénice.

Vers 1965, après l'organisation des deux écoles d'été de chimie quantique, se fit sentir la nécessité de fonder un groupe de quanto-chimistes susceptibles de contribuer à l'établissement d'une structure internationale au sein de cette discipline.

En 1967, naquit ainsi l'Académie Internationale des Sciences Moléculaires Quantiques qui rassemble les grands pionniers de ce domaine et de plus jeunes mais non moins éminents chercheurs qui ont su créer d'importantes écoles dans leurs pays respectifs.

En conformité avec l'article premier de ses statuts cette Compagnie s'efforce de favoriser et d'encourager le développement des applications de la Mécanique Ondulatoire à l'étude des phénomènes moléculaires.

C'est pourquoi elle attribue chaque année un prix à un jeune chercheur venant de produire des concepts, des théories ou des techniques particulièrement utiles. Les Professeurs Kołos, Levine, Hoffmann, Dalgarno et Jortner furent les premiers lauréats de cette Académie et nous aurons le privilège d'écouter trois de ces éminents collègues au cours des symposiums qui vont s'ouvrir dès demain matin.

C'est aussi pourquoi l'Académie Internationale des Sciences Moléculaires Quantiques a décidé d'organiser des Congrès dont le premier est celui qui nous réunit aujourd'hui.

Un grand nombre d'éthnies possédant leur langue et leurs idéologies contribuent dans le monde entier à la production de la connaissance physico-chimique. L'analyse du langage à l'aide des grammaires formelles met clairement en évidence les structures qu'une langue particulière impose à la pensée. Ces structures constituent en même temps un support utile au développement de la connaissance et des contraintes incitant à attaquer un problème selon des processus spécifiques. La vitalité, la richesse de la production scientifique reposent sur la multiplicité de ces modes de pensée. Et c'est pourquoi notre Académie, loin de vouloir imposer une langue, une idéologie, une théorie, s'intéresse à tous les groupes, à toutes les éthnies, mais souhaite les réunir afin que chacun apporte ses trouvailles et que le choc de ces problématiques entretienne cette évolution du savoir qui nous permet d'exercer une action de plus en plus étendue, de plus en plus réfléchie sur l'Univers qui nous englobe.

Le cinquantième anniversaire de la découverte par Louis de Broglie de l'onde associée à l'électron nous a fourni une excellent prétexte pour organiser ce Premier Congrès International de Chimie Quantique.

Nous espérons que cette réunion ne sera que le premier maillon d'une chaîne qui regroupera tous les trois ans en des lieux célèbres les quanto-chimistes du monde entier. La Nouvelle Orléans en 1976, Kyoto en 1979 seront peut-être des coordonnées spatiotemporelles sur nos lignes d'Univers.

As forty countries are represented I hope many languages will be heard. There is no official language. English speaking participants are numerous (about one hundred). It is gratifying to notice that eighty of them are over ocean visitors.

I must also acknowledge Doctors Price, Chissik and Ravensdale, who edited a book to celebrate the 50th anniversary of the discovery of wave mechanics. The foreword to this book written by Prime Minister Edward Heath underlines the importance of that discovery.

Más de un tercio de los participantes habla una lengua romance (español, francés, italiano, portugués, rumano). Por ello hemos creído oportuno aprovechar este Congreso, para celebrar una asamblea entre los quimícos cuánticos de expresión latina, con el fin de estudiar la organización de la proxima reunión de este grupo que tendrá lugar en Mexico en 1974.

Questo grupo è aperto a tutti i chimici teorici che terrebero a esprimersi in una lingua latina independentemente dalla lora nazionalità.

Ich begrüsse die Delegationen des deutschen Sprachgebietes, denke dabei besonders an den eminenten Wissenschaftler Schrödinger, der die Dissertationsarbeit von Louis de Broglie in so geniale Weise weiterzuentwickeln wusste; ich denke an die Herren Professoren Hund und Hückel, die unserer Akademie angehören und an der Teilnahme der Festsitzung verhindert sind, und an Herrn Professor Hartmann, Gründer der bewährten Zeitschrift *Theoretica Chimica Acta*.

私は 日本からおいでになった方々に、日本語で
御挨拶したいと思います。
　何故ならば、第一に 日本は遠く離れたアジア
の国です。そして量子化学に非常に重要な貢献
をいたしました。さらに 私の三冊の本は日本語に
翻訳されています。

J'aurai aimé saluer chaque délégation dans sa langue propre. Mes connaissances linguistiques ne me le permettant pas, c'est Mademoiselle Michèle Rabier de l'Opéra de Stuttgart qui le fera dans sa langue universelle : la chorégraphie.

En qualité de Président de l'Académie Internationale des Sciences Moléculaires Quantiques je ne saurais conclure sans remercier très chaleureusement tous ceux qui ont contribué à l'organisation de ce Congrès et en particulier Monsieur le Sénateur Maire Francis Palméro, Monsieur le Recteur Mayer et tous leurs collaborateurs.

SCIENTIFIC PROGRAM

Organized by the International Academy of Quantum Molecular Science this Congress has been held in Menton in the Palais de l'Europe put at the disposal of the participants by M. le Sénateur Francis Palméro, Maire de Menton.

The Scientific Committee of the Congress was constituted by the members of the Academy: Mme A. Pullman*, and MM. L. de Broglie (Nobel Prize), C. A. Coulson*, R. Daudel*, H. Eyring, V. Fock, K. Fukui*, J, O. Hirschfelder*, W. Heitler, E. Hückel, F. Hund, M. Karplus, M. Kotani, J. Koutecký*, H. C. Longuet-Higgins, P. O. Löwdin*, H. M. McConnell, R. S. Mulliken (Nobel Prize), R. G. Parr*, L. Pauling (Nobel Prize), R. Pauncz*, J. A. Pople*, B. Pullman*, C. C. J. Roothaan, E. Scrocco*, J. C. Slater, J. H. van Vleck and B. Wilson.

The names followed by an asterisk formed the Organizing Committee.

The Executive Committee was composed of the French members of the Organizing Committee.

The mornings (9 a.m-noon) were devoted to symposia consisting of general presentations followed by large discussions. The following topics were chosen:

   I. Methods of Quantum Chemistry: Löwdin, Chairman (Uppsala), Slater (Gainesville), Davidson (Seattle), Kołos (Warsaw).
  II. Electronic Structure and Conformation of Molecules: Koutecký, Chairman (Berlin), Pople (Pittsburgh), B. Pullman (Paris).
 III. General Theory of Chemical Reactivity: Daudel, Chairman (Paris), Karplus (Harvard), Fukui (Kyoto).
 IV. Formation and Evolution of Molecular Excited States: Parr, Chairman (Baltimore), Jortner and Mukamel (Tel-Aviv), Heilbronner (Basel).
  V. Effects of Environment on the Behaviour of Molecules: A. Pullman, Chairman (Paris), Buckingham (Cambridge), Sinanoğlu (New Haven), Morokuma, Iwata, and Lathan (Rochester).

The afternoons (3–7 p.m.) were reserved for studies in sections where the participants have presented communications. The following topics have been considered:

1. Methods in quantum chemistry. Use of computers.
2. Atomic and molecular spectroscopy.
3. Electronic structures and properties of chemical species (inorganic, organic, and coordination chemistry).
4. Chemical reactions. Collisions. Reaction paths. Chemical equilibrium. Catalysis. Application of quantum chemistry in industry.

5. Excitation and relaxation processes. Phototochemical reactions. Mass Spectroscopy.

6. Quantum biochemistry and quantum pharmacology.

7. Intermolecular forces, role of environment (solvent, lattice, matrix).

8. Quantum chemistry and its relationships with the physics of liquids and solids.

About four hundred scientists representing forty countries participated in the Congress.

The Executive Committee

A. DAUDEL, A. PULLMAN, B. PULLMAN

PART I

METHODS OF QUANTUM CHEMISTRY

Chairman: Per-Olov Löwdin

Quantum Chemistry Group, Uppsala University, Sweden

An Academy session in the Villa Maria Serena at Menton.

THE HISTORY OF THE $X\alpha$ METHOD

J. C. SLATER

Quantum Theory Project, University of Florida, Gainesville, Fla., U.S.A.

Abstract. The $X\alpha$ method for handling the quantum theory of atoms, molecules, and solids, has origins which go back 50 years, to the earliest days of wave mechanics. Most of the fundamental ideas were well formed by the 1930's. After the delay of World War II, and after the development of electronic digital computers during the 1950's, the method has been of practical value since the 1960's. Various recent developments, including the multiple scattering $X\alpha$ method, and the concept of the transition state, have been available only during the past several years, and they have led to a rapid rate of expansion in the use of the method, which is adaptable to a great variety of problems in the structure of matter.

1. The 1920's and 1930's

The talk which I wish to give is about the method of approximation to quantum mechanics for atoms, molecules, and solids which we are calling the $X\alpha$ method. This method has achieved great success during the past few years. But at this congress we are celebrating the 50 years which have elapsed since the discovery of wave mechanics. Even in those days 50 years ago, in 1923, I was concerned with much the same questions which we are talking about today. I believe it will be an interesting example of the continuity of scientific thought, even in a time of great change, if I organize my remarks around some of the events which I have been concerned with during these past 50 years, which have a direct bearing on the theories that we are developing at present.

In June, 1923, just 50 years ago, I received my doctor's degree at Harvard, having worked with P. W. Bridgman on the compressibility of the alkali halides. This work was partly experimental, partly theoretical, exploring the applicability of the theories of Born and von Kármán to interionic forces. I had convinced myself that no simple extension of the formulation of quantum theory then extant could lead to a real explanation of the repulsive forces between closed shell ions which formed an essential part of this problem. Through a Sheldon Fellowship from Harvard, I was able to spend the year 1923-24 in Europe, and I hoped to devote myself to thinking about quantum theory in a sufficiently fundamental manner to give some hint as to the new forms of quantum mechanics which I was convinced would appear in the near future. I passed the summer in France and Italy; my first visit to the beautiful Riviera was in September 1923, on my way back from Italy to England, where I attended the British Association meeting in Liverpool, and there met many of the leading European physicists. The fall months were spent at the Cavendish Laboratory at Cambridge, working with R. H. Fowler, attending lectures of Rutherford, Aston, and others. In December, just before Christmas, I crossed the North Sea to Copenhagen, where I spent the spring with Bohr and Kramers, before returning to the United States and to a position at Harvard in June 1924.

The year 1923 was the one when the Compton effect had attracted the attention of

R. Daudel and B. Pullman (eds.), The World of Quantum Chemistry, 3–15. All Rights Reserved
Copyright © 1974 by D. Reidel Publishing Company, Dordrecht-Holland

the leading physicists of the world. The contrast between the wave nature of X-ray diffraction patterns, and the corpuscular nature as suggested by Compton's explanation of his scattering effects, was striking, and by the end of the year I had convinced myself that we had to be prepared for the coexistence of both forms of theory. This view was stated clearly in a letter which I wrote from England on December 19, 1923, to E. C. Kemble of Harvard, with whom I had studied quantum mechanics:

I have come to a good deal more definite conclusions than I did at Harvard. I have, for one thing, flopped rather definitely to Lichtquanten of the most extreme kind. And the reason is sufficiently strange – I decided that that was the only possible way of explaining interference! The way I would do that is roughly something like this: have an electromagnetic *field*, determined by charges which move with the frequencies of quantum lines, and amplitudes to be determined by correspondence principle from the actual amplitudes of the motion; and use this field to regulate the motion of quanta, letting them travel roughly along Poynting's vector, giving up Maxwell's energy density except as an average, and giving up Einstein's straight line paths. Apparently it is possible to get a decidedly consistent scheme on those lines, with somewhat surprising results in fields where you wouldn't expect it.

One of the things which came from this theory was the possibility that the electromagnetic radiation field should be continuously emitted during stationary states, so that the breadth of spectral lines could be correlated with the lifetime in the stationary states. But I was not satisfied with these applications to radiation alone. I wished to make some connection with the theory of atomic structure and stationary states, and I described to Kemble some very preliminary results on the excited states of two-electron atoms, in which I considered the interactions of the oscillators representing the inner and outer electrons, and tried to derive the Rydberg states from such a picture. Here we had an early attempt to try to describe a self consistent field: each of the electrons was to move in the field of the other. I was of course familiar with the work which Schrödinger (1921), Hartree (1923), and others had done in the preceding few years in trying to describe the sodium atom along similar lines.

When I got to Copenhagen at Christmas time, 1923, Bohr and Kramers were very enthusiastic about my idea of having the electromagnetic field emitted by 'virtual oscillators', as we started to call them, during the stationary states, but unalterably opposed to the real existence of photons. The paper of Bohr et al. (1924a, b) was the outcome of these discussions. By refusing to admit the existence of point photons, and of their simultaneous existence with the light waves which accompanied them, the Copenhagen school lost the chance of making the synthesis between this picture and the similar coexistence of material particles and the wave functions which accompany them, which was the feature of the brilliant and simultaneous work of de Broglie. Instead, the Copenhagen school followed the lead which came out of the idea of virtual oscillators and their interactions within the atom, resulting in Heisenberg's matrix mechanics. The de Broglie waves, the Schrödinger wave equation, and Born's statistical explanation of the relation between the wave function and the motion of particles, were foreign to the Copenhagen thinking of the time, though they were entirely in line with the general attitude which I had had before I went to Copenhagen, and to which I returned as soon as I left Bohr (Slater, 1924, 1925a, b).

From the time I returned to Harvard in 1924, until the fall of 1926, I worked partly on radiation theory (Slater, 1925b), partly on the theory of complex spectra (Slater, 1926). The spinning electron, and Russell-Saunders coupling, were much in the air, particularly since Saunders was a Harvard professor, and I became thoroughly familiar with the theory of many-electron atoms as it could be worked out by the older quantum theory, supplemented by the exclusion principle. As soon as Schrödinger's monumental series of papers came out in 1926, I started asking myself how to apply wave mechanics to atomic problems. By the spring of 1927, I sent in my first paper on the helium atom (Slater, 1927), following the ideas which I had outlined to Kemble in 1923, but now expressed in terms of wave mechanics. I followed this with a second more elaborate paper (Slater, 1928), in which I considered in detail the nature of the correlation hole in a two-electron atom. But by then I felt it was time to apply wave mechanics to the complex spectrum theory, which I had been working out in early 1926 on the basis of the older quantum theory, and in 1929 (Slater, 1929) I published the paper on the determinantal method, in which linear combinations of determinantal functions were used to describe atoms with open shells of electrons.

The techniques which I worked out in this 1929 paper obviously were applicable to problems of molecules and solids as well as to atoms, and I hastened to go as far with that as I could. I had another travelling fellowship, this time a Guggenheim, to spend part of 1929 with Heisenberg and Hund in Leipzig, and while there I worked out many of the applications of the determinantal method, which I incorporated in a paper on cohesion in monovalent metals (Slater, 1930b). One characteristic of this paper, as of all my later work, was the use of the same techniques both for molecular and solid-state problems. I would have followed this by a paper on ferromagnetism, the analog of multiplets to crystals, but in the summer of 1929 I had shown a preprint of the 1929 determinantal paper to Felix Bloch, and he had been able already to make his application of the method to the ferromagnetism of a free-electron model of a metal (Bloch, 1929). This was the first paper in which spin polarization was taken into account, and in which an exchange-correlation energy was written as a sum of two terms, one in the $\frac{4}{3}$ power of the spin-up charge, one in the $\frac{4}{3}$ power of the spin-down charge.

Since Bloch had already made this beautiful application of the determinantal method, I proceeded instead to work further on the molecular problem (Slater, 1931b), showing how a superposition of determinantal functions formed from molecular orbitals could give a complete description of molecular multiplets. At the same time, after I had moved from Harvard to MIT, I was writing numerous papers about analytic atomic wave functions (Slater, 1930c), directed valence in polyatomic molecules (Slater 1931a), van der Waals forces (Slater and Kirkwood, 1931), and various other problems related to atomic and molecular theory. Hartree had published his first papers on the self-consistent field by means of wave mechanics, and in 1930 (Slater, 1930a) I pointed out how his method could be derived from variation methods, leading in the case of determinantal functions to what we now know as the Hartree–Fock method.

By 1932 or 1933 we thus had available the techniques which have formed the basis of most of the standard treatments of molecular theory: the analytic atomic wave functions, molecular orbitals formed as linear combinations of atomic orbitals, determinantal functions formed from these molecular orbitals, and linear combinations of these determinants to get approximations to the complete molecular wave functions. But I had already made a mental calculation of the magnitude of the computational problem which would be involved in applying these methods to large molecules, and its was clear to me that we would have to look for different and more powerful techniques if we hoped to be successful with solids, or with molecules as large as we really wished to be able to handle. I felt that the most promising approach was to make as good an application of the self-consistent field as we could work out for molecular systems. I had promised some time earlier to write a review paper on the electronic structure of metals, for *Reviews of Modern Physics*, and this paper (Slater, 1934b) gave me an opportunity to think through the nature of the self-consistent field.

During the preceding year I had been considering what was the nature of the potential field in which an individual electron would move in the self-consistent field. In the summer of 1933 I gave an invited paper, at a symposium on the application of quantum mechanics to chemistry, held at the meeting of the American Physical Society in Chicago, in which I emphasized this point of view. The other speakers at this symposium were Pauling, Eyring, and Mulliken. My general point of view was that an electron was acted on, in an N-electron system, by only N-1 other electrons, and that the electron which was removed from the total charge density of N electrons, to form the effective charge of N-1 electrons, must be removed from a sphere of influence surrounding the electron which we were considering. As I pointed out in my 1934 paper, (Slater, 1934b), this sphere of influence was largely that of an electron of the same spin as the one which we were considering. The charge distribution could be found from the type of arguments which Bloch (1929) had used in discussing ferromagnetism, and which Dirac (1930) had applied to the problem of exchange in the Fermi–Thomas atom. By 1934, when my paper came out, Wigner and Seitz (1933, 1934; Wigner, 1934) had started their very valuable set of papers on the cellular method for handling crystalline self-consistent-field problems, and their point of view on the potential was very similar to that which I had been working out.

As soon as the Wigner-Seitz work came out, I started applying their method to a more realistic model of the energy bands of sodium than they had used (Slater, 1934a). I saw at once that my approach could equally well be used for transition elements, and I was very much interested in getting into the problem of ferromagnetism, using realistic energy bands. One of my students, Krutter (1935), worked out energy bands for copper using the Wigner–Seitz method, and I proceeded (Slater, 1936) to extrapolate these energy bands, using the rigid-band model, back to the preceding element, nickel. This permitted a treatment of the ferromagnetism of nickel, using the general ideas which Bloch (1929) had developed, but with real nickel energy bands, and it was clear that the spin-polarized energy bands resulting from a self-consistent field allowed one to get a real explanation of the ferromagnetism of the $3d$ transition elements. It was obvious to me, however, that the simple energy-band theory was not adequate to describe ferromagnetism, since it did not directly lead to anything like the Heisenberg exchange integral. Work which I did with my student Shockley (Slater and Shockley, 1936) convinced me that optical excitations in insulating crystals were generally localized, and in 1937 I applied (Slater, 1937b) a similar idea to localized magnetic excitations in a magnetic crystal, showing that here we found the explanation of the Heisenberg exchange integral.

This 1937 work was done while visiting the Institute for Advanced Study at Princeton, and I had sufficient free time there, with an escape from my administrative duties as head of the physics department at MIT, to carry out several other pieces of work. For one thing, Wannier was there at the time, working out his ideas concerning excitons (Wannier, 1937), and I was able to use his methods for the problem of magnetic excitations. But also I had convinced myself that though the cellular method of Wigner and Seitz had certain valuable features for computing energy bands in crystals, it was inherently a rather unsatisfactory scheme. It involved solving Schrödinger's equation inside a polyhedral cell surrounding each atom, and fitting these solutions together over the bounding surfaces between cells. This could not be handled in any general and rigorous way. It seemed to me that a much more tractable starting point was the use of a potential which was spherical inside a sphere inscribed in such a cell, and was constant in the region between. With this potential, which we now call the muffin-tin potential, one could satisfy boundary conditions over the surface of each sphere, and use a different method of expansion of the wave function inside and outside the spheres. I worked through a scheme (Slater, 1937a) by which one would expand in plane waves in the region outside the spheres, and this method, now called the Augmented Plane Wave (APW) method, has proved to be a very useful and manageable computational technique. One of my students, Chodorow (1939), made a first exploration of the possibilities of this method just before the war, but it remained for the post-war period before we could make real use of it.

One other line of work during the 1930's has proved to make close connection with what we have been doing since then. I had been conscious for many years of the valuable possibilities in the virial theorem, and in preparing my talk for the symposium in 1933, which I mentioned earlier, I considered its application to interatomic forces in molecules. I wrote a paper (Slater, 1933) pointing out these applications, and they were discussed again in my 1934 paper (Slater, 1934b). It seemed to me, however, that there must be additional relations between quantum mechanics and interatomic forces which had not been explored. I noted that if one had the exact wave function of the electrons in a molecular system, there were two possible methods for getting the energy as a function of nuclear position. First, one could solve Schrödinger's equation for each set of nuclear positions. Secondly, one could take the charge density of electrons, use electrostatics to find the force on each nucleus arising classically from this charge, and integrate the forces to get the energy. I asked myself, would these results agree? I put this question to my student Feynman, who proved (Feynman, 1939) that in fact they would agree. We did not know at the time that Hellmann had proved this result, which is now known as the Hellmann-Feynman theorem, shortly before, but in any case it is a useful addition to our available techniques.

2. The 1950's and 1960's

The decade of the 1940's was almost a total loss as far as the theory of molecules and solids was concerned, on account of the Second World War and its aftermath. I was working on radar, and once that work was finished, I felt an obligation to write up what had been done, and then to help with the organization of various interdepartmental laboratories at MIT, where I was still head of the physics department. But one of the results of the war, and in particular of radar work, was that the theory of solids in general, of energy bands in particular, which had been of only academic importance in the 1930's, became the basis of the technology of solid-state electronics. The invention of the transistor, by Bardeen, Brattain, and my former student Shockley, made a revolution in the whole electronic field which cannot be overestimated. It was obvious that we ought to return to this field, and see if it was not possible to calculate energy bands properly, instead of relying on only the roughest semiempirical descriptions of their nature.

Two other theoretical developments of the latter 1940's showed that progress in solid-state theory had not stopped. Korringa (1947) developed the multiple-scattering method of handling the solution of Schrödinger's equation in a muffin-tin potential. In this method, instead of expanding the solution in the region between the spheres in plane waves, as I had done in my work of 1937 (Slater, 1937a), he expanded in spherical wavelets, by analogy with the scattering theory of X-rays, which had been worked out 30 years before. I at once felt that this method had potential advantages over the augmented plane wave method, and was anxious to investigate their respective merits. Then in addition, Löwdin (1947, 1948) had studied the cohesive energy of alkali halide crystals by use of determinantal wave functions, and had shown very convincingly that one could get a very good account of this problem from first principles. This particularly pleased me, since it formed the answer to the problem which had first turned me in the direction of quantum theory, in 1923.

In 1951, I shifted from department head to an Institute Professorship at MIT, an appointment giving me more time for research. I was then able to get back to the solid-state and molecular field, and proceeded to organize a Solid-State and Molecular Theory Group, to try to make a concerted attack on these problems. I realized that the young students coming along were totally ignorant of pre-war work, and my first step was to bring together some of my thoughts on the type of potential to be used in a self-consistent-field approach. These papers (Slater 1951a, b; 1953a) have formed the basis of most of what has been done since on the Xα method, but they were direct outgrowths of pre-war work, particularly as described in my 1934 *Reviews of Modern Physics* paper. One of my students, Pratt (1952), very shortly applied the method to the copper atom, and showed that it was quite usable.

It was not long after these papers that Gaspar (1954) pointed out that there was a real question whether the coefficient by which I had multiplied the $\frac{1}{3}$ power of the charge density to get the exchange-correlation energy, was properly chosen. He pointed out that if one sets up an expression for total energy, on the analogy of the Thomas-

Fermi-Dirac method, and varies the spin-orbitals in it to minimize the energy, one gets a coefficient only $\frac{2}{3}$ as large as I had advocated. At the time this attracted little attention, though I was in agreement with Gaspar that his method was more reasonable than mine. It was rediscovered by Kohn and Sham (1965) more than a decade later, and it was only in the middle 1960's that theorists became really aware of the two possible values of the coefficient. It is since that time that we have used a parameter α in the coefficient, equal to unity for my original suggestion, $\frac{2}{3}$ for that of Gaspar, Kohn, and Sham, and that we have taken to referring to the $X\alpha$ method as one which uses one of these values of α, or an intermediate value, in computing the exchange-correlation potential U_X.

The reason why it took more than ten years for this point of Gaspar's to have practical effect was that we were working all this time on methods of getting usable solutions of Schrödinger's equation for the muffin-tin potential. As soon as we started the Solid-State and Molecular Theory Group at MIT in 1951, I proposed to the students that they try to implement either my 1937 augmented plane wave method or Korringa's 1947 scattered-wave method, and use them in connection with my $X\alpha$ potential to make accurate calculations of energy bands. It took until 1960 to get a practical method into operation.

One must remember that in 1951 we were still using desk computers. The wartime development of electronic digital computers was just beginning to be made available for peacetime use. The students in the Soid-State and Molecular Theory Group – Pratt, Meckler, and others – quite on their own initiative started out to investigate the various computers under development at MIT, to see if there were not some that could be used for our calculations. Pratt's work (1952) on the copper atom had been done on a very primitive IBM computer which the registrar's office used to handle student grades. But a very sophisticated set of computers was under development in the electrical engineering department, Whirlwind I and II, under Jay Forrester, and Meckler and the others got permission to use these after hours. They really got our computing going. We (Saffren and Slater, 1953) investigated their use for the augmented plane wave calculations, concluding that a modified version which I had suggested shortly before (Slater, 1953b) was not as practical as the original 1937 scheme.

Finally in 1960 Wood (1960, 1962) got his first energy bands for iron using the APW method and an IBM digital computer, and applications of that method began to follow in profusion. In the meantime, a variation of the Korringa method suggested by Kohn and Rostoker (1954) had been programmed at the General Electric Research Laboratory, and at my suggestion, Segall (1961, 1962) of GE and Burdick (1961, 1963) of MIT undertook to intercompare the results of the APW and KKR (Korringa–Kohn–Rostoker) methods using identical potentials, in this case potentials for the copper crystal. They verified the fact that the final results were identical by both methods. Only after this intercomparison, in 1961 and following, were we really satisfied that we could get accurate solutions of the muffin-tin potential problem. Until that point, it was hardly worth while considering the details of the $X\alpha$ potential. But once the method was available and proved out, energy bands for a great many substances were computed, and compared with experiment. Almost all of these calculations were made for $\alpha = 1$, which appeared to work very satisfactorily. About the same time, Herman and Skillman (1963) published $X\alpha$ calculations of all the isolated atoms, also with $\alpha = 1$, with eigenvalues agreeing well with experiment as determined from X-ray measurements.

While all this development of computer technique was going on, there were a number of improvements in fundamental theory. Koster and I (Koster and Slater, 1954) worked out a scheme for handling localized excitations of crystals. Löwdin (1955), who had paid numerous visits to our solid-state and molecular theory group, had looked in a very thorough way at the whole problem of the self-consistent field, and in particular had shown that the local potential which would best approximate the exact Hartree-Fock solutions of the many-body problem was of the general form which I had proposed earlier (Slater, 1951a, b; 1953a). Finally, Hartree (1958), who over the years had frequently visited us at MIT, made what proved to be a final visit in 1957, the year before his death. He used results which he had had available for many years for the self-consistent field in copper, to derive potentials for the various orbitals in that atom, which would best represent the Hartree–Fock problem. These results some years later proved invaluable in further studies of the accuracy of $X\alpha$ approximations to the Hartree–Fock equation (Slater *et al.*, 1969).

In 1964, I moved my main base of operations from MIT to the University of Florida. In Gainesville, with new colleagues and new students, I hoped to develop the $X\alpha$ method in several directions. First, I hoped to go further with the problems of ferromagnetism on which I had made a beginning by the cellular method in 1936. Connolly (1967), then a graduate student at Gainesville, took on the problem of the ferromagnetism of nickel, and showed that by use of the $X\alpha$ method and the APW solution of Schrödinger's equation, it was possible to make the method really quantitative. It was a great satisfaction that his results in general verified the much cruder calculations of the 1930's. Cho (1967, 1970) made similar calculations on ferromagnetic EuS, and Wilson (1968, 1969, 1970; Wilson *et al.*, 1970) started work on the antiferromagnetic oxides, particularly MnO. Only at this point, with the very accurate work which was then possible, did we begin to see that the choice of α was quite crucial for getting a proper evaluation of magnetic properties. What we found was that values of α nearer to $\frac{2}{3}$ than to unity were required, to lead to correct determination of the magnetic behavior. In Japan, Wakoh and Yamashita (1966), who were carrying out very similar calculations, came to essentially the same conclusions.

For several years there was a very confusing situation regarding the proper choice of α. It appeared that for getting the eigenvalues which were appropriate for computing optical transitions, the value $\alpha = 1$ was more accurate, whereas for deciding which orbitals were occupied, $\alpha = \frac{2}{3}$ was better. The lower value also gave better values for the orbitals themselves. This puzzle was resolved, when I observed (Slater and Wood, 1971) that the eigenvalues of the $X\alpha$ method were given by an essentially different formula from that holding for the Hartree–Fock method. In the $X\alpha$ method, there is a rigorous theorem, which does not hold for the Hartree–Fock method, stating that the Fermi statistics holds exactly, so that the electrons must go to the states of lowest one-electron energy to give the lowest value for the total energy. A different procedure must be used for finding optical transitions, and this was handled by setting up what we call the transition state, (Slater and Johnson, 1972), in which the occupation numbers are half way between those of the initial and final state. By solving the self-consistent-field problem for this transition state, but using the smaller values of α, we prove to be taking account to a high approximation of the rearrangement of the electronic structure in going from one energy level to another.

While these things were going on, I also was very anxious to start work along a line which had appealed to me for several years before leaving MIT. I was convinced that the combination of the $X\alpha$ potential and the muffin-tin approximation, which we were using for solid-state problems, was capable of application to molecules as well as solids. I felt it more promising than the use of approximate molecular orbitals formed as linear combinations of analytic atomic orbitals, plus a configuration interaction of determinantal functions, which the quantum chemists persisted in using. I have already stated that I had become convinced of this point as early as 1933, but the chemists did not agree with me. Even in the Solid-State and Molecular Theory Group, able workers such as M. P. Barnett, I. G. Csizmadia, and J. W. Moskowitz, were continuing to use these standard methods, developing some of the earliest usable computer programs for polyatomic molecules. But I had been able to compare their monumental use of computer time with the much more modest requirements of the work by the $X\alpha$ method and the APW solution of the muffin-tin problem which the energy-band workers in the same group were using. I became convinced then that it was necessary to convince the chemists that our $X\alpha$ methods could be adapted to chemical problems.

As soon as I reached Florida, I started trying to interest members of my group there in this possibility (Slater, 1965). The one who took up my suggestion was K. H. Johnson (1966, 1967), who was a postdoctoral worker with our research group from 1965 to 1967. Johnson had made earlier use of the $X\alpha$ method for crystals, using the KKR or scattered-wave method, and it at once occurred to him that that method could be readily adapted to setting up computer programs for applying the $X\alpha$ muffin-tin approximations to molecules. F. C. Smith, Jr., who followed Johnson as a post-doctoral worker at Gainesville, assisted with these programming problems. Since 1967, when Johnson went to the department of Metallurgy and Materials Science at MIT, followed by Smith in 1969, they have collaborated on the development of what we now call the MSXα method, multiple-scattering Xα (Slater and Johnson, 1972; Johnson and Smith, 1972). This method has produced usable programs capable of handling molecules, or molecular clusters in crystals, containing more and heavier atoms than can be treated by the more conventional methods. These approximations are admittedly not perfect, but they are very convenient and practical to use, comparable in computer speed with the most simplified of the semiempirical approximation methods in use by the chemists, and yet often superior in accuracy to good Hartree–

Fock calculations. The development of this procedure during the past two or three years has been remarkable, and I shall say something of it, and of probable directions of progress, in the next section.

3. Developments in the 1970's

The literature which has appeared describing results of the method, since 1970, has been too extensive to list in detail in this paper. Instead, we give references to review articles by the writer and by Johnson (Slater, 1972d; Johnson, 1973), a book in press by the writer (Slater, 1974), and a few recent papers (Averill, 1972; Conklin et al., 1972; Connolly, 1972; Connolly and Sabin, 1972; Danese, 1972; Johnson, 1972; Johnson et al., 1973a, b); Johnson and Wahlgren, 1972; Konowalow et al., 1972; Phillips et al., 1972; Slater, 1972a, b, c; Trickey et al., 1972; Wahlgren and Johnson, 1972) which together contain references to much of the most recent work. In this section I shall describe a few of the directions in which the method is developing.

The effort of the first papers was to give one-electron energies of molecular orbitals, and the related transition-state excitation energies, for a variety of molecules and clusters of atoms. Among those which have been studied, by the groups at MIT and Gainesville, as well as by associated workers in a number of other institutions, are the $(MnO_4)^-$, $(SO_4)^{-2}$, and $(ClO_4)^-$ ions, SF_6, C_2H_6, N_2, CO, NO, CH_4, H_2O, H_2 and Li_2, CO_2, C_3O_2, N_2O, C_4F_4S, C_6H_6, NH_3, H_2O_2, P_4, and P_8. Many of these have been studied experimentally elsewhere by the electron photoemission method, and the comparisons of the calculated and observed excitation energies are very good. Among clusters of atoms from crystals which have been studied are $TiCl_4$, $(NiF_6)^{-4}$, $(CuCl_4)^{-2}$, $(PtCl_4)^{-2}$, $Ni(CO)_4$, $Fe(CO)_5$, and parts of the guanidinium and the phosphate ions, as they join to form guanidinium phosphate.

In addition to the one-electron energies and optical excitations, the method gives an account of cohesive energy. We can get at this for a crystal by studying the total energy as a function of lattice spacing, and for a molecule by finding total energy as a function of nuclear coordinates. Work has been done on many examples of both crystals and molecules. For crystals, using the $X\alpha$ method with APW calculations of energy bands, we find good agreement between theory and experiment for such varied substances as alkali halides, alkali metals, transition group metals (vanadium, copper), and inert gas crystals (solid argon). The muffin-tin approximation would not be expected to be seriously inaccurate for these close-packed crystals. For small molecules, however, we find much less satisfactory agreement between theory and experiment, and very recent unpublished work at Gainesville by Danese and Connolly is indicating that it is the muffin-tin approximation, rather than the $X\alpha$ method which is to blame.

It is not too hard to use the molecular orbitals arising from the muffin-tin method as unperturbed wave functions in carrying through a perturbation treatment, regarding the departures from the muffin-tin potential as a perturbation. When this is done, the results are greatly improved. Thus, the simple muffin-tin results on C_2 gave no binding, whereas when the perturbation calculation is carried through, we not only find binding, but in tolerably good agreement with the best configuration-interaction calculations carried through by the standard method. In the interaction of two neon atoms, the muffin-tin results gave no binding, in spite of the binding which we found in solid argon for the crystalline application of the method. However, we find that when the muffin-tin corrections are included, we get a slight binding, of the order of magnitude of that experimentally observed. Danese and Connolly expect to continue work along these lines, with more complicated molecules.

The corrections to the muffin-tin potential are expensive in computer time, and our expectation is that for diatomic molecules and similar cases where the corrections are vital, the method we are using will not be as useful as the standard methods. However, the environment of an atom does not have to be very much more nearly spherical than this for the uncorrected muffin-tin method to be fairly satisfactory. Thus, Danese (1972) has found that for methane, the uncorrected muffin-tin approximation gives a fairly satisfactory result. Our hope is that we shall be able to evaluate the general magnitude of the corrections, so that they will be used only in the few cases where they are really required.

Another direction in which we are making progress and planning further work is in the field of magnetism. Classical magnetic theory has been almost entirely based on a few concepts, such as the Heisenberg exchange integral and the crystal field, which had only minimal direct theoretical justification. These concepts have been used to suggest semiempirical formulas, by use of which the necessary constants could be found from experiment. It has been my hope ever since the 1930's to change this situation, and to show that the results could be found directly from fundamental theory. The difficulty of course is the small size of the energy terms involved. I have already described how Connolly (1967), using the $X\alpha$ method, was able to get results for nickel, Cho (1967, 1970) for EuS, and Wilson (1968, 1969, 1970; Wilson *et al.*, 1970) for MnO. More recently, Hattox (Conklin *et al.*, 1972) has been able to follow the crystal of vanadium through from a large internuclear separation, in which a ferromagnetic state would have a lower energy than the nonmagnetic form, to the observed internuclear distance, where the magnetism has disappeared. This type of phenomenon had been expected theoretically since the 1930's, but has never been derived before from a fundamental theory. There are many more detailed magnetic results which we now see our way to attack, and my hope is that the next few years may see much progress in the direction of relating magnetic theory to the $X\alpha$ method.

These are only a few of the many directions in which we are making, or are hoping to make progress with this versatile method. Its strength is that it is so close to a real solution of the many-electron problem that it can be used to discuss many types of phenomena which previously could only be handled by semiempirical methods. Each year we understand it better and have many new results to report, and the many young colleagues who are working on it, many of whom I have not been able to name, are sure to lead to great progress in the years to come.

Acknowledgement

This work was supported in part by the National Science Foundation, in Grant GH-32006.

References

Averill, F. W.: 1972, *Phys. Rev.* **B6**, 3637.
Bloch, F.: 1929, *Z. Phys.* **57**, 545.
Bohr, N., Kramers, H. A., and Slater, J. C.: 1924a, *Phil. Mag.* **47**, 785.
Bohr, N., Kramers, H. A., and Slater, J. C.: 1924b, *Z. Phys.* **24**, 69.
Burdick, G. A.: 1961, *Phys. Rev. Letters* **7**, 156.
Burdick, G. A.: 1963, *Phys. Rev.* **129**, 138.
Cho, S. J.: 1967, *Phys. Rev.* **157**, 632.
Cho, S. J.: 1970, *Phys. Rev.* **B1**, 4589.
Chodorow, M. I.: 1939, *Phys. Rev.* **55**, 675.
Conklin, J. B., Jr., Averill, F. W., and Hattox, T. M.: 1972, *J. Phys. (Paris)* **33**, C3-213.
Connolly, J. W. D.: 1967, *Phys. Rev.* **159**, 415.
Connolly, J. W. D.: 1972, *Int. J. Quant. Chem.* **6S**, 201.
Connolly, J. W. D. and Sabin, J. R.: 1972, *J. Chem. Phys.* **56**, 5529.
Danese, J. B.: 1972, *Int. J. Quant. Chem.* **6S**, 209.
Dirac, P. A. M.: 1930, *Proc. Cambridge Phil. Soc.* **26**, 376.
Feynman, R. P.: 1939, *Phys. Rev.* **56**, 340.
Gaspar, R.: 1954, *Acta Phys. Acad. Sci. Hung.* **3**, 263.
Hartree, D. R.: 1923, *Proc. Cambridge Phil. Soc.* **21**, 625.
Hartree, D. R.: 1958, *Phys. Rev.* **109**, 840.
Herman, F. and Skillman, S.: 1963, *Atomic Structure Calculations*, Prentice-Hall, Englewood Cliffs, N.J., U.S.A.
Johnson, K. H.: 1966, *J. Chem. Phys.* **45**, 3085.
Johnson, K. H.: 1967, *Int. J. Quant. Chem.* **IS**, 361.
Johnson, K. H.: 1972, *J. Phys. (Paris)* **33**, C3-195.
Johnson, K. H.: 1973, in *Advances in Quantum Chemistry* (ed. by Per-Olov Löwdin), Vol. 7, Academic Press, New York, p. 143.
Johnson, K. H. and Smith, F. C., Jr.: 1972, *Phys. Rev.* **B5**, 831.
Johnson, K. H. and Wahlgren. U.: 1972, *Int. J. Quant. Chem.* **6S**, 243.
Johnson, K. H., Messmer, R. P., and Connolly, J. W. D.: 1973a, *Solid State Comm.* **12**, 313.
Johnson, K. H., Norman, J. G., Jr., and Connolly, J. W. D.: 1973b, in *Computational Methods for Large Molecules and Localized States in Solids* (ed. by F. Herman *et al.*), Plenum Press, New York, p. 161.
Kohn, W. and Rostoker, J.: 1954, *Phys. Rev.* **94** (1111).
Kohn, W. and Sham, L. J.: 1965, *Phys. Rev.* **140**, A1133.
Konowalow, D. D., Weinberger, P., Calais, J. L., and Connolly, J. W. D.: 1972, *Chem. Phys. Letters* **16**, 81.
Korringa, J.: 1947, *Physica* **13**, 392.
Koster, G. F. and Slater, J. C.: 1954, *Phys. Rev.* **95**, 1167.
Krutter, H. M.: 1935, *Phys. Rev.* **48**, 664.
Löwdin, P.-O.: 1947, *Arkiv Mat. Astron. Fis.* **A35**, 9.
Löwdin, P.-O.: 1948, *Arkiv Mat. Astron. Fis.* **A35**, 30.
Löwdin, P.-O.: 1955, *Phys. Rev.* **97**, 1474.
Phillips, E. W., Connolly, J. W. D., and Trickey, S. B.: 1972, *Chem. Phys. Letters* **17**, 203.
Pratt, G. W., Jr.: 1952, *Phys. Rev.* **88**, 1217.
Saffren, M. M. and Slater, J. C.: 1953, *Phys. Rev.* **92**, 1126.
Schrödinger, E.: 1921, *Z. Phys.* **4**, 347.
Segall, B.: 1961, *Phys. Rev. Letters* **7**, 154.
Segall, B.: 1962, *Phys. Rev.* **125**, 109.
Slater, J. C.: 1924, *Nature* **113**, 307.

Slater, J. C.: 1925a, *Nature* **116**, 278.
Slater, J. C.: 1925b, *Phys. Rev.* **25**, 395.
Slater, J. C.: 1926, *Phys. Rev.* **28**, 291.
Slater, J. C.: 1927, *Proc. Nat. Acad. Sci.* **13**, 423.
Slater, J. C.: 1928, *Phys. Rev.* **32**, 349.
Slater, J. C.: 1929, *Phys. Rev.* **34**, 1293.
Slater, J. C.: 1930a, *Phys. Rev.* **35**, 210.
Slater, J. C.: 1930b, *Phys. Rev.* **35**, 509.
Slater, J. C.: 1930c, *Phys. Rev.* **36**, 57.
Slater, J. C.: 1931a, *Phys. Rev.* **37**, 481.
Slater, J. C.: 1931b, *Phys. Rev.* **38**, 1109.
Slater, J. C.: 1933, *J. Chem. Phys.* **1**, 687.
Slater, J. C.: 1934a, *Phys. Rev.* **45**, 794.
Slater, J. C.: 1934b, *Rev. Mod. Phys.* **6**, 209.
Slater, J. C.: 1936, *Phys. Rev.* **49**, 537.
Slater, J. C.: 1937a, *Phys. Rev.* **51**, 846.
Slater, J. C.: 1937b, *Phys. Rev.* **52**, 198.
Slater, J. C.: 1951a, *Phys. Rev.* **81**, 385.
Slater, J. C.: 1951b, *Phys. Rev.* **82**, 538.
Slater, J. C.: 1953a, *Phys. Rev.* **91**, 528.
Slater, J. C.: 1953b, *Phys. Rev.* **92**, 603.
Slater, J. C.: 1965, *J. Chem. Phys.* **43**, S228.
Slater, J. C.: 1972a, *J. Chem. Phys.* **57**, 2389.
Slater, J. C.: 1972b, *J. Phys. (Paris)* **33**, C3-1.
Slater, J. C.: 1972c, *J. Phys. (Paris)* **33**, C3-7.
Slater, J. C.: 1972d, in *Advances in Quantum Chemistry* (ed. by Per-Olov Löwdin), Vol. 6, Academic Press, New York, p. 1.
Slater, J. C.: 1974, *The Self-Consistent Field for Molecules and Solids*, Vol. 4 of *Quantum Theory of Molecules and Solids*, McGraw-Hill, New York.
Slater, J. C. and Johnson, K. H.: 1972, *Phys. Rev.* **B5**, 844.
Slater, J. C. and Kirkwood, J. G.: 1931, *Phys. Rev.* **37**, 682.
Slater, J. C. and Shockley, W.: 1936, *Phys. Rev.* **50**, 705.
Slater, J. C. and Wood, J. H.: 1971, *Int. J. Quant. Chem.* **4S**, 3.
Slater, J. C., Wilson, T. M., and Wood, J. H.: 1969, *Phys. Rev.* **179**, 28.
Trickey, S. B., Averill, F. W., and Green, F. R.: 1972, *Phys. Letters* **41a**, 385.
Wahlgren, U. and Johnson, K. H.: 1972, *J. Chem. Phys.* **56**, 3715.
Wakoh, S. and Yamashita, J.: 1966, *J. Phys. Soc. Japan* **21**, 1712.
Wannier, G. H.: 1937, *Phys. Rev.* **52**, 191.
Wigner, E.: 1934, *Phys. Rev.* **46**, 1002.
Wigner, E. and Seitz, F.: 1933, *Phys. Rev.* **43**, 804.
Wigner, E. and Seitz, F.: 1934, *Phys. Rev.* **46**, 509.
Wilson, T. M.: 1968, *Int. J. Quant. Chem.* **IIS**, 269.
Wilson, T. M.: 1969, *J. Appl. Phys.* **40**, 1588.
Wilson, T. M.: 1970, *Int. J. Quant. Chem.* **IIIS**, 757.
Wilson, T. M., Wood, J. H., and Slater, J. C.: 1970, *Phys. Rev.* **A2**, 620.
Wood, J. H.: 1960, *Phys. Rev.* **117**, 714.
Wood, J. H.: 1962, *Phys. Rev.* **126**, 517.

CONFIGURATION INTERACTION DESCRIPTION
OF ELECTRON CORRELATION

ERNEST R. DAVIDSON

Chemistry Dept. BG-10, University of Washington, Seattle, Wash. 98195, U.S.A.
Laureate of the International Academy of Molecular Sciences

The fundamental goal of quantum chemistry is the development of a qualitatively correct set of concepts for describing the chemical and physical properties of molecules. When this goal is achieved, a well-trained chemist should be able to correlate and predict the behavior of most chemical systems without extensive calculation. In order to achieve this goal it is necessary to obtain high-accuracy wave functions for a few typical chemical systems. Further it is necessary to understand these wave functions on two levels. First one must understand what types of terms and effects must be included in the mathematical description of the wave function. This level of understanding has probably now been attained as a result of twenty years of large-scale calculations. On the second level one must understand the relationship between the physical and chemical properties of the constituent parts of the system and the physical and chemical properties exhibited by the system itself. This level of understanding has proven frustratingly difficult to achieve by the direct *ab-initio* approach.

A secondary goal of quantum chemistry, which has recently become possible, is the routine tabulation of physical properties of chemical systems from the results of high-accuracy calculations. Such tabulations, whether obtained from experiment or theory, are of value only if the properties tabulated are useful in furthering our understanding of nature. Tabulation of a few potential surfaces for chemical reactions, for example, is of value because such surfaces are needed in the development of better theories of chemical reactions.

During the 50 years that quantum chemistry has existed there have been many methods proposed for computing wave functions. Since the advent of high-speed computers there has been an emphasis on developing a feasible, high-accuracy procedure for polyatomic molecules. Most of the major obstacles to this development have been overcome in the last 20 years so that today modest accuracy for small polyatomic systems is possible. Since many aspects of the present method are fairly new, further technological refinements can be expected to lower the computing cost considerably.

Expansion of the wave function as a linear combination of configurations seems inevitable because of the close connection of linear vector spaces with the basic postulates of quantum mechanics (Dirac, 1958). Calculation of the wave function then involves evaluation of matrix elements of the hamiltonian between configurations and the calculation of the expansion coefficients. The expansion coefficients can be found by perturbation theory provided the configurations are all approximately eigenfunctions of the hamiltonian. Otherwise the coefficients must be found by the linear variation method from a secular equation for a truncated, discrete set of configurations.

R. Daudel and B. Pullman (eds.), The World of Quantum Chemistry, 17–30. All Rights Reserved
Copyright © 1974 by D. Reidel Publishing Company, Dordrecht-Holland

Because of the difficulty in computing or interpreting millions of coefficients, the configurations used in the linear variation method must be selected to give a rapidly convergent expansion of the wave function. At the same time, however, the matrix elements of the Hamiltonian must be easy to evaluate. Finding a set of configurations which meet these two contradictory conditions has been one of the major bottlenecks of quantum chemistry.

Configurations which involved interelectronic distances explicitly gave excellent results for very simple molecules (Hyleraas, 1930; James and Coolidge, 1933). It does not seem feasible, however, to form matrix elements for such configurations for polyatomic molecules. Configurations which are built from products of orbitals, properly spin adapted and antisymmetrized, are the simplest conceivable expansion functions and at the same time, the most complicated functions for which the matrix elements of the hamiltonian can still be readily formed. Configuration interaction calculations using antsymmetrized products of nonorthogonal spin-orbitals were tried for small molecules (Ebbing, 1962; Davidson, 1961; Harris and Michels, 1967a). These results were discouraging both in terms of the number of configurations required for an accurate energy and the time required to form the Hamiltonian matrix for several electrons.

Expansion of the wave functions for atoms in Slater determinants built from orthonormal orbitals had been discussed by Slater as early as 1929. But these early discussions were mainly concerned with removing near degeneracies among physically meaningful configurations and not with systematic calculation of the correlation energy (Condon and Shortley, 1935). Description of molecular wave functions in terms of Slater determinants built from molecular orbitals which were themselves linear combinations of atomic orbitals is also a very old idea (Mulliken, 1928). Removal of near degeneracies by inclusion of a small amount of configuration interaction was a common qualitative concept in the discussion of potential curves and spectra (Mulliken, 1937).

During the period from 1950 to 1970 quantum chemistry moved from a state of great uncertainty about how to find the wave functions for molecules to the present state of being able to systematically produce results of any desired degree of accuracy given sufficient computer time. Because of the simplicity of the formulas for matrix elements between Slater determinants, provided a list of certain basic integrals is available (Boys and Cook, 1960), configuration interaction for many electron systems was possible in 1950. Preliminary results of such CI calculations were discouraging, however (Grimaldi, 1965). Many conceptual and technological difficulties have had to be overcome before reasonable results could be produced. The technological problems included rapid generation of electron-repulsion integrals for a reasonable basis set, rapid transformation of these integrals to a molecular orbital basis set, and rapid solution of the CI problem for large numbers of configurations. As difficult as these technological problems were, the conceptual problems were in some ways more difficult to solve since they required discarding many preconceptions and then, by numerical experimentation aided by perturbation theory analysis, trying to discover systematic procedures for selecting basis functions, for selecting transformation coefficients from basis functions to molecular orbitals, and for selecting configurations for inclusion in the wave function.

Rapid progress has been made in the last few years in the technology of integral computation. Integrals involving Slater-type orbitals can now be generated at not too unreasonable a cost (McLean, 1971; Harris and Michels, 1967b). But contracted gaussian orbitals based on extensions of the work of Boys (1950) and Whitten (1966) allow significantly larger basis sets to be used for polyatomic molecules. Probably contracted Gaussian orbitals will be the preferred basis set during the next decade.

Transformations from integrals over atomic orbitals to integrals over molecular orbitals requires a time proportional to K^5 (where K is the number of basis functions) to transform K^4 integrals. Using a modification of the random access procedure suggested by McLean (1971) the transformation requires less time than computing the basic integrals over contracted Gaussian functions for K less than 50. Even for 100 basis functions the transformation time is less than double the time needed to form the integrals.

Rapid solution of large CI problems has been made possible by several developments. First was the development by Nesbet (1965) and Shavitt (1970) of relaxation methods for generating eigenvectors for a very large, sparse matrix provided good initial guesses to the eigenvectors were available. Use of a formula tape when the same set of configurations are used several times reduces the time for finding several points on a potential surface or for repeating a calculation with slightly different molecular orbitals. Further progress in reducing the cost of constructing the Hamiltonian can be expected.

For example, suppose $\psi_{v,I}$,

$$\psi_{v,I} = \sum_{p=1}^{T_I} c_{p,v,I} D_{p,I} \quad v = 1, \ldots, R_I$$

are spin-coupled combinations of a set of Slater determinants $D_{p,I}$ which differ only in the spin. Suppose further that these determinants have been ordered so that the spin-projections of the first R_I of them are linearly independent. Then

$$\langle \psi_{v,I} | H | \psi_{\mu,J} \rangle = \sum_{p=1}^{T_I} \sum_{q=1}^{T_J} c_{p,v,I} c_{q,\mu,J} \langle D_{p,I} | H | D_{q,J} \rangle$$

can always be written as

$$\langle \psi_{v,I} | H | \psi_{\mu,J} \rangle = \sum_{p=1}^{R_I} \sum_{q=1}^{T_J} b_{p,v,I} c_{q,\mu,J} \langle D_{p,I} | H | D_{q,J} \rangle,$$

where if

$$\mathbf{c} = \begin{pmatrix} c_{1,1} \cdots c_{1,R} \\ \vdots \quad\quad \vdots \\ c_{R,1} \cdots c_{R,R} \\ \text{-------} \\ \vdots \quad\quad \vdots \\ c_{T,1} \quad\quad c_{T,R} \end{pmatrix} = \left(-\frac{B}{A} - \right),$$

then $\mathbf{b} = (B^T)^{-1}$. Hence, no matter how $\mathbf{c}$ is chosen, the problem can be reduced from considering $T_I T_J$ matrix elements to considering only $R_I T_J$. This means that for any choice of $\mathbf{c}$ the advantages claimed for orthogonalized spin-projected determinants (Löwdin, 1962) can be obtained and the best choice of $\mathbf{c}$ is a conceptual rather than a technical problem.

The conceptual problems were solved mainly by use of natural orbitals as a tool for analyzing and understanding the description of correlation effects. Natural spin orbitals were originally defined by Löwdin (1955). Given any wave function Ψ,

$$\varrho(x, x') = N \int \Psi(x, x_2, ..., x_N) \, \Psi^*(x', x_2, ..., x_N) \, dx_2 ... dx_N,$$

where $x = (\mathbf{r}, \xi)$ is the combined position and spin coordinate of the particle. If ϱ is regarded as the kernel of an integral operator, then it is Hermitian, positive definite, and has trace N. Further, if Ψ is a CI wave function built from a finite basis set, then ϱ has a finite matrix representation in that same basis set. The natural spin orbitals, defined as eigenfunctions of ϱ

$$\int \varrho(x, x') \chi_i(x') \, dx' = \lambda_i \chi_i(x),$$

give a diagonal representation of ϱ,

$$\varrho = \sum \lambda_i \chi_i(x) \chi_i^*(x'),$$

which is the best approximation to ϱ in the sense that it involves the fewest possible terms and term-by-term it minimizes the least-squares residual error. Further, it can be shown (Coleman, 1963) that a set of Slater determinants built from these χ_i give the best description of Ψ in certain restricted senses. The λ_i, called occupation numbers, lie between 0 and 1 and total to N. Further each λ_i equals the sum of the squares of coefficients from every Slater determinant which involves χ_i.

If Ψ is not degenerate due to spin or point-group symmetry, the χ_i are spin-equivalent orbitals of pure symmetry (Bingel, 1970). If Ψ is degenerate, however, the χ_i lose their spin and point-group equivalence and symmetry properties. Since large scale CI can only be carried out efficiently with spin and symmetry restricted orbitals, it is common to use symmetry-constrained natural orbitals to discuss open-shell problems. Throughout this discussion 'natural orbital' will be used to refer to these restricted space-orbitals which diagonalize the ensemble average charge density defined as

$$\bar{\varrho}(\mathbf{r}, \mathbf{r}') = g^{-1} \sum_{\nu=1}^{g} \sum_{\xi} \varrho_\nu(x, x')\big|_{\xi = \xi'},$$

where the average includes all the ϱ_ν which are degenerate by symmetry. It can be shown that $\bar{\varrho}$ is the totally symmetric part of each ϱ_ν.

Two-particle functions take a particularly simple form in terms of natural spin-orbitals (Löwdin and Shull, 1956). For each NSO, χ_i, there is a normalized conjugate function ψ_i defined by

$$\mu_i\psi_i(x_2) = \int \chi_i^*(x_1)\, \Psi(x_1, x_2)\, dx_1,$$

which is also a natural orbital with $\lambda = |\mu_i|^2 = \lambda_i$ (Carlson and Keller, 1961) (although, if λ_i is degenerate, ψ_i may not be one of the χ_i). The wave function then has a generalized Fourier series expansion

$$\Psi(x_1, x_2) = \sum \mu_i \chi_i(x_1)\, \psi_i(x_2),$$

which is diagonal in the sense that the number of terms equals the number of spin-orbitals rather than the square of that number as it would for a complete CI calculation with any other choice of orbitals.

For the ground state of H_2 the natural expansion takes the form (Davidson and Jones, 1962)

$$\Psi = 0.996\, \sigma_g\sigma_g - 0.099\, \sigma_u\sigma_u - 0.046\,(p_xp_x + p_yp_y). \cdots,$$

where the σ_g orbital is very nearly the SCF orbital for H_2, σ_u resembles the antibonding LCAO orbital and p_x is a π-orbital the same size as the σ_g orbital. The pair distribution $\Gamma(r_1, r_2)$ then is given approximately as (Davidson, 1972a)

$$\Gamma(r_1, r_2) = \varrho(1)\varrho(2)\, \gamma_x\gamma_y\gamma_z,$$

where the correlation factors   are given as

$$\gamma_z = 1 - 2(0.099)(\sigma_u/\sigma_g)_1\,(\sigma_u/\sigma_g)_2,$$
$$\gamma_x = 1 - 2(0.046)(p_x/\sigma_g)_1\,(p_x/\sigma_g)_2,$$
$$\gamma_y = 1 - 2(0.046)(p_y/\sigma_g)_1\,(p_y/\sigma_g)_2.$$

Thus γ_z is greater than 1 whenever the electrons are at opposite ends of the molecule and less than 1 when they are at the same end. Similarly γ_x or γ_y increase the probability of finding the electrons on opposite sides of the bond axis. Because σ_u, p_x, and p_y are large where σ_g is large, the corrections are large where ϱ is large but the corrections make very little change in $\varrho(r)$. The kinetic energy is greatly changed, however, since σ_u, p_x, and p_y have additional nodes and hence much higher kinetic energy than σ_g. Although the σ_u orbital is very nearly a spectroscopic orbital, the p_x orbital is nearly orthogonal to each of the p_x Rydberg orbitals.

For many-electron wave functions the form of the expansion in Slater determinants is not simplified by natural orbitals although the number of important configurations is usually greatly reduced. The dominant configuration for most medium-size molecules near their equilibrium geometry has a coefficient c_0 larger than 0.95. Further, this first natural configuration has a very large overlap with the SCF wave function. This does not mean that the natural orbitals appearing in the first natural configuration resemble the canonical SCF orbitals, however, since they can, and generally do, differ by a unitary transformation which strongly mixes occupied orbitals of the same symmetry which are close together in energy (Davidson, 1972b).

The natural orbitals which do not appear in the first natural configuration are all tightly localized in the same region of space as the SCF charge density because the correlation error in the SCF wave function is largest where the SCF charge density is large. As a consequence the SCF charge density is even better than the Brillouin theorem would indicate while the SCF momentum density is significantly in error for large momenta. This trivial observation that the correlation error is localized greatly simplifies the calculation. For example, the basis set for the ground state of a molecule need not include Rydberg or continuum orbitals. Instead localized functions of the same general size and shape as the SCF orbitals but with more nodes are needed to describe correlation effects. Previous discussion about the large contribution from continuum orbitals (Shull and Löwdin, 1955) is very misleading because these continuum orbitals really combine into quite localized natural orbitals through extensive cancellation at large distances from the molecule. Essentially because of the large gaps between shells in a Coulomb potential, there is room in each shell for many additional functions without orthogonality problems. But since those additional orbitals do not naturally occur in the spectrum of a Coulomb potential, they can only be represented in Coulomb orbital expansions as linear combinations of the inner tails of Rydberg or continuum orbitals for which the large outer parts have cancelled.

Thus for example, Hirschfelder and Löwdin (1959, 1965) were able to obtain the van der Waals energy of H_2 at large internuclear distance using only a $1s$ and one set of p orbitals instead of the usual perturbation expansion. The p orbital they used, xe^{-r}, had the same orbital exponent as the $1s$ orbital and hence has very small overlap with the hydrogenic excited state p orbitals which all have much smaller orbital exponents.

As a second example, the correlation between the $1s^2$ pair of electrons in beryllium gives rise to a set of natural orbitals with more radial or angular nodes but of the same size as the $1s$ orbital (Chan and Davidson, 1968). All of these orbitals are so small that they feel little effect from the $2s$ and $2p$ valence shell orbitals.

These examples show that the basis sets usually used to expand the SCF orbitals are quite appropriate for describing correlation effects. Usually over 50% of the correlation energy can be obtained with the integrals used for a double zeta plus polarization calculation of the SCF energy. At the level of triple zeta plus double zeta for the next higher value of l and single zeta for an l value higher by 2 than the minimum atomic orbital basis, about 90% of the correlation energy can be recovered. Since this requires about 70 basis functions for the water molecule, calculations at this level of accuracy are not presently feasible for larger polyatomic molecules. This is a clear reminder of the fact that, even at best, an orbital description of electron correlation is very cumbersome.

For a wave function in which there is a dominant configuration with a coefficient in excess of 0.95 the largest contributions from other configurations represent the correlation between pairs of electrons. Based on results for Be and LiH it was hoped at first that only intrapair correlations would matter. But it has now been decisively demonstrated that interpair correlations, in aggregate, are far more important than intrapair correlations for explaining the correlation energy correction to spectral and dissociation energies (Bender and Davidson, 1969; Nesbet, 1968; Jungen and Ahlrichs, 1970). For example, the largest single correction to the dissociation energy of the hydrogen fluoride molecule comes from the correlation between the pi electrons of fluorine and the extra electron in the bonding orbital (Bender and Davidson, 1967). Thus, for purposes of discussing correlation energy one must think in terms of $N(N-1)/2$ pairs rather than the chemists' notion of $N/2$ pairs of electrons.

The fact that double excitations of pairs of electrons would give the largest contribution to the correlation energy was anticipated from perturbation theory (Sinanoğlu, 1962). The major advantage of natural orbitals is providing an efficient set of virtual orbitals so that only a few terms are required to describe each electron pair. Just as for H_2, the effect of an individual double excitation $ij \rightarrow ab$ may be interpreted as providing a correlation factor

$$\gamma_{ij}^{ab} = 1 - 2c_{ij \rightarrow ab} (\psi_a/\psi_i)_1 (\psi_b/\psi_j)_2$$

for the electron pair (i, j). Even if c_0 is as large as 0.95 many of the remaining coefficients may be as large as 0.05. Since the ratio (ψ_a/ψ_i) may be as large as 2 or 3 near the region where ψ_i is largest, the correlation factor may provide a 25% correction to the pair distribution with seemingly small coefficients $c_{ij \rightarrow ab}$. Thus, for the HF molecule the correlation between the π-electrons and the bonding electrons is provided by $2p\pi$, $\sigma \rightarrow 3p\pi$, σ^*, where σ^* is the antibonding orbital with which all chemists are familiar and $3p\pi$ is a π-orbital with a radial mode through the region of maximum density of the $2p\pi$ orbital. This configuration causes the bonding electrons to be shifted toward the hydrogen atom if a π-electron is nearer than average to the fluorine atom. It is interesting to note, in this interpretation, why the Dewar-type (Dewar and Wulfman, 1958) vertical π-correlation does not occur. To describe this effect would require an excitation $(2p\pi)^2 \rightarrow (2s)^2$, but the $2s$ orbital is usually already occupied with another pair of electrons. Thus the Pauli exclusion principle prevents this type of correlation. Instead the π-electrons in a conjugated system tend to be strongly correlated horizontally through excitations into the antibonding π-orbitals. The small vertical correlation is mainly of the type $(2p\pi)^2 \rightarrow (3p\pi)^2$ which keeps the electrons in the same lobe away from each other.

For systems such as CO and H_2CO the antibonding π orbital is responsible for a very large part of the correlation energy of the valence shell (Siu and Davidson, 1970; Langhoff et al., 1973). Since the bonding π-orbital is polarized toward oxygen, the antibonding orbital is polarized toward carbon. This antibonding orbital is used by several pairs of electrons as a dominant correlation orbital. For example, for H_2CO the dominant correlation effect is $\pi^2 \rightarrow \pi^{*2}$ which describes the left-right correlation of the π^2-pair. The coefficient for this configuration is more than twice as large as for any other correlation effect. The next most important configuration $\pi, n \rightarrow \pi^*, n^*$ describes the correlation between the left-right position of the π electrons and the in-out position of the nonbonding electron pair.

In terms of natural orbitals, the contribution from single excitations are generally

even smaller than for SCF orbitals. One exception to the negligibility of single excitations are the spin-polarization excitations,

$$\sigma\alpha\sigma\beta\pi\alpha \rightarrow \sigma\sigma^*\pi \, (2\alpha\alpha\beta - \alpha\beta\alpha - \beta\alpha\alpha),$$

which are essential for explaining many spin-dependent properties of planar aromatic
radicals when spin-equivalent orbitals are used (Chang *et al.*, 1968). These excitations,
which make a negligible contribution to the energy, introduce a correlation between
the $\sigma\beta$ electron and the $\pi\alpha$ electron. Since the $\sigma\alpha\pi\alpha$ pair is partially correlated by the
Fermi hole in the SCF wave function their Coulomb hole correlation is less pronounc-
ed. As a consequence the σ^2-pair is spin-polarized with the $\sigma\beta$ electron shifted away
from the $\pi\alpha$ electron (and the $\sigma\alpha$ electron shifted toward the $\pi\alpha$ electron in order to
keep the charge density unchanged). This effect is essential to understanding the
ESR spectra of planar aromatic radicals. For methyl radicals (Chang *et al.*, 1970a) the
spin density at the carbon nucleus was found to be sensitive to both the spin polarization of the valence shell and the core electrons. Minimum basis sets are unable to
describe core correlation effects and hence seriously overestimate the spin density at the
carbon nucleus. A similar effect has been noticed in the calculation of spin-spin dipole
coupling constants for formaldehyde and methylane. In the $n \rightarrow \pi^*$ triplet state of
formaldehyde the net unpaired spin induced in the CH bond by the unpaired electron
in the nonbonding oxygen orbital interacts with the spin of the π^* electron through
spin-dipole coupling and changes the spin-dipole coupling constants by 50% of their
SCF value.

Triple excitations from the first natural configuration are generally negligible.
Quadruple excitations, as Sinanoğlu (1962) and others have explained, are important
as the result of simultaneous pair correlations. The fraction of the correlation energy
which comes from quadruple excitations is roughly $1 - c_0^2$ where c_0 is the coefficient
of the first natural configuration. For a molecule such as N_2 this amounts to about 7%
of the correlation energy and is not negligible. As shown by Sinanoğlu (1962) and
Nesbet (1968), the effect of quadruple excitations is approximately included in the
sum of independent-electron-pair correlation energies obtained by performing variational calculations on individual pairs of electrons. Unfortunately, for semidisjoint
pairs of electrons such as $2p_x^2$ and $2p_x 2p_y$ in neon, the pair-pair interaction may be as
large as 15% of the correlation energy (Barr and Davidson, 1970). Thus, if all orbitals
were very localized and not interpenetrating the Nesbet independent-electron-pair
method would be expected to give very good results. For a molecule such as N_2 with
six strongly penetrating molecular orbitals the situation approaches an electron gas
where independent pairs will not work at all. Since most molecules consist of inter-
locking octets, it must be expected that the Nesbet approach will be in error by about
20% as it is for CO, HF, etc. But neglecting quadrupole or higher excitations will also
give a sizeable relative error which increases as the size of the molecules is increased.
For a hydrocarbon chain longer than about 30 carbons most of the correlation energy
should come from quadruple or higher excitations.

In conclusion, for small polyatomic molecules such as urea and formaldehyde where we can now do accurate calculations, the wave function is best described by a dominant configuration plus selected single, double and quadruple excitations. For still larger molecules this approximation will probably prove inadequate. None of the methods such as independent-electron-pairs, which can handle the unlinked cluster problem conveniently, seem able to handle the pair-pair interactions present in the chemical octets of polyatomic molecules. Many-body-perturbation theory is promising but has not yet produced reliable results for a molecule containing more than one non-hydrogenic atom.

Another aspect of the selection of configurations is the selection of proper spin-couplings. If configurations are selected on the basis of their probable importance in the wave function then not all spin-couplings from a given product of space orbitals will be equally important. Bunge (1970) has shown that the selection of those spin-projected single determinants which have non-zero matrix elements with the dominant configuration give most of the correlation energy. If this idea is carried to its extreme, and integral dependent coupling coefficients are allowed, then only one configuration

$$\sum_p \langle D_{p,I}| H |\Psi_0\rangle D_{p,I}$$

out of the set of R_I possible orthonormal couplings has a nonvanishing matrix element with Ψ_0. Clearly, if a fixed formula tape is to be used to form the H matrix for several similar sets of molecular orbitals then an integral independent coupling would be preferable. This can still be achieved by the same reasoning, although more than one coupling may contribute. For example if the $D_{p,I}$ differ from ψ_0 by a double substitution of space orbitals, then only two distinct electron repulsion integrals, I_1 and I_2 are involved in all of the $\langle D_{p,I}| H |\Psi_0\rangle$. Hence

$$\sum_p \langle D_{p,I}| H |\Psi_0\rangle D_{p,I} = I_1 \sum_p c_{p,I} D_{p,I} + I_2 \sum_p d_{p,I} D_{p,I}.$$

But then any couplings orthogonal to the two functions

$$\Psi_{1,I} = \sum c_{p,I} D_{p,I} \quad \text{and} \quad \Psi_{2,I} = \sum d_{p,I} D_{p,I}$$

will have zero connecting elements with Ψ_0. This may be seen by examining one such coupling $\sum x_{p,I} D_{p,I}$ with

$$\sum x_{p,I} c_{p,I} = 0,$$

and

$$\sum x_{p,I} d_{p,I} = 0.$$

Then

$$\langle \sum x_{p,I} D_{p,I}| H |\Psi_0\rangle = \sum_p x_{p,I} \langle D_{p,I}| H |\Psi_0\rangle,$$
$$= \sum_p x_{p,I} (I_1 c_{p,I} + I_2 d_{p,I}),$$
$$= 0.$$

Hence the two functions $\psi_{1,I}$ and $\psi_{2,I}$ span the subspace which interacts with ψ_0 and their coefficients are integral independent. For most purposes only these two spin-couplings would be needed to provide an accuracy comparable to the inherent accuracy of the basis set.

The conceptual problem of finding the proper set of molecular orbitals to use for description of correlation effects was solved by the introduction of natural orbitals. Experience has shown that while spectroscopic orbitals fall into the categories of core orbitals, bonding orbitals, antibonding orbitals, Rydberg orbitals and continuum orbitals, these categories do not describe natural orbitals. Instead of the Rydberg and continuum orbitals, natural orbitals include correlation orbitals the same size as core orbitals or bonding orbitals but with additional nodes. Some of these additional orbitals more or less coincide with the antibonding orbitals but most of them require basis functions in their description which do not correspond to any spectroscopic atomic orbitals.

There are, however, several technical difficulties in constructing such a set of natural orbitals because they do not appear as the eigenfunctions of any simply constructed operator except the density matrix, and the density matrix can only be constructed after the wave function has been found. These technical difficulties have been solved by several techniques. To obtain an initial guess to the first natural configuration for an open shell molecule, the analytic expansion SCF equations can be solved. For an open shell system these take the form

$$2F_0\Psi_i = \sum_j \lambda_{ji}\Psi_j \quad \text{for} \quad n_i = 2,$$

$$F_i\Psi_i = \sum_j \lambda_{ji}\Psi_j \quad \text{for} \quad n_i = 1,$$

$$F_0 = h + J - \tfrac{1}{2}K,$$

$$F_i = F_0 + \tfrac{1}{2}\sum_j \beta_{ji}K_j.$$

The off-diagonal Lagrangian multipliers may be eliminated to give the easily solved equations

$$H_0\Psi_i = \varepsilon_i\Psi_i \quad \text{for} \quad n_i = 2,$$

$$G_i\Psi_i = \varepsilon_i\Psi_i \quad \text{for} \quad n_i = 1,$$

where

$$G_i = P_iH_iP_i,$$

$$P_i = 1 - \sum_{j<i} |\Psi_j\rangle\langle\Psi_j|,$$

$$H_i = F_i - \sum_{j>i,\,n_j=1} n_i^{-1}\{|\Psi_j\rangle\langle\Psi_j| F_j + F_j |\Psi_j\rangle\langle\Psi_j|\}.$$

While these canonical SCF orbitals give a good approximation to the first natural configuration, it should be remembered that they generally differ from the leading natural orbitals by a unitary transformation.

An approximate set of correlation orbitals may then be constructed by (a) diagonalizing the exchange operator in the virtual space, (Bender and Davidson, 1966; Chan and Davidson, 1968), (b) solving the Kutzelnigg equations in the uncoupled pair approximation (Kutzelnigg, 1964), (c) doing a Nesbet independent-pair calculation for the important pairs and finding the natural orbitals from each pair (Edmiston and Krauss, 1966; Bender and Davidson, 1967), (d) forming a first-order perturbation theory density matrix using the virtual orbitals from a Fock operator with one repulsion term removed or the orbitals from the internally consistent SCF equations or (e) solving the MCSCF equations (Liu, 1973). Meyer (1971) has shown that the natural orbitals from each pair can be used as a nonorthogonal set to give a very compact wave function for small molecules. The advantage of this comes from the fact that correlation orbitals from different pairs are often similar enough that orthogonalization increases the number of configurations required. Most workers, however, prefer the orthogonal formalism and Schmidt orthogonalize the set of independent pair correlation orbitals (Edmiston and Krauss, 1966).

For some purposes these sets of approximate natural orbitals may be good enough if a large configuration interaction calculation is done. For example, if the SCF configuration plus all single and double excitations are included, the wave function is independent of the choice of virtual orbitals. For large enough basis sets some selection of configurations is essential. In this case even the approximate natural orbitals may not be good enough. One can, however, iterate the calculation by using the natural orbitals from a fairly good wave function to construct a still better one.

The first results produced using the iterative natural orbital scheme with relatively few configurations gave significantly improved accuracy (Bender and Davidson, 1966). More recently, however Schaefer (1972) has noticed that, although the accuracy improves for a few iterations, the iterative results eventually diverge. There are two observations to be made about this divergence. First of all it comes about because the first natural configuration drifts away from the SCF wave function. This can be prevented by use of frozen natural orbitals (Barr and Davidson, 1970) which are obtained by suppressing the density matrix elements coupling the virtual and occupied orbitals before diagonalizing ϱ. This still allows the occupied orbitals to change by a unitary transformation among themselves. There are many conceptual advantages to the frozen natural orbitals but they do have the disadvantage that single excitations are needed to get good molecular properties (Bender and Davidson, 1968).

The second observation is that part of the divergence problem that Schaefer has encountered is caused by using a very limited, fixed set of configurations. Because the form of the important configurations may change when they are re-expressed in terms of natural orbitals, the original iterative natural orbital work selected a new list of configurations for each iteration based on their second order perturbation theory contribution to the energy. An extreme example of this problem occurs in the $\pi \rightarrow \pi^*$ $^1A_1$ excited state of formaldehyde (Elbert et al., unpublished), where the π and π^* electrons may be described (with optimized orbitals) as

$$2^{-1/2}(\pi\pi^* + \pi^*\pi)(1 + \langle \pi \mid \pi^* \rangle^2)^{-1/2},$$

where the π and π^* orbitals have an overlap of 0.355, or as

$$0.942(\pi\pi^* + \pi^*\pi)2^{1/2} + 0.237\,\pi\pi + 0.237\,\pi^*\pi^*,$$

where π and π^* are orthogonal, or by natural orbitals as

$$0.9034u^2 - 0.4287v^2,$$

where u and v look like $\pi \pm \pi^*$ (if the orthogonal π, π^* orbitals are used). Clearly any procedure which tried to optimize either π, π^* form by iteratively substituting its u, v natural orbitals for π and π^* would give nonsense. Although Schaefer's procedure is less extreme than this, any fixed functional form must be carefully examined to see if it is invariant when re-expressed in its own natural orbitals if convergence is expected, Allowing the list of configurations to change on each iteration is only slightly more expensive and is less likely to cause divergence.

By way of summary let us consider a state-of-the-art calculation of the spin-dipole parameters of methylene (Langhoff and Davidson, 1973). The basis set used consisted of contracted gaussian lobe functions with three s-type contractions on each hydrogen, four s and two p on the carbon and four lobes in a cloverleaf pattern in the middle of each bond. The contractions were obtained from using Husinga's (1965) exponents with our own contraction scheme. This basis set gives the SCF energy of carbon and hydrogen separately to within 0.002 Hartress and seems to be within 0.005 of the SCF limit for CH_2. This is slightly better than what is usually obtained from a double-zeta plus polarization wave function. The time required to solve the open-shell SCF equations was negligible compared with the time to do the integrals. The SCF orbitals show the familiar pattern (Chang et al., 1970b) of incomplete orbital following with an effective HCH angle of 146° when the geometrical HCH angle is 135°.

The correlation energy was calculated with the same basis set. Approximately 50% of the estimated correlation energy was obtained. Iteration to frozen natural orbitals improved the energy slightly while reducing the number of important configurations by 17%. Since the correlation energy was largest at 180°, this calculation increased the SCF bond angle by 2° and lowered the SCF barrier by 30%. Any exact statement about the molecular geometry is difficult since the missing correlation energy is more than 10 times larger than the height of the barrier. Probably the equilibrium bond angle is about 134° and the barrier is about 5 kcal. This is sufficient to bind two or three vibrational levels.

The spin-dipole contribution to the zero field splitting parameters was predicted to be about 0.79 cm^{-1} for D and 0.05 cm^{-1} for E. When this is combined with the best estimate of the spin-orbit contribution to D, 0.01 cm^{-1}, the result is about 0.04 cm^{-1} above the current experimental value (Wasserman et al., 1971). The effect of CI on D is small near 135° but larger at both 180° and at smaller bond angles. Above 135° most of the correlation affect on D comes from spin-polarization of the CH bond. This is also reflected in a negative spin density at the protons.

It would be fairly easy to devise a basis set which would give 90% of the correlation energy of CH_2. This has not been done since we were interested in finding a basis set small enough to be useful for formaldehyde and urea. We have now calculated many properties of formaldehyde in the ground state, in the $^1A_1(\pi \to \pi^*)$ state, in the $^3A_1(\pi \to \pi^*)$ state and in the $^3A_2(n \to \pi^*)$ state. Among the more interesting results are that the $\pi \to \pi^*$ excitation energy $(^1A_1 \to {}^1A_1)$ is found to be 11.3 eV in good agreement with Peyerimhoff *et al.* (1971). Also the D value for the $^3A_1 (\pi \to \pi^*)$ state is predicted to be -0.6 cm^{-1} and for the $^3A_2 (n \to \pi^*)$ state is predicted to be $+0.5$ cm^{-1}.

References

Barr, T. L. and Davidson, E. R.: 1970, *Phys. Rev.* **A1**, 644.

Bender, C. F. and Davidson, E. R.: 1966, *J. Phys. Chem.* **70**, 2675.

Bender, C. F. and Davidson, E. R.: 1967, *J. Chem. Phys.* **47**, 360.

Bender, C. F. and Davidson, E. R.: 1968, *J. Chem. Phys.* **49**, 4222.

Bender, C. F. and Davidson, E. R.: 1969, *Phys. Rev.* **183**, 23.

Bingel, W. A.: 1970, *Theor. Chim. Acta* **16**, 319.

Boys, S. F.: 1950, *Proc. Roy. Soc.* (*London*) **A200**, 542.

Boys, S. F. and Cook, G. B.: 1960, *Rev. Mod. Phys.* **32**, 285.

Bunge, A.: 1970, *J. Chem. Phys.* **53**, 20.

Carlson, B. C. and Keller, J. M.: 1961, *Phys. Rev.* **121**, 659.

Chan, A. C. H. and Davidson, E. R.: 1968, *J. Chem. Phys.* **49**, 727.

Chang, S. Y., Davidson, E. R., and Vincow, G.: 1968, *J. Chem. Phys.* **49**, 529.

Chang, S. Y., Davidson, E. R., and Vincow, G.: 1970a, *J. Chem. Phys.* **52**, 1740.

Chang, S. Y., Davidson, E. R., and Vincow, G.: 1970b, *J. Chem. Phys.* **52**, 5596.

Coleman, A. J.: 1963, *Rev. Med. Phys.* **35**, 668.

Condon, E. U. and Shortley, G. H.: 1935, *The Theory of Atomic Spectra*, Cambridge University Press.

Davidson, E. R.: 1961, *J. Chem. Phys.* **35**, 1189.

Davidson, E. R.: 1972a, *Rev. Mod. Phys.* **44**, 451.

Davidson, E. R.: 1972b, *Adv. Quant. Chem.* **6**, 235.

Davidson, E. R. and Jones L. L.: 1962, *J. Chem. Phys.* **37**, 2966.

Dewar, M. J. S. and Wulfman, C. E.: 1958, *J. Chem. Phys.* **29**, 158.

Dirac, P. A. M.: 1958, *The Principles of Quantum Mechanics*, Oxford University Press.

Ebbing, D. D.: 1962, *J. Chem. Phys.* **36**, 1361.

Edmiston, C. E. and Krauss, M.: 1966, *J. Chem. Phys.* **45**, 1833.

Elbert, S. T., Langhoff, S. R., and Davidson, E. R.: unpublished.

Grimaldi, F.: 1965, *J. Chem. Phys.* **43**, S59.

Harris, F. E. and Michels, H. H.: 1967a, *Int. J. Quant. Chem.* **1S**, 329.

Harris, F. E. and Michels, H. H.: 1967b, *Adv. Chem. Phys.* **13**, 205.

Hirschfelder, J. O. and Löwdin, P.-O.: 1959, *Mol. Phys.* **2**, 229.

Hirschfelder, J. O. and Löwdin, P.-O.: 1965, *Mol. Phys.* **9**, 491.

Huzinaga, S.: 1965, *J. Chem. Phys.* **42**, 1293.

Hyleraas, E. A.: 1930, *Z. Phys.* **65**, 209.

James, H. M. and Coolidge, A. S.: 1933, *J. Chem. Phys.* **1**, 825.

Hungen, M. and Ahlrichs, R.: 1970, *Theor. Chim. Acta* **17**, 339.

Kutzelnigg, W.: 1964, *J. Chem. Phys.* **40**, 3640.

Langhoff, S. R. and Davidson, E. R.: 1973, *Int. J. Quant. Chem.* **7**, 759.

Langhoff, S. R., Elbert, S. T., and Davidson, E. R.: 1973, *Int. J. Quant. Chem.* **7**, 999.

Liu, B.: 1973, *J. Chem. Phys.* **58**, 1925.

Löwdin, P.-O.: 1955, *Phys. Rev.* **97**, 1474.

Löwdin, P.-O.: 1962, *Rev. Mod. Phys.* **34**, 520.

Löwdin, P.-O. and Shull, H.: 1956, *Phys. Rev.* **101**, 1730.

McLean, A. D.: 1971, in *Proceedings of the Conference on Potential Energy Surfaces in Chemistry*, IBM Research Laboratory, San Jose, Calif., pp. 87–112.

Meyer, W.: 1971, *Int. J. Quant. Chem.* **5S**, 341.

Mulliken, R. S.: 1928, *Phys. Rev.* **32**, 186.

Mulliken, R. S.: 1937, *Phys. Rev.* **51**, 310.

Nesbet, R. K.: 1965, *J. Chem. Phys.* **43**, 311.

Nesbet, R. K.: 1968, *Phys. Rev.* **175**, 2.

Peyerimhoff, S. D., Buenker, R. J., Kammer, W. E., and Hsu, H.: 1971, *Chem. Phys. Letters* **8**, 129.

Schaefer, H. F.: 1972, *The Electronic Structure of Atoms and Molecules*, Addison-Wesley, Cambridge, Mass., U.S.A.

Shavitt, I.: 1970, *J. Comput. Phys.* **6**, 124.

Shull, H. and Löwdin, P.-O.: 1955, *J. Chem. Phys.* **23**, 1362.

Sinanoğlu, O.: 1962, *J. Chem. Phys.* **36**, 706.

Siu, A. and Davidson, E. R.: 1970, *Int. J, Quant. Chem.* **4**, 223.

Slater, J. C.: 1929, *Phys. Rev.* **34**, 1293.

Wasserman, E., Hutton, R. S., Kuck, V. J., and Yager, W. A.: 1971. *J. Chem. Phys.* **55**, 2593.

Whitten, J. L.: 1966, *J. Chem. Phys.* **44**, 359.

SOME PROBLEMS OF THE THEORY OF INTERMOLECULAR INTERACTIONS

WŁODZIMIERZ KOŁOS

Quantum Chemistry Group, University of Warsaw, Warsaw 22, Poland
Laureate of the International Academy of Sciences

1. Introduction

In 1965 at the Sanibel Island Symposium, in honour of Professor R. S. Mulliken, Professor J. O. Hirschfelder (1965) started his lecture on intermolecular forces with the following remark: "it is surprising how little is known about intermolecular forces when you consider their importance in a wide variety of problems." It seems to be even more surprising that in spite of some progress in this field the above statement is still valid in 1973.

There are very well known difficulties which one encounters when applying the standard approaches of quantum chemistry to the study of intermolecular interactions. The Rayleigh–Schrödinger perturbation theory adequately describes the interaction at large separations but fails to give the van der Waals minimum since in this region exchange is already not negligible. In the SCF method one neglects completely the electron correlation which is known to play a very important role in some problems of intermolecular interactions. On the other hand, in a variational approach with correlation a very high accuracy is needed to get reliable results for the weak intermolecular interaction energy defined as a difference of two large quantities.

In the following we shall discuss some recent developments and current problems of the theory of intermolecular interactions.

2. Perturbation Theory Approach

Let us consider two interacting atoms or molecules a and b, with n_a and n_b electrons, respectively. The total spin-independent n-electron Hamiltonian H of the interacting systems $(n = n_a + n_b)$ commutes with a group G of transformations R_i

$$R_i H = H R_i, \tag{1}$$

where G is the direct product of the symmetry group F of the nuclear framework and of the group S_n of permutations of all n electrons of the total system

$$G = F \times S_n. \tag{2}$$

The total Hamiltonian can be expressed as

$$H = H_0 + V_{ab}, \tag{3}$$

R. Daudel and B. Pullman (eds.), The World of Quantum Chemistry, 31–42. All Rights Reserved
Copyright © 1974 by D. Reidel Publishing Company, Dordrecht-Holland

where $H_0 = H_a + H_b$ is the Hamiltonian of the noninteracting systems, and V_{ab} represents their interaction. The Hamiltonian H_0 commutes with a group G^0

$$G^0 = F \times S_{n_a} \times S_{n_b}, \tag{4}$$

which does not include permutations of electrons between a and b. It is seen that G^0 is a subgroup of G. Hence the perturbation theory expansion

$$\psi = \sum_{i=0} \psi^{(i)} \tag{5}$$

with
$$\psi^{(i)} = \sum_{nm} c_{nm}^{(i)} \psi_{nm}^{(0)}, \tag{6}$$

$$\psi_{nm}^{(0)} = \psi_{an} \psi_{bm}, \tag{7}$$

where the basis functions $\psi_{nm}^{(0)}$ span the Hilbert space of H_0, need not converge (Claverie, 1971) to the physically admissible eigenfunctions of H.

By using the basis (7) one introduces the polarization approximation which yields the polarization energy, $E_{pol}^{(i)}$, in each order of the perturbation theory.

The symmetry adapted basis ψ_{nm} can be obtained from $\psi_{nm}^{(0)}$ by employing a projection operator A_j to get

$$^j\psi_{nm}^{(0)} = A_j \psi_{nm}^{(0)}, \tag{8}$$

where j denotes the j+th irreducible representation of G. When considering the interaction of two one-electron atoms, j denotes the singlet or triplet state of the total system. In this case it is sometimes useful to define the Coulomb and exchange energy

$$Q^{(i)} = \tfrac{1}{2}(E_{sing}^{(i)} + E_{trip}^{(i)}),$$
$$K^{(i)} = \tfrac{1}{2}(E_{sing}^{(i)} - E_{trip}^{(i)}). \tag{9}$$

Usually the exchange energy is more conveniently defined with reference to the polarization energy

$$^jE_{exch}^{(i)} = {}^jE^{(i)} - E_{pol}^{(i)}, \tag{10}$$

where $E_{pol}^{(i)}$ is the energy calculated in the ith order of the perturbation theory using the basis (7). Note that $E_{pol}^{(i)}$ is not identical with $Q^{(i)}$ and the exchange contribution in $Q^{(i)}$ can be defined as

$$Q_{exch}^{(i)} = Q^{(i)} - E_{pol}^{(i)}. \tag{11}$$

Thus, the total interaction energy can be expressed as

$$^jE_{int} = \sum_{i=1} {}^jE^{(i)} = \sum_{i=1} (E_{pol}^{(i)} + {}^jE_{exch}^{(i)}). \tag{12}$$

If overlap is neglected and the multipole expansion is used for the interaction potential V_{ab}, the polarization energy, $E_{pol}^{(1)} + E_{pol}^{(2)}$, is the well-known van der Waals energy. $E_{pol}^{(1)}$ represents the familiar electrostatic energy and $E_{pol}^{(2)}$ the sum of the induction and dispersion energies which result when single or double excitations, respectively, are taken into account in the expansion (6) for $\psi^{(1)}$. The above distinction between the induction and dispersion energy can be retained if overlap is not neglected and the unexpanded interaction potential is used in the second-order perturbation theory. In this case even for the interaction of neutral atoms one gets non-zero values of $E^{(1)}$ and $E^{(2)}_{\text{ind}}$.

Several symmetry adapted perturbation theories have been developed (for references see Chipman *et al.*, 1973) to calculate $^j E^{(2)}_{\text{exch}}$. However, applications of the formalisms have been confined practically to one- and two-electron systems.

In the case of the $^2\Sigma_u$ state of H_2^+ new results have been recently obtained (Chałasiński and Jeziorski, 1973a) using the concept of the exchange polarization energy which in the second order has been shown (Murrell *et al.*, 1965) to differ from $E^{(2)}_{\text{exch}}$ by only small terms of the order of S^3, and which can be relatively easily calculated from the first-order polarization function defined in Equations (6) and (7). In H_2^+ there is only the exchange induction energy which can be evaluated analytically (Chałasiński and Jeziorski, 1973a). For large R it goes like $-(1/18)Re^{-R}$ and represents only 80% of $E_{\text{exch}} - E^{(1)}_{\text{exch}}$. This shows that in contrast to the Coulomb energy it is not possible, in the second-order perturbation theory, to get the proper asymptotic behaviour of the exchange energy. Higher-order terms are clearly needed for this purpose.

The exchange induction energy for H_2^+ can also be evaluated (Chałasiński and Jeziorski, 1973b) using the polarization function resulting if the interaction potential is expanded in the well-known multipole series. The convergence of the expansion is fairly good. Foor H_2, however, the main contribution to the exchange polarization energy comes from the exchange dispersion energy which has been found (Andzelm *et al.*, 1973) to converge very slowly if the multipole expansion of the potential is used to calculate $\psi^{(1)}$.

When considering H_2 special attention should be paid to the $b^3\Sigma_u^+$ state of the system which plays an important role in the theory of intermolecular interactions since the interaction of hydrogen atoms in this case is analogous to the interaction of two closed-shell atoms. Therefore, the various contributions to the interaction energy in the $^3\Sigma_u^+$ state of H_2 are in many respects much more representative than those for H_2^+. For $R=6$ and $R=8$ a.u. They are given (in cm^{-1}) in Table I.

The first-order energy $^3E^{(1)} = E^{(1)}_{\text{pol}} + {}^3E^{(1)}_{\text{exch}}$ is easily obtained from the Heitler–London wave function. The third-order polarization energy $E^{(3)}_{\text{pol}}$ has been calculated (Bowman, 1971) using the proper not expanded interaction potential. Note that the leading term of the expanded potential (Chan and Dalgarno, 1968), which for $R \geqslant 8$

TABLE I

Contributions to the interaction energy, E_{int}, of two hydrogen atoms in the $^3\Sigma_u^+$ state (in cm^{-1})

R	$^3E_{\text{int}}$	$^3E^{(1)}$	$E^{(2)}_{\text{pol}}$	$^3E^{(2)}_{\text{exch}}$	$E^{(3)}_{\text{pol}}$
6	41.1	86.0	-56.4	13.3	-1.8
8	-4.3	3.0	-7.9	0.5	0.1

yields reasonable values of $E_{pol}^{(3)}$, for $R=6$ results in $E_{pol}^{(3)} = +2.4$ cm^{-1} in contrast to the correct value -1.8 cm^{-1}. The third-order exchange energy may be assumed to be negligible.

The second-order Coulomb and exchange energies, $Q^{(2)}$ and $K^{(2)}$ respectively, can be obtained from Equations (9) using the accurate variational energies for the singlet and triplet state (Kolos and Wolniewicz, 1965, 1973), and substracting off the first and third-order corrections. $Q^{(2)}$ differs from $E_{pol}^{(2)}$ by a small exchange correction, defined in Equation (11), which for $R=6$ amounts to $Q_{exch}^{(2)}=0.3$ cm^{-1} and for $R=8$ is negligible (Bowman, 1971). In the first order the analogous correction at $R=6$ amounts to $Q_{exch}^{(1)}=0.2$ cm^{-1}. When this is taken into account one gets the values of $E_{pol}^{(2)}$ and $^3E_{exch}^{(2)}$ listed in Table I. For $R=8$ they are in a fair agreement with those obtained directly from the perturbation theory (Bowman, 1971; Certain et al., (1968). For $R=6$ small discrepancies exist which may be due to insufficient accuracy of the results of the perturbation theory or to the effect of higher-order corrections.

An important conclusion follows from Table I, viz., it seems that in the vicinity of the van der Waals minimum the interaction energy may be fairly well approximated by

$$^3E = {}^3E^{(1)} + E_{pol}^{(2)} \tag{13}$$

neglecting $^3E_{exch}^{(2)}$ and all higher-order corrections. If $E_{pol}^{(2)}$ is calculated using the multipole expansion for the interaction Hamiltonian one gets $E_{pol}^{(2)} = -8.3$ cm^{-1} for $R=8$ whereas the result obtained with only the dipole-dipole term in the potential is $E_{pol}^{(2)} = -5.4$, i.e., much worse. This indicates that the usual calculations of $E_{pol}^{(2)}$ from the atomic or molecular polarizabilities may be not very reliable. It has already been mentioned that $E_{pol}^{(2)}$ can be separated into the dispersion and induction energy

$$E_{pol}^{(2)} = E_{disp}^{(2)} + E_{ind}^{(2)}. \tag{14}$$

If the expanded potential is used one gets $E_{ind}^{(2)}=0$. With the not expanded potential (Chałasiński and Jeziorski, 1974) $E_{ind}^{(2)} = -2.5$ cm^{-1} for $R=6$ and, since it decreases exponentially, its value for $R=8$ is negligible. Therefore at large distances $E_{disp}^{(2)}$ predominates. Its value can also be calculated directly (Kreek and Meath, 1969). For $R=8$ the result agrees fairly well with the value of $E_{pol}^{(2)}$ given in Table I; for $R=6$ apparently the direct computation has not reached its limit.

3. Variational Approaches

3.1. VERY ACCURATE CALCULATIONS

The only system for which at present very accurate variational calculation of the interaction energy can be carried out are two hydrogen atoms. However, even this case has not been closed and further work is desirable. Small discrepancies still exist between experimental and theoretical energies, especially for excited electronic states, and an analysis of the adiabatic potential energy curves shows that they are open for improvements. Even for the electronic ground state there are discrepancies between the experimental and theoretical vibrational energies (Kołos and Wolniewicz, 1968; Herzberg,

1971). Partly they are obviously due to the nonadiabatic effects. However, a future calculation of these effects will have to be based on very accurate adiabatic energies. Therefore the H_2 molecule has recently been reinvestigated and some improvement over the previous results for the long-range part of the ground state potential energy curve has been obtained. A sample of preliminary results is shown in Table II.

TABLE II

Long-range dissociation energies for
the ground state of H_2

R (a.u.)	D (cm^{-1})	
	a	b
6	178.9	182.7
8	11.6	12.1

[a] Kołos and Wolniewicz (1965).
[b] Kołos and Wolniewicz (1973).

3.2. SCF CALCULATIONS

The well-known success of the SCF method in treating diatomic and medium size polyatomic molecules has stimulated applications of this method to the study of intermolecular interactions. In this case the SCF approach is particularly attractive. For simplicity, let us consider the interaction of two closed-shell atoms a and b. In such a case the SCF method is capable of describing properly the dissociation of the SCF molecule into two SCF atoms. If the molecular orbitals for the total system are built up from frozen atomic orbitals of a and b the one-determinant wave function is identical with the zeroth-order wave function in the language of the perturbation theory. If the molecular orbitals are subject to variation they comprise part of the first-order correction to the wave function, viz., this part which is due to the polarization or inductive effects. Hence, the SCF method gives the total first-order and part of the second-order interaction energy. It does not give, however, any dispersion energy which is a pure interatomic correlation effect. In addition the computed values may be in error due to the neglect of the intra-atomic electron correlation.

Hence an SCF calculation cannot give a realistic interaction energy of two rare-gas atoms, where the attractive contribution to the interaction energy is entirely due to the dispersion interaction. However, one may expect the SCF method to describe fairly well the interaction between polar or ionic systems where the main attractive contribution to the interaction energy is due to the electrostatic and inductive effects.

This is confirmed by numerical calculations of the H_2O–H_2O or H_2O–ion interaction energies as shown in Table III, where the results of the most extended SCF calculations are compared with experiment.

The agreement between the theoretical and experimental interaction energy of two water molecules, listed in Table III, is astonishingly good. This is very surprising indeed

TABLE III

Comparison of theoretical binding energies and of experimental ΔH values (in kcal/mol) for some molecular complexes

Molecular complex	D_{theor}	ΔH_{exper}
$H_2O \ldots H_2O$	4.72[a] 4.84[b]	5.0[h]
$Li^+ \ldots H_2O$	35.3[c] 36.0[d]	34[i]
$Na^+ \ldots H_2O$	25.2[d] 25.2[e]	24[i]
$F^- \ldots H_2O$	24.1[f] 23.5[g]	23.3[i]
$Cl^- \ldots H_2O$	11.9[g]	13.1[i]

[a] Hankins et al. (1970).
[b] Diercksen (1971).
[c] Clementi and Popkie (1972).
[d] Diercksen and Kraemer (1972).
[e] Kistenmacher et al. (1973a).
[f] Diercksen and Kraemer (1970).
[g] Kistenmacher et al. (1973b).
[h] Pimentel and McClellan (1960).
[i] Džidič and Kebarle (1970).

since the dispersion interaction of two water molecules has been estimated (Coulson, 1959; Hofacker, 1958) to be 2–3 kcal/mol. If one accepts this value the good agreement shown in Table III can only be rationalized by assuming that the dispersion interaction energy is cancelled by the changes in intramolecular correlation energy. For this reason both effects, i.e. the intermolecular correlation (dispersion) energy and the changes of intramolecular correlation energy caused by the environment seem to belong to the most important problems of the present day ab initio theory of intermolecular interactions.

3.3. CI CALCULATIONS

The effect of interatomic electron correlation on the interaction energy has recently been studied very estensively for two helium atoms. Calculations of the pure interatomic correlation effect require localization of the orbitals and have been carried out using a slightly modified CI method (Schaefer et al., 1970; McLaughlin and Schaefer, 1971) or the MC SCF approach (Bertoncini and Wahl, 1970). In both cases a potential minimum of about 12 K was obtained, i.e. somewhat deeper than the experimental values which range from 10.4 to 11.2 K. From the known effect of electron correlation on atomic polarizabilities one might expect that the neglect of the intra-atomic electron correlation is responsible for the too low potential minimum. This has recently been confirmed by direct calculation carried out by Bertoncini and Wahl (1973) using the MC SCF method. A 20-configuration wave function was used in which 16 terms accounted for 85% of the intra-atomic correlation energy, whereas the additional four configurations were the most important ones to account for the interatomic correlation. This raised the potential minimum by about 10% thus bringing the theoretical value into good agreement with the experiment. The above raising of the potential conforms to the asymptotic result (Davison, 1966) obtained by calculating the influence of electron correlation on the value of C_6 in the asymptotic potential $V = - C_6/R^6$.

In the vicinity of the potential minimum the full MC SCF calculation for two helium atoms (Bertoncini and Wahl, 1973) has given results similar to those obtained previously by Das and Wahl (1971) for the H ... He interaction. By using the MC SCF method and taking into account the intra-atomic correlation in the helium atom they decreased the previously computed potential depth by about 10%.

New light on the problem under consideration is shed by recent calculation by Kutzelnigg *et al.* (1973) who studied the interaction between the Li^+ ion and the H_2 molecule. They used the independent pair approximation based on direct determination of pair-natural-orbitals, and performed computations for various geometrical configurations of the system. In the case of the equilibrium internuclear separation in H_2 and linear H_2Li^+ complex the interaction energy as a function of the distance between Li^+ and the geometrical center of H_2 is shown in Figure 1. The curve labeled SCF denotes the SCF result and the curve CORR has been obtained by taking into account the intra- and intermolecular electron correlation. However, of particular interest is the breakdown of the correlation effect into various contributions. At the potential minimum, which is mainly due to the polarization effect, the intermolecular correlation energy (between H_2 and Li^+) is small which is consistent with a small polarizability of Li^+. The change of the intra-atomic correlation energy in Li^+ is also small which is understandable in view of a small perturbation of Li^+ by the hydrogen molecule. Hence, as shown by direct computation, the main contribution to the correlation effect in the interaction energy comes from the intramolecular correlation in H_2. In the field of the Li^+ ion the H_2 molecule is strongly polarized and, especially for large H ... H distances, it resembles the H^+H^- system, hence the considerable change in the correlation energy. The numerical results are shown in Figure 2 where ΔE_{corr} denotes the total change of correlation energy and ΔE_{corr} (H_2) the change of correlation energy in H_2 as a function of the H_2 ... Li^+ separation.

The above numerical results are not sufficiently accurate to yield definite conclusions however, some conclusions seem to be quite firm. Electron correlation has certainly a non-negligible effect on the interaction energy. In the case of rare gas atoms the attractive part of the potential curve is entirely due to interatomic electron correlation. Because of a small perturbation of the atomic wavefunctions (at the van der Waals minimum) the change of the intra-atomic correlation energy is relatively small but

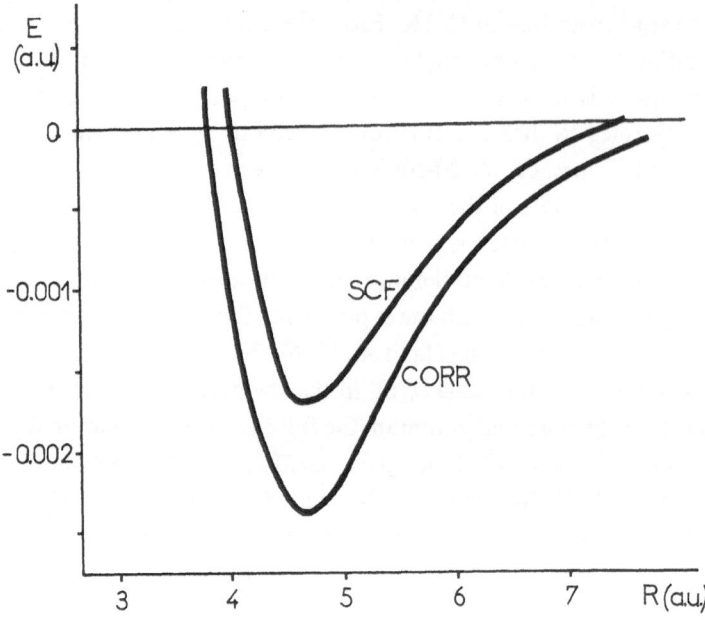

Fig. 1. Interaction energy for linear $H_2 \ldots Li^+$ complex as a function of the distance between Li^+ and the geometrical center of H_2, for internuclear distance in H_2; $R = 1.4$ a.u.

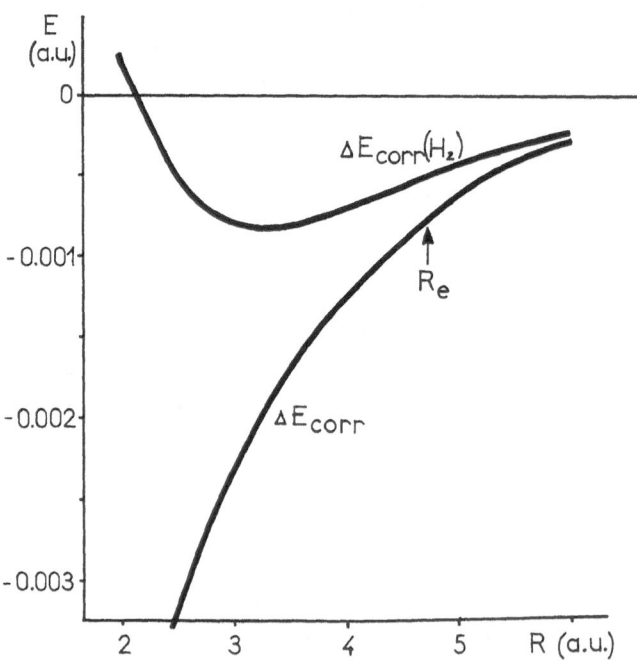

Fig. 2. Change of correlation energy in the linear $H_2 \ldots Li^+$ complex as a function of the distance between Li^+ and the geometrical center of H_2, for internuclear distance in H_2; $R = 1.4$ a.u.

seems to be not negligible. In the case of ionic (polar) systems the main part of the interaction energy is due to the first order and inductive effects and is taken care off within the SCF model. However, even crude estimates of the intermolecular correlation, based on polarizabilities, suggest that in some cases the dispersion interaction may give a considerable contribution. In addition the calculation by Kutzelnigg *et al.* (1973) shows that the change of intramolecular correlation caused by mutual polarization of the molecules may also be very important. Thus accurate theoretical interaction energies seem to be not attainable without explicitly taking into account the correlation of all the electrons of the interacting systems.

4. Many-Body Effects

Now problems arise in the theory of intermolecular interactions when there are aggregates of $n(n>2)$ interacting systems. In this case it is convenient to express the interaction energy in the form of the multibody expansion (Margenau and Stampler, 1967)

$$E(n) = \sum_{m=2}^{n} E(m, n), \tag{15}$$

where $E(m, n)$ denotes the m-body contribution to the interaction energy of n atoms.

The familiar second-order dispersion energy calculated in the polarization approximation is known to be additive. The lowest-order nonadditive dispersion-type contribution to the interaction energy is represented by the well-known 'triple-dipole' effect resulting in the third-order perturbation theory (Axilrod and Teller, 1943; Axilrod, 1951). The induction energy is non-additive in the second order, and if the wave function is properly symmetrized one gets nonadditive effects in each order of the perturbation theory (Jansen, 1965).

Let us consider the interaction of n hydrogen atoms with all spins parallel. If their $1s$ orbitals are denoted by $\chi_a(1), \chi_b(2), \ldots \chi_x(n)$ where $a, b, \ldots x$ refer to the nuclei and $1, 2, \ldots . n$ to the electrons, the zeroth-order wave function has the form

$$\Psi^{(0)} = (n!)^{-1/2} A\psi(1, 2, \ldots n), \tag{16}$$

where

$$\psi(1, 2, \ldots n) = \chi_a(1) \chi_b(2) \ldots \chi_x(n) \tag{17}$$

and A denotes the antisymmetrizer.

The first-order interaction energy may be expressed as

$$E^{(1)}(n) = \frac{\int \psi(1, 2, \ldots, n) A [H'\psi(1, 2, \quad , n)] \, d\tau}{\int \psi(1, 2, \ldots, n) A\psi(1, 2, \ldots, n) \, d\tau}, \tag{18}$$

where H' is the interaction Hamiltonian. Since H' may be represented as a sum of two-body interactions

$$H' = \sum_{\alpha > \beta = 1}^{n} V_{\alpha\beta}, \tag{19}$$

we have

$$E^{(1)}(n) = \sum_{\alpha > \beta = 1}^{n} v_{\alpha\beta}(n), \tag{20}$$

where

$$v_{ab}(n) = \frac{\int \psi(1, 2, ..., n) A [V_{ab}(1, 2) \psi(1, 2, ..., n)] \, d\tau}{\int \psi(1, 2, ..., n) A \psi(1, 2, ..., n) \, d\tau}. \tag{21}$$

By expanding the denominator in (21) and ordering the terms one gets the multibody expansion for v_{ab}

$$v_{ab}(n) = \sum_{m=1}^{n} v_{ab}(m, n), \tag{22}$$

where, e.g.,

$$v_{ab}(2, n) =$$
$$\langle ab \mid V_{ab} \mid ab \rangle - \langle ab \mid V_{ab} \mid ba \rangle + S_{ab} \langle ab \mid V_{ab} \mid ab \rangle + O(S_{ab}^4). \tag{23}$$

The leading term in (23) can easily be estimated to be of the order of S_{ab}^2/R_{ab}. Thus the two-body contribution to the interaction energy is of the order of $\sum_{\alpha < \beta = 1}^{n} S_{\alpha\beta}^2/R_{\alpha\beta}$.

By using the Mulliken approximation one can show that the three-body contribution to the interaction energy of, say, three atoms a, b, c contains terms of the order of $-S_{ab}S_{bc}S_{ca}$ and $+kS_{ac}^2 S_{bc}^2$, where k is probably always positive. For an isosceles triangle the former is larger than the latter, whereas for a linear configuration acb one gets $S_{ab} < S_{ac}S_{bc}$ and the positive contribution predominates being, however, in absolute value smaller than the negative one for the isosceles triangle. Thus the three-body contribution, depending on the geometry, can either increase or decrease binding of three atoms.

In the case of four atoms an analogous estimation suggests that the four-body contribution should be roughly the same for both the tetrahedral and square configurations, whereas the three-body contribution for the square should be smaller than for the tetrahedron.

All the above conclusions with regard to the first-order interactions have been confirmed by numerical computations for three and four hydrogen atoms (Kołos and Leś, 1972a).

The total nonadditive contribution to the interaction energy of three atoms can be expressed as

$$E(3, 3) = \sum_{i=1} E^{(i)}(3, 3) = \sum_{i=1} [E_{pol}^{(i)}(3, 3) + E_{exch}^{(i)}(3, 3)]. \tag{24}$$

$E_{pol}^{(2)}(m, n)$ can be separated into $E_{ind}^{(2)}(m, n)$ and $E_{disp}^{(2)}(m, n)$, however, one can show that the latter is additive, i.e. $E^{(2)}_{disp}(m, n) = 0$ for $m > 2$. Values of some three-body energies for three hydrogen atoms in the $^4A'_2$ state, forming an isosceles triangle, are listed in Table IV.

At $R = 8$ and 6 a.u. the value of $E^{(1)}(3, 3)$ represents about 1% and 10%, respectively of the total two-body interaction energy. It seems that from the nonadditive contributions missing in Table IV only $E^{(2)}_{exch}(3, 3)$ may be of some importance. If this were

TABLE IV

Some nonadditive effects in the interaction energy of three hydrogen atoms in the $^4A'_2$ state (in a.u.)

R	$E^{(1)}(3, 3)$[a]	$E^{(2)}_{ind}(3, 3)$[b]	$E^{(3)}_{pol}(3, 3)$[c]
6	-0.5837×10^{-4}	-0.6152×10^{-6}	-0.2953×10^{-5}
8	-0.5252×10^{-6}	-0.2015×10^{-8}	-0.2217×10^{-6}

[a] Kołos and Leś (1972a).
[b] Jeziorski (1973).
[c] Axilrod (1951).

indeed the case, one could conclude that in the vicinity of the van der Waals minimum the nonadditive correction to the interaction energy of three hydrogen atoms is small and is mainly due to the first-order and triple-dipole effects. For three helium atoms SCF calculations have been carried out (Novaro and Beltran-Lopez, 1972; Block et al., 1971; Kołos and Leś, 1972b) and for the van der Waals minimum the results do not differ significantly from those obtained in the first-order perturbation theory. This suggests that in the second order the exchange induction energy is small. However, the exchange dispersion energy is not taken into account in the SCF method, and therefore its magnitude for three helium atoms is not known. In addition, approximate calculations carried out by Jansen (1965) suggest that at least in some cases the second-order nonadditive corrections may be larger than those obtained in the first order. More work on nonadditive second-order effects is certainly desirable.

One should also mention a very interesting SCF calculation of the nonadditive three-body effects in the interaction of three water molecules (Hankins et al., 1970). The effects are large in magnitude and their sign depends on the geometry of the trimer. In both aqueous liquids and solids their net effect amounts to increased bonding. However, it has already been pointed out that electron correlation may play a fairly important role in these interactions and it is not included in the SCF approach. Within the SCF method interactions of several molecules have been studied recently by Clementi (1973) and Diercksen (Kraemer and Diercksen, 1972; Diercksen, 1973). The many-body effects are also likely to play an important role in biophysical problems (Sinanoğlu et al., 1964) however, little work has been carried out in this direction.

Concluding this brief review let me express hope that at the Second International Congress of Quantum Chemistry considerable progress in the theory of intermolecular interactions will be reported and we shall get a deeper insight into the phenomena which are of great importance in physics, chemistry and biology.

The author is indebted to Mr B. Jeziorski for interesting discussions.

References

Andzelm, J., Chałasiński, G., and Jeziorski, B.: 1973, unpublished results.
Axilrod, B. M.: 1951, *J. Chem. Phys.* **19**, 724.
Axilrod, B. M. and Teller, E.: 1943, *J. Chem. Phys.* **11**, 299.
Bertoncini, P. and Wahl, A. C.: 1970, *Phys. Rev. Letters* **25**, 991.
Bertoncini, P. and Wahl, A. C.: 1973, *J. Chem. Phys.* **58**, 1259.
Block, R., Roël, R., and Ter Maten, G.: 1971, *Chem. Phys. Letters* **11**, 425.
Bowman, J. D.: 1971, Ph. D. Thesis, University of Wisconsin, Theoretical Chemistry Institute, Report No. WIS-TCI-463.
Certain, P. R., Hirschfelder, J. O., Kołos, W., and Wolniewicz, L.: 1968, *J. Chem. Phys.* **49**, 24.
Chałasiński, G. and Jeziorski, B.: 1973a, *Int. J. Quant. Chem.* **7**, 63.
Chałasiński, G. and Jeziorski, B.: 1973b, *Int. J. Quant. Chem.* **7**, 745.
Chałasiński, G. and Jeziorski, B.: 1974, *Mol. Phys.*, to be published.
Chan, Y. M. and Dalgarno, A.: 1968, *Mol. Phys.* **14**, 101.
Chipman, D. M., Bowman, J. D., and Hirschfelder, J. O.: 1973, *J. Chem. Phys.*, to be published.
Claverie, P.: 1971, *Int. J. Quant. Chem.* **5**, 273.
Clementi, E.: 1973, private communication.
Clementi, E. and Popkie, H.: 1972, *J. Chem. Phys.* **57**, 1077.
Coulson, C. A.: 1959, in *Hydrogen Bonding* (ed. by D. Hadzi), Pergamon Press.
Das, G. and Wahl, A. C.: 1971, *Phys. Rev.* **A4**, 825.
Davidson, W. D.: 1966, *Proc. Phys. Soc. (London)* **87**, 133.
Diercksen, G. H. F.: 1971, *Theor. Chim. Acta* **21**, 335.
Diercksen, G. H. F.: 1973, private communication.
Diercksen, G. H. F. and Kraemer, W.: 1970, *Chem. Phys. Letters* **5**, 570.
Diercksen, G. H. F. and Kraemer, W.: 1972, *Theor. Chim. Acta* **23**, 387.
Džidič, I. and Kebarle, P.: 1970, *J. Phys. Chem.* **74**, 1466.
Hankins, D., Moskowitz, J. W., and Stillinger, F. H.: 1970: *J. Chem. Phys.* **53**, 4544.
Herzberg, G.: 1971, Nobel Lecture.
Hirschfelder, J. O.: 1965, *J. Chem. Phys.* **43**, S199.
Hofacker, L.: 1958, *Z. Naturforsch.* **13a**, 1044.
Jansen, L.: 1965, *Adv. Quant. Chem.* **2**, 119.
Jeziorski, B.: 1973, unpublished results.
Kistenmacher, H., Popkie, H., and Clementi, E.: 1973a, *J. Chem. Phys.* **58**, 1689.
Kistenmacher, H., Popkie, H., and Clementi, E.: 1973b, *J. Chem. Phys.*, in press.
Kołos, W. and Leś, A.: 1972a, *Chem. Phys. Letters* **14**, 167.
Kołos, W. and Leś, A.: 1972b, *Int. J. Quant. Chem.* **6**, 1101.
Kołos, W. and Wolniewicz, L.: 1965, *J. Chem. Phys.* **43**, 2429.
Kołos, W. and Wolniewicz, L.: 1968, *J. Chem. Phys.* **49**, 404.
Kołos, W. and Wolniewicz, L.: 1973, unpublished results.
Kraemer, W. P. and Diercksen, G. H. F.: 1972, *Theor. Chim. Acta* **23**, 393.
Kreek, H. and Meath, W. J.: 1969, *J. Chem. Phys.* **50**, 2289.
Kutzelnigg, W., Staemmler, V., and Hoheisel, C.: 1973, *Chem. Phys.* **1**, 27.
Margenau, H. and Stamper, J.: 1967, *Adv. Quant. Chem.* **3**, 129.
McLaughlin, D. R. and Schaefer, H. F.: 1971, *Chem. Phys. Letters* **12**, 244.
Murrell, J. N., Randić, M., and Williams, D. R.: 1965, *Proc. Roy. Soc. (London)* **A284**, 566.
Novaro, O. A. and Beltran-Lopez, V.: 1972, *J. Chem. Phys.* **56**, 815.
Pimentel, G. C. and McClellan, A. D.: 1960, *The Hydrogen Bond*, W. A. Freeman, San Francisco, Calif., U.S.A.
Schaefer, H. F., McLaughlin, D. R., Harris, F. E., and Alder, B. J.: 1970, *Phys. Rev. Letters* **25**, 988.
Sinanoğlu, O., Abdulnur, S., and Kestner, N. R.: 1964, in *Electronic Aspects of Biochemistry* (ed. by B. Pullman), Academic Press, New York.

PART II

ELECTRONIC STRUCTURE AND CONFORMATION OF MOLECULES

Chairman: Jaroslav Koutecký

Miss Michèle Rabier of Stuttgart Opera in a choreographic creation conveying the evolution of the electron concept (after an argument of Raymond Daudel).

REMARKS ON CALCULATIONS OF MOLECULAR ELECTRONIC STRUCTURE

JAROSLAV KOUTECKÝ

Institut für Physikalische Chemie der Freien Universität Berlin,
1 Berlin 33, Thielallee 63–67, Germany

A number of calculation schemes, ranging from semi-empirical methods on various levels of sophistication to ab initio one-electron methods with various sizes of basis sets developed in recent years, have been remarkably successful in statements on steric properties of molecules. Predictions of probably stable conformations of simple and quite complicated molecules, estimates of rotation barriers, Walsh's rules and the qualitative description of the initial and final situation for molecular systems undergoing chemical reaction given by Woodward–Hoffman rules are examples of mentioned successes.

The common feature of all these methods is the one-electron approach and LCAO assumption, even when some methods try to go beyond one-electron approximation. In spite of very basic differences in level and character of the particular approximation the overall qualitative agreement indicates existence of some common basic physical property which controls the geometrical arrangements of the molecules and their possible changes. One could assume that behind the success of LCAO schemes there could be something like orbital type control, which is basically determined by the form of orbitals localized on individual atoms.

The success of various quantum-chemical methods in describing molecular conformations can be understood as due to the fact that there is no creating and breaking of chemical bonds during the conformation changes. It means that the interaction between relatively distant parts of the molecule might be described as interaction of approximately closed-shell electronic systems. It is not necessary to consider any new free valences. Therefore, from this point of view the closed shell Slater determinant can be considered as a reasonable approximation to the state of the electronic systems. The approximate concept of interactions of the closed shell subsystems is certainly possible to apply on the molecules yielding according to one-electron approach all occupied MO's indicating only bondings within the individual parts of the molecule among which interaction is considered. Such quite general and qualitative concept near to ideas of Sovers *et al.* [1] and Lowe [2] should serve only as an example of the possible kind of explanations of the successes of the various computational schemes.

On the other hand, the interaction of the closed-shell systems for large distances, and therefore also the interaction between two distant parts of a bigger molecule is difficult to describe by one-electron approximation due to dispersion forces. We are caught between two kinds of difficulties of the one-electron description: the first difficulty is due to possible changes in distribution of chemical bonds which can arise from more close interactions; the second difficulty represents the existence of disper-

R. Daudel and B. Pullman (eds.), The World of Quantum Chemistry, 45–47. All Rights Reserved
Copyright © 1974 by D. Reidel Publishing Company, Dordrecht-Holland

sion forces for larger distances between two parts of the molecular system. The additional complication can arise when the energies of two conformers do not differ sufficiently. In this case, in principle the 0-point energy of molecules vibrations should be considered too.

In view of these shortcomings of one-electron approximation the mentioned successes of the one-electron related methods should not be taken as completely satisfacory but these results should serve also as a challenge to find explanations why these methods work.

The whole sequence of widely used simple molecular orbital methods is connected with the name of John Pople, the first speaker of today's symposium. All these methods which reflect the thinking about molecules in chemical terms can be considered as step by step improving the level of sophistication of the same general idea. This basic idea is the belief that the chemical properties of the molecules can be described for purpose of chemists more or less by LCAO self-consistent one-electron approximation, which must be also invariable under transformations leaving the physical properties unchanged. The steps of the level of sophistications are characterized (1) by the definition of the collective of electrons which are explicitly considered and (2) by assumptions upon molecular integrals. The PPP method [3] treats only π-electrons and assumes zero differential overlap, CNDO/2 and INDO [4] takes into consideration all valence electrons with different kinds of neglecting differential overlap. The *ab initio* method with small basis sets will be presented in the contribution by Pople today. These small basis set *ab initio* method is a further step in improving the level of sophistication of the method potentially useful for conformation studies of relatively large molecules.

One of the main contributors to the application of quantum-chemical methods in the field of biophysics and biology is Bernard Pullman, the second speaker of today's Symposium. The determination of the conformations of biologically interesting molecules involves necessarily consideration of very large molecular systems and therefore the application of relatively simple methods is inevitable. At the Institut de Biologie Physico-Chimique in Paris the PCILO method [s] (Perturbative Configuration Interaction Using Localized Orbitals) has been developed. The PCILO method starts with the Slater determinant describing the localized bonds according to the chemical formula and takes into account the interaction among these orbitals by Rayleigh–Schrödinger perturbation theory. Pullman and coworkers have carried out a systematic investigation of the conformations of important biological systems using PCILO method with parametrization similar to that of CNDO/2. The comparison of PCILO, CNDO/2 and extended Hückel method has been made also for a number of biological systems.

A large amount of effort has been invested in the *ab initio* calculations on various levels for rotational barriers and the barriers between the stable and unstable conformation states for various, relatively small molecules. This type of calculations is important mainly for investigation of conformation of molecules of special interest. On the other hand, these calculations give deeper insight into the general nature of the conformation problem. The number of investigators is too large that the short time does not allow me to mention their merits.

Even if successes of all methods mentioned above have been numerous, the limitation of applicability of particular methods is often met. One point can be taken as sure: if the creating and breaking of chemical bonds occurs during the geometrical rearrangement of a molecule, one-electron approximation is no longer suitable. Border situations can also arise between the following cases, namely (1) when the interaction of the atomic orbitals can be classified as a chemical bond and (2) when interactions of other kinds occur. In such cases the one-electron approach should be handled with special care. I personally believe that the success of simple methods, including semi empirical methods, have justification in the physics of the problem. This justification must be related with the fact that chemistry is a science and not a mosaic of noncoherent facts.

Less effort than in the concrete calculations has been certainly invested in the systematic analysis of the internal structure of the individual methods for solving electronic structures and related conformation problems. Therefore, it is not fully understood why the individual methods in particular cases work and why others fail. In my opinion, parallel to wide application of the quantum-chemical methods, investigation of the physical meaning of the individual models behind the calculation schemes should be emphasized and supported. Even if such analysis looks unproductive, the better understanding of the physics behind can lead to new methods and more sure application of the existing methods in the range of their applicability.

References

[1] Sovers, D. J., Kern, C. W., Pitzer, R. I., and Karplus, E.: 1968, *J. Chem. Phys.* **49**, 2592.
[2] Lowe, J. P.: 1973, *Science* **179**, 527.
[3] Parr, R. G.: 1963, *Quantum Theory of Molecular Electronic Structure*, Benjamin, New York.
[4] Pople, J. A. and Beveridge, D. L.: 1970, *Approximate Molecular Orbital Theory*, McGraw-Hill, New York.
[5] Diner, S., Malrieu, J. P., Jordan, F., and Gilbert, M.: 1969, *Theoret. Chim. Acta* **15**, 100.

MOLECULAR ORBITAL THEORY OF THE
CONFORMATION OF SMALL ORGANIC MOLECULES

J. POPLE

Mellon Institute, Pittsburgh, Pa., U.S.A.

1. Introduction

The changes in molecular energy that occur when rotations take place about single bonds are of central importance in determining conformations and other structural details of organic molecules. Experimental work on these potential functions has recently been supplemented by a wide range of theoretical studies using molecular orbital techniques [1, 2]. These have proved quite promising and indicate that currently available theoretical methods may have considerable predictive value in stereochemistry as well as affording insight into the controlling electronic factors. In this review, some examples of such applications will be given, based on recent *ab initio* molecular orbital studies in this laboratory.

It is useful to begin a discussion of molecular conformations by considering the simplest molecules containing bonds about which rotation may occur. These are the parent hydrides. One can next consider the effects of replacing one or more hydrogens in these molecules by one or more monovalent substituents to investigate how internal rotation is influenced by substitution. Then the third step is to consider replacement of a hydrogen atom by a second rotor group, leading to a class of molecules with two coupled rotors. This is an area where experimental data is still rather sparse and theory may be useful. We shall follow these general lines of development.

2. Quantum-Mechanical Methods

Molecular orbital theory has developed rapidly in the last decade and a number of techniques at various levels of sophistication are now available for widespread application. A given mathematical procedure which can be applied at a uniform level to an arbitrary three-dimensional molecule in any configuration is usefully regarded as a model chemistry [3] which can be explored in any detail, subject only to limitations of computing facilities.

The simpler molecular orbital methods are of the semiempirical kind where major mathematical approximations are made and certain parameters are obtained by appeal to experimental data rather than by computation. These include Extended Hückel Theory (EHT) [4], Complete Neglect of Differential Overlap (CNDO [5], Intermediate Neglect of Differential Overlap (INDO) [6] and modifications of these [7]. Although relatively inexpensive to perform (see Table I), such models are known to have certain limitations when applied to problems of internal rotation and the detailed consequences of the approximations made are hard to assess. For small organic molecules, a

R. Daudel and B. Pullman (eds.), The World of Quantum Chemistry, 49–59. All Rights Reserved
Copyright © 1974 by D. Reidel Publishing Company, Dordrecht-Holland

more detailed study using *ab initio* methods is now possible and this review will be mostly concerned with studies of this sort.

For the diamagnetic molecules considered here, the wave function in simple molecular orbital theory is represented as a single determinant of doubly occupied molecular orbitals ψ_i, each of which is expressed as a linear combination of basis functions ϕ_μ

$$\psi_i = \sum_\mu c_{\mu i} \phi_\mu. \tag{1}$$

The coefficients $c_{\mu i}$ in these expansions are chosen to minimize the total calculated energy, leading to the well-known equations of Roothaan [8]. If the basis functions ϕ_μ are centered on the atomic nuclei, a basis set for each atomic number leads directly to a theoretical model chemistry. In our laboratory, we have made use of a number of such bases, each aiming to lead to a simple and widely applicable model with certain definite features [3].

The simplest type of basis set is minimal – that is the ϕ functions are only just sufficient in number to describe the ground states of the atoms ($1s$ only for hydrogen, $1s$, $2s$, $2px$, $2py$, $2pz$ for first row atoms such as carbon and oxygen). At this level we use a basis set STO-3G in which each member of a set of Slater-type (exponential) atomic orbitals is replaced by a least-squares fitted sum of three Gaussian functions [9].

The second type of basis set is described as split-valence and contains two sets of basis functions for the valence shell of an atom. One of these (ϕ') is for the inner part of the valence shell region and the other (ϕ'') for the outer part. Thus there are nine functions ϕ_μ for an atom such as carbon ($1s$, $2s'$, $2px'$, $2py'$, $2pz'$, $2s''$, $2px''$, $2py''$, $2pz''$) and two for hydrogen ($1s'$, $1s''$). The basis set we have used is 4-31G indicating a four-Gaussian inner shell function and inner and outer valence parts containing three and one Gaussians respectively [10]. The split-valence has the advantage of increased 'in-out' flexibility to describe local expansion or contraction of the electronic structure. Basis sets with a split inner shell also are sometimes described as 'double zeta' and have similar properties.

Further improvements to the basis set can be made by the addition of basis functions with higher angular quantum numbers than any of the orbitals occupied in the atomic ground state. This means d-functions for carbon, nitrogen, . . . and p-functions for hydrogen. A useful basis for conformational studies on the smallest organic molecules which we have used is 6-31G^* which is a split-valence 6-31G basis to which d-functions on heavy atoms are added [11].

Table I gives approximate relative computing times for the various theoretical models mentioned above. At the highest of these levels (6-31G^*), the results appear to reflect those at the Hartree–Fock limit (infinitely flexible basis set ϕ_μ). As we shall see, there is considerable evidence that most features of internal rotation and stereochemistry are well described at this level, that is without appeal to configuration interaction or many-determinant wave functions.

Given a molecular orbital wave function, the interpretation of changes in electronic

TABLE I

Relative times for molecular orbital computations

Method or basis	Type	Relative time[a]
EHT	Semiempirical; independent electrons	1
CNDO		
INDO	Semiempirical; electron interaction with zero differential overlap	5
MINDO		
STO-3G	Minimal Slater-type basis	150
4-31G	Split-valence basis	1000
6-31G*	Split-valence basis plus d-polarization functions	6000

[a] The absolute times are approximately those (in seconds) for a single calculation (using a Univac 1108 computer) on a molecule the size of ethanol.

structure is sometimes aided by examining the Mulliken populations [12]. For a given atomic orbital (or basis function), a gross population q_μ is defined by

$$P_{\mu\nu} = 2 \sum_i^{occ} c_{\mu i} c_{\nu i}, \tag{2}$$

$$q_\mu = \sum_\nu P_{\mu\nu} S_{\mu\nu}, \tag{3}$$

where $S_{\mu\nu}$ is the overlap matrix. The overlap populations are defined as $2P_{\mu\nu}S_{\mu\nu}$. By partial summation of these populations it is possible to get gross atom populations or atom–atom overlap populations.

3. Rotation About Single Bonds in Simple Hydrides

For the stereochemistry of molecules with carbon, nitrogen and oxygen, the most fundamental information concerns the rotational potential for the C—C, C—N, C—O N—N, N—O and O—O bonds in the parent hydrides. These are the molecules ethane, methylamine, methanol, hydrazine, hydroxylamine and hydrogen peroxide, all of which have been the subject of many theoretical studies. We shall discuss these in terms of the *ab initio* molecular orbital methods described in the previous section.

Internal rotation in ethane, methylamine and methanol involves a three-fold potential which may be adequately described by a single Fourier term,

$$V(\phi) = \tfrac{1}{2}V_3(1 - \cos 3\phi), \tag{4}$$

where ϕ is a dihedral angle relative to an eclipsed configuration. The simplest proce- dure for calculating a potential barrier such as V_3 is to assume 'rigid rotation' – that is vary only the angle ϕ, keeping the bond lengths and other angles constant. A more satisfactory procedure is 'flexible rotation' in which the surface is more fully explored and all other geometrical parameters are reoptimized for each value of the angle ϕ.

For the carbon–carbon bond in ethane, there have been very many theoretical studies both with rigid and flexible rotation. Virtually all of these studies find the staggered form to be the more stable (V_3 negative and minima at $\phi = \pm 60°$, $180°$) in agreement with experimental observation. Table II lists the results for the three basis sets mentioned above and compared with the most extensive treatment yet published (for single determinant wave functions). The rigid rotor results (for STO-3G, 4-31G and 6-31G*) here and elsewhere are based on bond lengths and angles specified in a standard geometrical model. The figures in Table II show that a barrier of the correct order of magnitude is given at all levels, there being a slight reduction when flexibility is taken into account.

TABLE II

Rotation barriers in ethane (kcal/mole)

Method	Rigid rotor	Flexible rotor
STO-3G	3.3	2.9
4-31G	3.3	–
6-31G*	3.5	–
Best MO[a]	3.4	3.2
Expt	–	2.9

[a] Clementi and H. Popkie: 1972, *J. Chem. Phys.* **57**, 4870.

The qualitative interpretation of the rotational barrier in ethane has also been the subject of numerous studies [13]. Here we shall only note that there appears to be a clear relationship with vicinal hydrogen–hydrogen interactions. Table III shows the atom–atom overlap populations. These indicate that eclipsed (*cis* or synperiplanar) vicinal hydrogens are antibonding (negative overlap populations) and that hydrogens with dihedral angles greater than 90° are bonding but the overlap populations are somewhat smaller in magnitude. The relatively strong antibonding interaction for $\phi = 0$ dominates in the eclipsed conformation.

For the three molecules ethane, methylamine and methanol, the experimental values of V_3 are approximately in the ratio $3:2:1$. This feature is well reproduced by 4-31G calculations [2] with rigid rotation as shown in Table IV. The vicinal overlap populations show the same type of dependence on dihedral angle as in ethane (Table III), so it seems reasonable to interpret these barriers in the same manner as in ethane, the reduction in magnitude being due to fewer hydrogen–hydrogen interactions.

For the remaining molecules in this series, the potential function is no longer three-fold and it is necessary to use a fuller Fourier series

$$V(\phi) = \tfrac{1}{2} \sum_{n=1}^{3} V_n (1 - \cos n\phi). \qquad (5)$$

Of the three molecules NH_2—NH_2, NH_2—OH and OH—OH, hydrogen peroxide has been studied most fully. Experimentally H_2O_2 is known to be a twisted molecule [14]

TABLE III

Vicinal H—H populations in ethane (6-31G^*)

ϕ	Eclipsed (D_{3h})	Staggered (D_{3d})
0	-0.016	–
60	–	-0.006
120	$+0.006$	–
180	–	$+0.008$

TABLE IV

Three-fold rotational barriers (V_3, kcal/mole)

Molecule	V_3 (4-31G)	V_3 (expt.)
CH_3—CH_3	3.3	2.9 [a]
CH_3—NH_2	2.1	2.0 [b]
CH_3—OH	1.1	1.1 [c]

[a] S. Weiss and G. E. Leroi: 1968, *J. Chem. Phys.* **48**, 962.
[b] D. R. Lide: 1957, *J. Chem. Phys.* **27**, 343.
[c] E. V. Ivash and D. M. Dennison: 1953, *J. Chem. Phys.* **21**, 1804.

(symmetry C_2) with an HOOH dihedral angle of about 120°. For this case, the theoretical results are found to depend quite significantly on the type of basis set used. For a minimal Slater-type basis [15–16] it is found that the potential curve is very flat with little energy difference between $\phi = 120°$ and the *trans* form ($\phi = 180°$). The split valence 4-31G basis gives a *trans* minimum if full geometry optimization is carried out [2]. It became evident several years ago [17] that the addition of polarization functions and geometry optimization was necessary to give an adequate account of H_2O_2. The 6-31G^* results do in fact give a twisted minimum with a small *trans* barrier and a much larger *cis* barrier in reasonable agreement with available experimental data. Figure 1 shows the 6-31G^* curves obtained both with rigid rotation and flexible rotation (with variable bond angles). These are in reasonable agreemenent with those obtained by a more extensive calculation of Veillard [17].

The interpretation of the electronic factors associated with the rotational potential curve for hydrogen peroxide is aided by splitting the total potential into the three Fourier components according to Equation (5). The corresponding values of V_n are listed in Table V. These may be discussed individually. To begin with, V_3 is small and is approximately one ninth of V_3 for ethane, consistent with a hydrogen–hydrogen interaction for this term. The main features of the potential curve are clearly dominated by V_1 and V_2. The negative value for V_1 implies a preference for *trans* over *cis* conformations which may be ascribed partly to hydrogen–hydrogen repulsions in the gas phase and partly to dipole–dipole interactions which favor the *trans* arrangement. Finally, the negative value of V_2 must be associated with an electronic mechanism

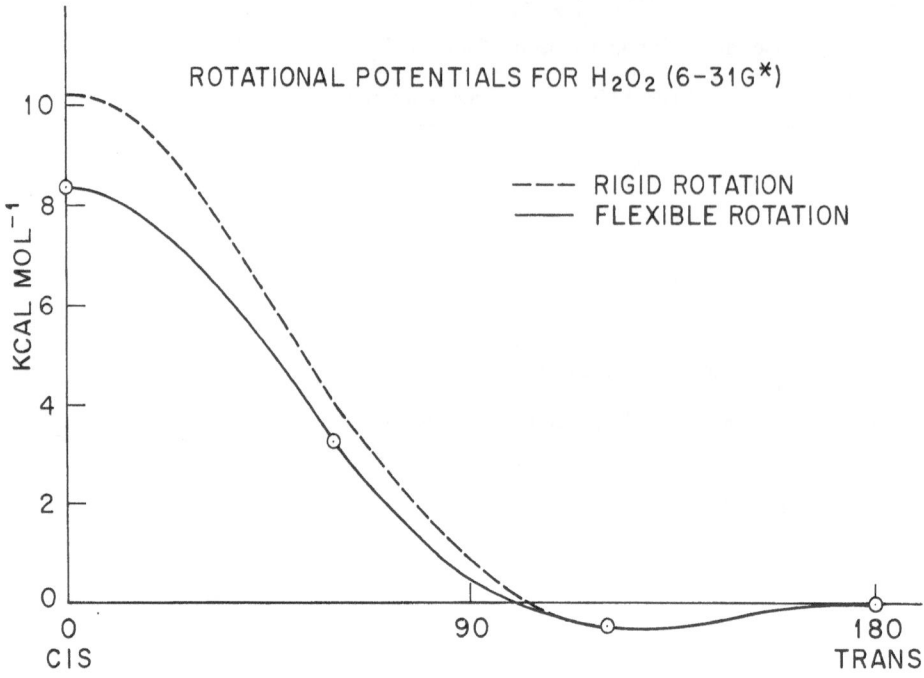

Fig. 1.   6-31G^* potential curves for hydrogen peroxide.

TABLE V

Fourier compounds of H_2O_2
rotation potential (6-31G^*)

	V_n (kcal/mole)
V_1	− 8.0
V_2	− 3.7
V_3	− 0.3

favoring the perpendicular form ($\phi = 90°$) over either planar form. Two such mechanisms are possible. One arises from the fact that in either planar form, there will be
four π-electrons, two occupying a π-bonding orbital and two a π-antibonding orbital.
The gross populations of the $p\pi$ atomic functions are then both 2.00. However, it is
generally found that the antibonding effect in such circumstances outweighs the
bonding effect leading to some net destabilization. This is reflected in a negative $2p\pi$
overlap population between the oxygen atoms in either of these planar forms. For the
perpendicular form, on the other hand, the repulsion is relieved since π-electrons may
transfer from the $p\pi$ lone pair atomic function in one oxygen atom into vacant antibonding σ^* orbitals on the other. This is illustrated in Figure 2. Indeed this electron
transfer can lead to some double bond character in the bond. However, the gross
populations for the $p\pi$ atomic functions on one oxygen are only reduced to 1.997 so
the effect is not very large.

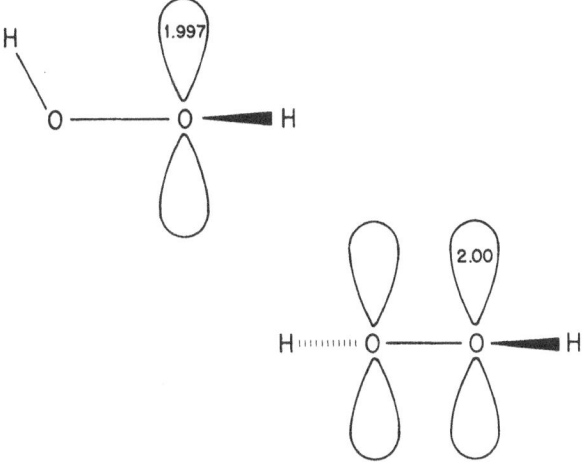

π-Orbital Populations in Hydrogen Peroxide (4–31G)

Fig. 2. $2p\pi$ gross populations for hydrogen peroxide (4-31G)

The remaining molecules hydrazine and hydroxylamine have also been examined by *ab initio* molecular orbital theory [4,18–19]. The 4-31G basis leads [2] to a twisted form for hydrazine and a C_s from with the OH bond *cis* to the nitrogen bond pair for hydroxylamine. It is interesting to note that the hydroxylamine result is different to that obtained by INDO theory [20], presumably because the latter neglects the dipolar interactions of directed lone pairs on neighboring atoms.

4. Effect of Polar Substituents on Rotational Barriers

If one of the hydrogen atoms in one of the parent molecules discussed in the previous section is replaced by a polar substituent, the symmetry of the potential function may be reduced and the energy changes may be altered. Theoretical studies on a number of such molecules indicate that polar substituents, particularly fluorine, modify the simple potential functions quite strongly.

Figure 3 shows the 4-31G rotational potential for fluoromethanol FCH$_2$—OH using standard bond lengths and angles [2, 21]. For methanol itself, the potential is a simple threefold function with a barrier of only about one kcal mole. The substitution of fluorine changes this drastically. The FCOH *trans* conformation turns out to be an unstable maximum instead of a local minimum. The minima correspond only to the FCOH *gauche* forms with a dihedral angle close to 60°. According to this potential function, the *trans* rotational barrier is 5.6 kcal/mole. Preliminary results with the 6-31G^* potential give a similar potential with a *trans* barrier of 4.6 kcal/mole [22]. Related results have been obtained by Wolfe *et al.* [23].

For fluoromethanol, the interpretation is again aided by breaking down the full rotational potential with Fourier components according to Equation (5) (ϕ being measured relative to the FCOH *cis* conformation). The values of V_1, V_2 and V_3 are

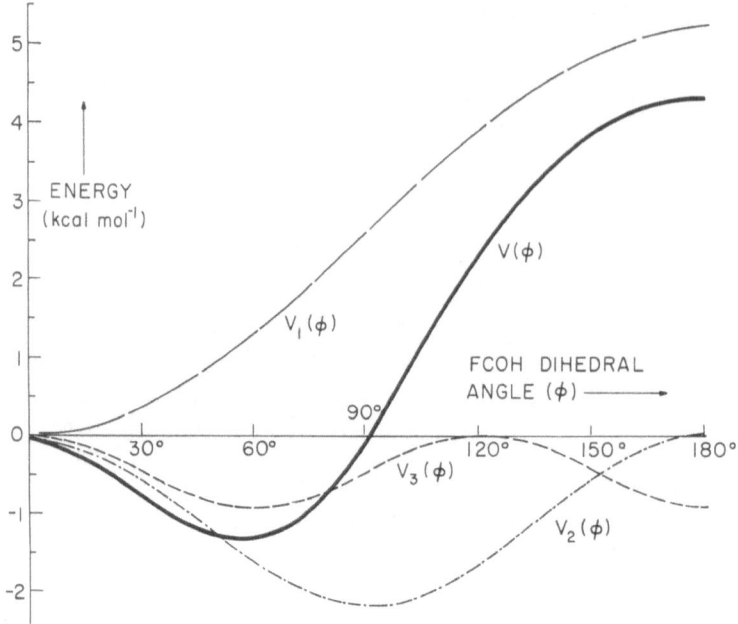

Fig. 3.   Rotation potential for fluoromethanol FCH_2—OH (4-31G)

given in Table VI. The value for V_3 is close to that in methanol. For V_1, there is a large positive value, corresponding to a higher (less stable) energy for FCOH *trans* compared with *cis*. This may be attributed largely to dipolar interaction between the polar C—F bond and the directed lone pairs on the oxygen atoms which will be in an attractive arrangement in the *cis* conformation. In addition there is a negative V_2 term which favors $\phi = \pm 90°$ over either *cis* or *trans* conformations. A plausible interpretation of the term involves back donation of π-lone pair electrons from the oxygen atom into the substituted methyl group. If the C—F bond is aligned approximately parallel to the axis of the $2p\pi$ oxygen lone pair atomic orbital (which occurs at $\phi = \pm 90°$) the back donation is accentuated. This shows up in the corresponding $2p\pi$ gross populations on oxygen (Figure 4). For the *trans* form, this population is close to the value in methanol but for the $\phi = 90°$ form, it is further reduced. In qualitative terms, this cor-

TABLE VI

Fourier components of
FCH₂OH rotation
potential (4-31G)

	V_n (kcal/mole)
V_1	+ 5.3
V_2	− 2.2
V_3	− 1.0

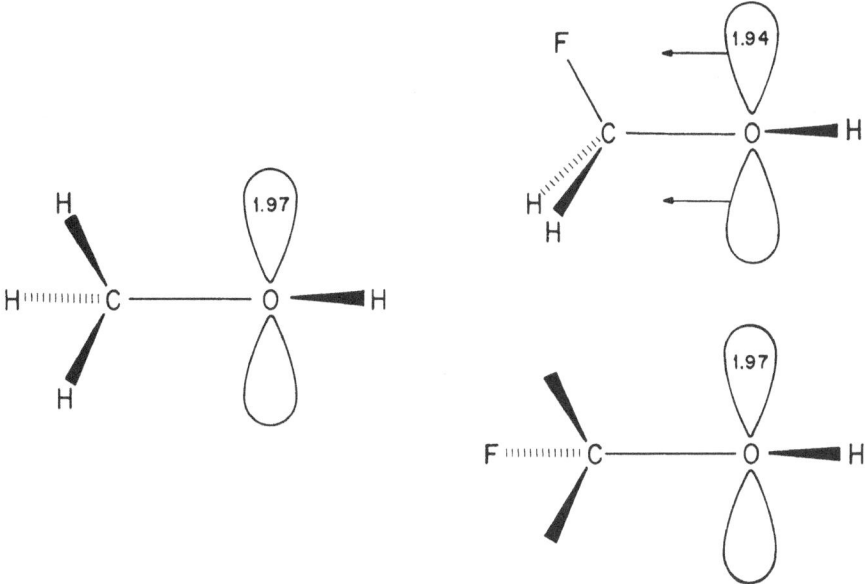

π – Orbital Populations in CH_3OH and FCH_2OH (4–31G)

Fig. 4. $2p\pi$ gross populations for fluoromethanol FCH_2—OH (4-31G)

responds to additional contributions of resonance structures such as

Fluorine substitution has also been found to produce substantial changes in other rotational potential. A 4-31G study [2] of fluoromethylamine FCN_2—NH_2 shows a deep minimum in a conformation with the C—F bond *trans* to the nitrogen lone pair direction. In this conformation, both the dipole–dipole interaction and the back-donation of lone pair electrons are stabilizing.

5. Double Rotor Molecules

If one of the hydrogens in one of the parent molecules is replaced by another rotor group, the resulting molecule has two coupled rotational degrees of freedom and a full treatment requires a two-dimensional potential surface rather than a single potential curve. Two rotational coordinates ϕ and ψ are required

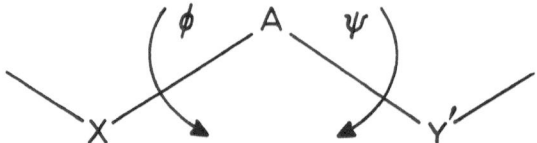

A number of molecules of this type have been partially examined with the 4-31G basis [24]. Here we shall discuss examples involving carbon–oxygen single bonds.

For the molecule ethanol, standard model calculations with the 4-31G basis give a C—O rotational potential with three distinct minima corresponding to two *gauche* and *trans* conformations [2]. The *trans* form is more stable by about 0.7 kcal/mole which is consistent with experimental observation. The barriers between these forms are of the order of 1 kcal/mole, so that the potential function is not very different from that of methanol. It is evident that the effect of a substituted methyl group on the C—O bond is much less than that of a fluorine atom. The rotational potential of the methyl group in ethanol is apparently similar to that in ethane.

For methyl ether, the two C—O bonds are equivalent, each giving a three-fold potential. 4-31G calculations with standard geometry [24] give a minimum energy for a staggered-type conformation (COCH *trans*). The barrier for a single CH_3 rotation is calculated (3.0 kcal/mole) to be considerably higher than in methanol (1.1 kcal/mole). This increase is found experimentally [25, 26] (2.7 kcal/mole in methyl ether vs. 1.1 kcal/mole in methanol). The origin of this increase is not entirely clear, although it may be associated with the close approach of hydrogens on different methyl groups.

One of the most interesting two-rotor problems is methanediol involving two equivalent coupled carbon–oxygen single bonds. Although the methanediol is not well characterized experimentally, it is an important model compound for the O—C—O group found in many carbohydrate systems. Since OH is itself a polar substituent somewhat similar to fluorine, it is expected to modify the rotational potential for the other bond. A 60° grid study of the two-dimensional 4-31G surface (using standard bond lengths and angles) has been published [24–25] and is reproduced in Table VII.

TABLE VII

Potential surface for methanediol (4-31G) (kcal/mole)[a]

θ ϕ	0	60	120	180
0	0	−1.8	−6.5	−6.8
60	−1.8	−3.7	−8.0	−7.4
120	−6.5	−8.0	−11.2	−6.4
180	−6.8	−7.4	−6.4	+0.5
240	−6.5	−6.0	−5.8	−6.4
300	−1.8	−1.8	−6.0	−7.4

[a] $\phi = \psi = 0$ corresponds to the all *trans* arrangement. Values ϕ greater than 180° are not listed as they are related to those given by symmetry.

It does indeed show a much wider variation than would be expected on the basis of two additive potentials. The (0, 0) conformation which corresponds to the all *trans* arrangement is a local *maximum* rather than a local minimum. This is analogous to the potential maximum in the *trans* form of fluoromethanol mentioned in the previous section. The gap between this structure and the lowest energy form ($+SC$, $+SC$)

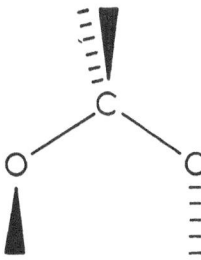

is more than 11 kcal/mole. This is somewhat reduced (to about 9 kcal/mole) at the $6-31G^*$ level [22], but the spread is evidently large. The implications of this potential surface for the stereochemistry of the anomeric center in carbohydrate chemistry are discussed elsewhere [25].

References

[1] Allen, L. C.: 1969, *Ann. Rev. Phys. Chem.* **20**, 315.
[2] Radom, L., Hehre, W. J., and Pople, J. A.: 1972, *J. Amer. Chem. Soc.* **94**, 2371.
[3] Pople, J. A.: 1972, 'Quantum Chemistry, Theory of Geometries and Energies of Small Molecules', in *Computational Methods for Large Molecules and Localized States in Solids* (ed. by F. Herman, A. D. McLean, and R. K. Nesbet), Plenum Press, New York, p. 11.
[4] Hoffman, R.: 1954, *J. Chem. Phys.* **22**, 571.
[5] Pople, J. A. and Segal, G. A.: 1965, *J. Chem. Phys.* **43**, S136.
[6] Pople, J. A., Beveridge, D. L., and Dobosh, P. A.: 1967, *J. Chem. Phys.* **47**, 2026.
[7] Dewar, M. J. S. and Haselbach, E.: 1970, *J. Amer. Chem. Soc.* **92**, 590.
[8] Roothaan, C. C. J.: 1951, *Rev. Mod. Phys.* **23**, 69.
[9] Hehre, W. J., Stesart, R. F., and Pople, J. A.: 1969, *J. Chem. Phys.* **51**, 2657.
[10] Ditchfield, R., Hehre, W. J., and Pople, J. A.: 1971, *J. Chem. Phys.* **54**, 724.
[11] Hariharan, P. C. and Pople, J. A.: 1973, *Theor. Chim. Acta* **28**, 213.
[12] Mulliken, R. S.: 1955, *J. Chem. Phys.* **23**, 1833.
[13] Lowe, J. P.: 1973, *Science* **175**, 527.
[14] Oelfke, W. C. and Gordy, W.: 1969, *J. Chem. Phys.* **51**, 5336.
[15] Newton, M. D., Lathan, W. A., Hehre, W. J., and Pople, J. A.: 1970, *J. Chem. Phys.* **52**, 4064.
[16] Stevens, R. M.: 1970, *J. Chem. Phys.* **52**, 1397.
[17] Veillard, A.: 1970, *Theor. Chim. Acta* **18**, 21.
[18] Pedersen, L. and Morokuma, K.: 1967, *J. Chem. Phys.* **46**, 3941.
[19] Fink, W. H., Pan, D. C., and Allen, L. C.: 1967, *J. Chem. Phys.* **47**, 895.
[20] Gordon, M. and Pople, J. A.: 1968, *J. Chem. Phys.* **49**, 4643.
[21] Pople, J. A. and Gordon, M.: 1967, *J. Am. Chem. Soc.* **89**, 4253.
[22] Hariharan, P. C. and Pople, J. A.: to be published.
[23] Wolfe, S., Rauk, A., Tel, L. M., and Csizmadia, I. G.: 1971, *J. Chem. Soc. B* 136.
[24] Radom, L., Lathan, W. A., Hehre, W. J., and Pople, J. A.: 1972, *Aust. J. Chem.* **25**, 1601.
[25] Jeffrey, G. A., Radom, L., and Pople, J. A.: 1972, *Carbohydrate Res.* **25**, 117.

CONFORMATIONAL STUDIES IN QUANTUM BIOCHEMISTRY

BERNARD PULLMAN

Institut de Biologie Physico-Chimique, Laboratoire de Biochimie Théorique associé au C.N.R.S.,
13, rue P. et M. Curie, Paris 5, France

1. Introduction

The studies I am going to describe here aim at the theoretical determination of *the conformational basis of molecular biology and pharmacology*. For a number of years the applications of quantum-mechanical theories and methods to biochemistry have been centered essentially on the electronic aspects of the structures and problems studied (B. Pullman and A. Pullman, 1963). This was a natural situation because of the importance of these aspects about which practically nothing was available and because of the tendency of these theories and methods to concentrate on such aspects.

Important as these contributions have been for the elucidation of a large number of biostructures and mechanisms, they left aside the second fundamental aspect pertaining to these problems, namely the *conformational* one. It is well known that the activity of biological molecules and, in particular, of biopolymers is strongly dependent upon their conformation. In fact, the conformational criterion is frequently a prerequisite for their functioning. The understanding of the factors governing conformational stability of biomolecules and the evaluation of the preferred conformers is therefore of utmost interest for the development of quantum biochemistry.

The need for such a promotion of quantum-mechanical studies appears the more necessary as there was during the last years a prominent development of what may be called 'empirical' studies in this field. These consist of partitioning the potential energy of the system into several discrete contributions, such as non-bonded and electrostatic interactions, barriers to internal rotations, hydrogen bonding, etc., which are then evaluated with the help of *empirical formulae* deduced from studies on model compounds of small molecular weight (for general reviews of such works, see Ramachandran and Sasisekharan, 1969; Scheraga, 1968). In the simplest approximation of these procedures (the 'hard sphere' approximation), due to Ramachandran and his collaborators and which practically inaugurated this area of research, the problem is even limited to the sole evaluation of allowed or forbidden contacts, with the help of van der Waals (or similar) radii.

Interesting as such attempts are, they suffer from two obvious drawbacks. First, whatever the practical justification for the partitioning of the total potential energy into a series of components, the procedure involves necessarily an element of arbitrariness and, possibly, incompleteness. Second, the fundamental formulae and parameters used to define the various components are far from being well established and differ, often appreciably, from one author to another. A more rigorous deal may therefore be expected from a quatum-mechanical approach.

R. Daudel and B. Pullman (eds.), The World of Quantum Chemistry, 61–89. All Rights Reserved
Copyright © 1974 by D. Reidel Publishing Company, Dordrecht-Holland

Possibilities of significant advances in this new direction have recently become clearly evident, due essentially to the elaboration of new methods of computation which *deal simultaneously with all valence, σ and π, or even all (including inner-shell) electrons.* They are therefore able to evaluate the total molecular energy corresponding to any given configuration of the constituent atoms and thus to choose the preferred ones. Among these methods, operating all within the general scheme of the molecular orbital method, the most prominent are: *the Extended Hückel Theory, the Iterative Extended Hückel Theory, the CNDO/2 and INDO methods, the PCILO method and* the *ab initio* or *nonempirical* procedure.

Occasionally, all of them have been used to investigate specific conformational problems of biochemistry or pharmacology. In our laboratory we have, since two or three years, developed a large programme of research corresponding to a *systematic* exploration of the conformational properties of fundamental biological and pharmacological compounds with the view of assessing the role of these properties in their behaviour and function. For this sake we have used essentially the PCILO method, and occasionally, when the dimensions of the system allowed it and the problem was of particular importance, the SCF *ab initio* procedure.

I cannot go here into details of the methods. (For a general presentation of them see, e.g., B. Pullman and A. Pullman, 1973.) I would just like to remind you of the general principles of the PCILO method (Diner *et al.*, 1969 and references therein), which is the one which we have most used, with which the majority of the complete available results have been obtained, and which is perhaps the less known one.

PCILO stands for *Perturbative Configuration Interaction using Localized Orbitals.* Its fundamental idea is to choose a set of reasonable bonding and antibonding orbitals localized on the chemical bonds. Such a set may be constructed on the basis of hybridized atomic orbitals (χ_i), the bond orbitals being obtained as linear combinations of distinct hybrids taken two by two, each bonding orbital Φ_i being associated with an orthogonal antibonding orbital Φ_i^*:

$$\Phi_i = C_{i1}\chi_{i1} + C_{i2}\chi_{i2},$$
$$\Phi^* = C_{i2}\chi_{i1} - C_{i1}\chi_{i2}.$$

A localized orbital representing a lone pair is described by a single hybrid orbital.

The bonding orbitals are then used to construct a fully localized Slater determinant. This determinant represents the zero-order wave function for the ground state of the system. The antibonding orbitals are utilized to build the excited states and a configuration interaction matrix is constructed on this basis. Then the lowest eigenvalue and eigenstate, i.e., the energy and the wave function of the ground state of the system, are obtained by a Rayleigh–Schrödinger perturbation expansion truncated after the third order.

As a technical simplification, the principal working hypotheses of the CNDO/2 procedure have been retained, in particular the hypothesis of complete neglect of differential overlap as well as the general parametrization of this procedure (Pople and

Segal, 1965, 1966). A more detailed description and the programme can be obtained from Q.C.P.E.

I would like to illustrate here the nature and the significance or the results obtained on three examples. The first two are concerned with the two fundamental biopolymers, proteins and nucleic acids. The third refers to a specific, small but very important molecule, acetylcholine, which, although a natural neurotransmitter, is a key compound in molecular pharmacology and whose study, because of abundance of information, experimental and theoretical, yields some conclusions of a wide general interest.

2. Proteins and Their Constituents

One of the main problems in the study of the conformation of proteins is the determination of the conformational possibilities of their 20 different constituent amino acid residues. This is done in general with the help of the so-called 'dipeptide' model, the definitions and conventions of which are illustrated in Figure 1 (Ramachandran and Sasisekharan, 1969; Pullman and Maigret, 1973; B. Pullman and A. Pullman, 1973).

The fundamental observation which is at the basis of this approximation is that because of the planarity of the peptide unit, the flexibility of the main backbone of the polypeptide chain originates essentially from the possibilities of rotation about the N—C^α and C^α—C' single bonds adjacent to the α-carbons. Consequently the conformation of, say, a pair of peptide units can be specified by giving the values of the two dihedral angles Φ and Ψ around the C^α carbon joining these units. The conformation of the backbone of the whole polypeptide chain may be described by indicating the sequence of these angles along the chain.

The second fundamental observation indicates that, while the values assigned to one angle of a pair, Φ_i, depend markedly on the values assigned to the other angle of the same pair, Ψ_i, the interactions associated with rotations of one such pair are largely independent of the angles assumed by the neighbouring pairs Φ_{i-1}, Ψ_{i-1} and Φ_{i+1}, Ψ_{i+1}. Otherwise speaking short-range interactions of an amino acid residue in a polypeptide chain which essentially involve the two peptide units directly connected with the residue represent the major determinant of its conformational possibilities. As a

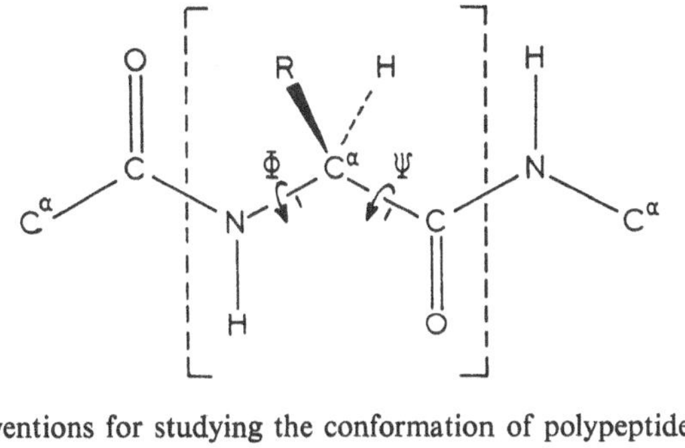

Fig. 1.   Standard conventions for studying the conformation of polypeptides (Edsall *et al.*, 1966). limits of a residue.

result of this state of affairs each pair of related rotational angles Φ and Ψ can, at least at first approximation, be treated separately from the others of the chain.

The possible values of each such pair will depend on the nature of the side chain attached to carbon α and the possible or preferred values of the side-chain rotational angles χ_1, χ_2 etc. The conformational possibilities of each of the 20 residues, related to the state of the backbone, will thus be characterized by a set of associated personal values of Φ and Ψ, depending themselves on the set of χ's of the side chain.

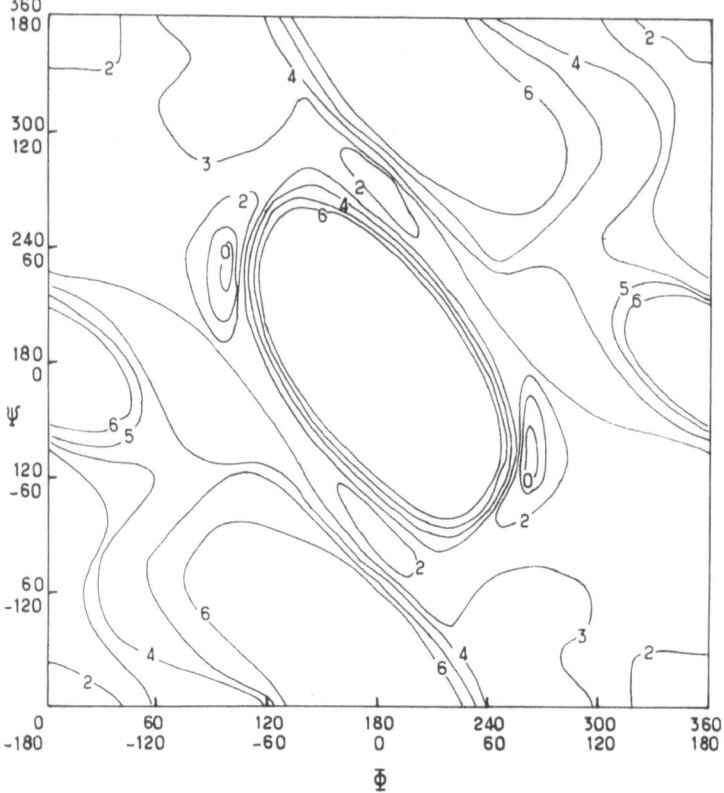

Fig. 2.   PCILO conformational energy map for the GLY residue. Isoenergy curves in kcal/mole with respect to the global minimum taken as energy zero.

Let us look at a few examples. Figure 2 presents the PCILO conformational energy map for the glycyl (GLY) residue, the simplest of all residues corresponding to $R = H$ (Maigret *et al.*, 1970). Indicated are iso-energy curves in kcal/mole with respect to the global energy minimum taken as energy zero. This energy zero (most stable conformation) is predicted for the combination $\Phi = 90°$, $\Psi = 240°$ and the symmetrical position $\Phi = 270°$, $\Psi = 120°$. Local energy minima are also seen, in particular one 2 kcal/mole above the gobal one at $\Phi = \Psi = 0°$. The isoenergy curves are limited on the figure to 6 kcal/mole above the minimum, considered as a limit for stable conformations.

Figure 3 presents similar results for the alanyl residue (I, $R = CH_3$), the simplest

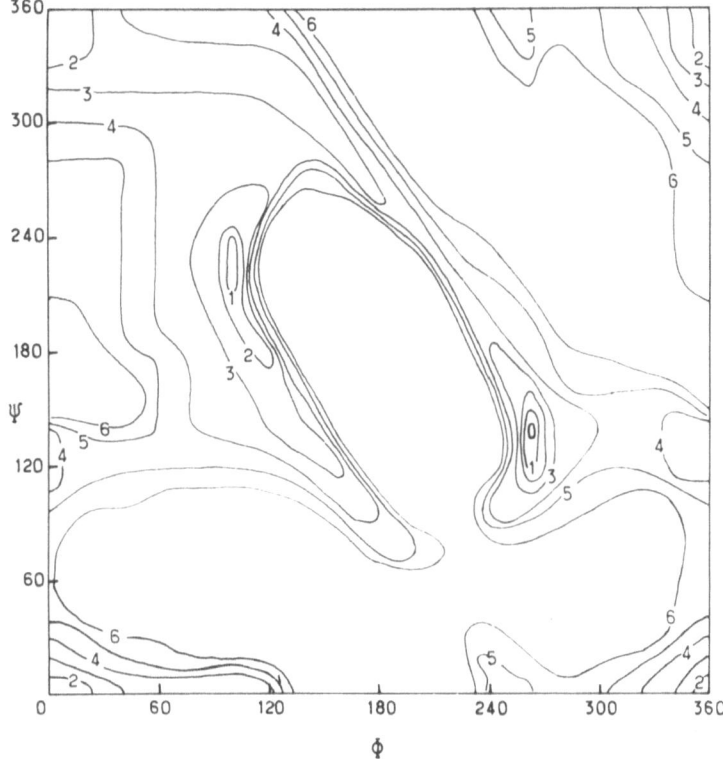

Fig. 3. PCILO conformational energy map for the ALA residue. Isoenergy curves in kcal/mole with respect to the global minimum takenas energy zero.

residue containing a β carbon (Maigret *et al.*, 1970). The most striking aspect of this map is the decrease of the conformational stability zone with respect to that of the GLY residue: otherwise, we observe the same position for the nearly doubly degenerate global minimum (in this case it is only *nearly* degenerate because the residue does not have the symmetry of the GLY residue) and the presence of the local energy minimum at $\Phi = \Psi = 0°$.

The immediate question is, of course, what is the practical significance of these results for the conformation of these residues in proteins? The answer to this question is nowadays relatively easy; it suffices to compare these theoretical indications with the experimentally observed conformations of these residues in globular proteins as given by crystal X-ray studies. A number of such proteins have been studied with sufficiently high resolution to provide information about the Φ and Ψ angles of all the constituent residues. Among such proteins are: lysozyme, myoglobine, α-chymotrypsin, carboxypeptidase, erythrocruorin, insulin, ribonuclease-S, rubrodoxin, subtilisin, oxyhaemoglobin, etc. They represent altogether over 2000 residues among which a large number are GLY and ALA residues. (For detailed references see Pullman, 1971; B. Pullman and A. Pullman, 1973.)

Figure 4 presents:

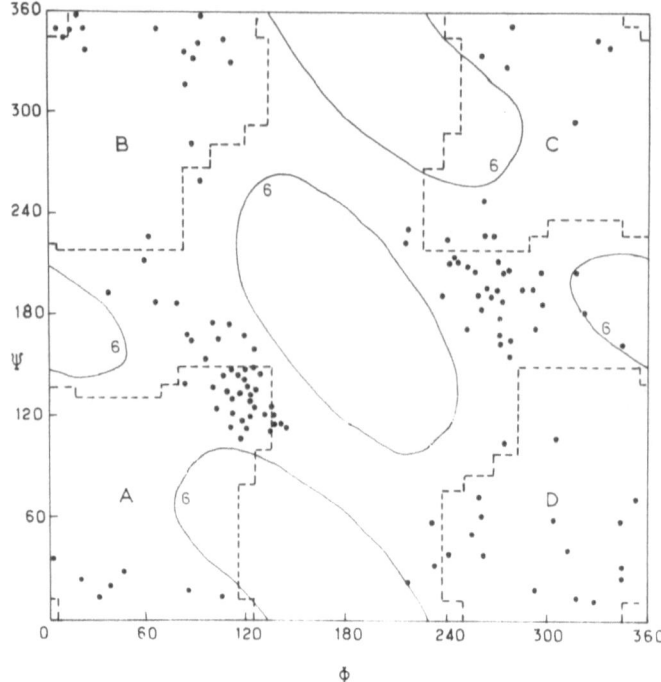

Fig. 4.   Conformations of the GLY residues in globular proteins. —— limits of the PCILO stable zone of the *GLY* residue at 6 kcal/mole above the global minimum. – – – limits of the stable zone in typical empirical computations.

(a) the contours of the PCILO computations for the GLY residue limited at the value of 6 kcal/mole above the deepest minimum;

(b) the limits of a typical 'allowed' conformational space as obtained in the 'hard sphere' approximation of the empirical methods (Ramachandran and Sasiseharan, 1969) and which is also typical of the results obtained in a large number of more developed empirical calculations using partitioned potential energy functions;

(c) the experimental conformations of the glycyl residues in the above globular proteins as determined by high resolution X-ray studies.

Figure 5 presents the same results for the ALA residue. The confrontation of the theoretical and experimental data leads immediately to two essential conclusions:

(1) It is seen that in both cases the quantum-mechanical calculations impose less restrictions on the allowed or preferred conformational space than do the empirical ones.

(2) The agreement between theory and experiment is much better with the quantum-mechanical calculations than with the empirical ones, many representative points lying in the space forbidden by the latter but allowed by the former.

This situation has an important conceptual consequence: it indicates that following the quantum-mechanical computations, the 'protein effect' apparently does not create 'extraordinary' conformations, which would correspond to high energy regions of the

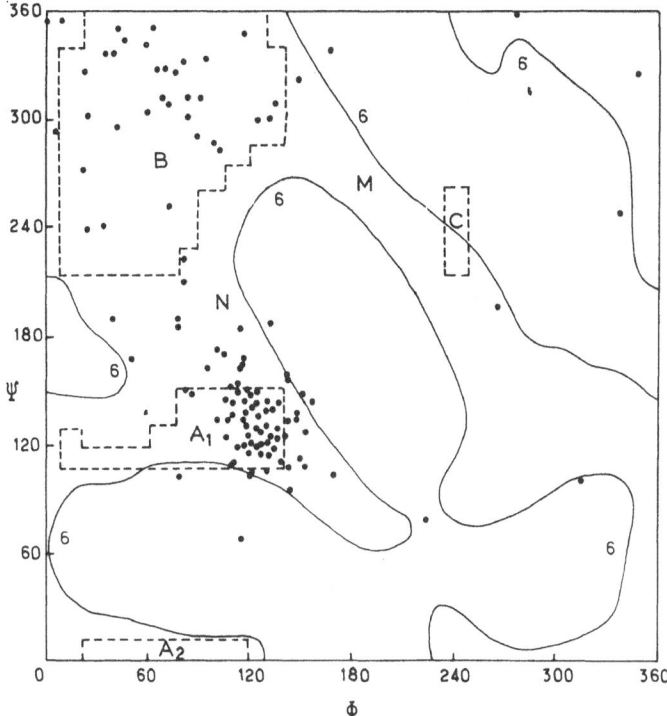

Fig. 5. Conformations of the ALA residues in globular proteins. ——— limits of the PCILO stable zone of the ALA residue at 6 kcal/mole above the global minimum. --- limits of the stable zone in typical empirical computations.

calculations for the 'dipeptides'. It just operates within the conformational stable zone of the individual residue. A different conclusion would have been drawn from the results of the empirical computations.

There is, however, also a third observation which can be made on the basis of the confrontation of the theoretical results with the conformation of the residues in globular proteins. While included within the predicted conformationally stable zone, the experimental conformations do not cluster especially around the energy minima of the PCILO results, in particular around the global one, but are widely scattered. The reasons for this situation are easy to determine, when one considers the nature of the predicted most stable conformations. They correspond to quite particular intra-molecularly hydrogen bonded structures, *highly specific for the model dipeptide studied*: the most stable form associated with the coordinates $\Phi = 90°$, $\Psi = 240°$ corresponds to a seven-membered ring (Figure 6a) called C_7, stabilized by an intramolecular hydrogen bond between $H^2 \ldots O^1$. The secondary energy minimum at $\Phi = \Psi = 0°$ corresponds to the fully extended (FE) form of the molecule stabilized by a weak intramolecular hydrogen bond formed between $O^2 \ldots H^1$ and leading to a cyclic pentagonal structure (called C_5) (Figure 6b). These conformations are specific for the model utilized and need not be of *particular* importance in proteins where other hydrogen bonds, say between more distant residues, have a large probability to be established.

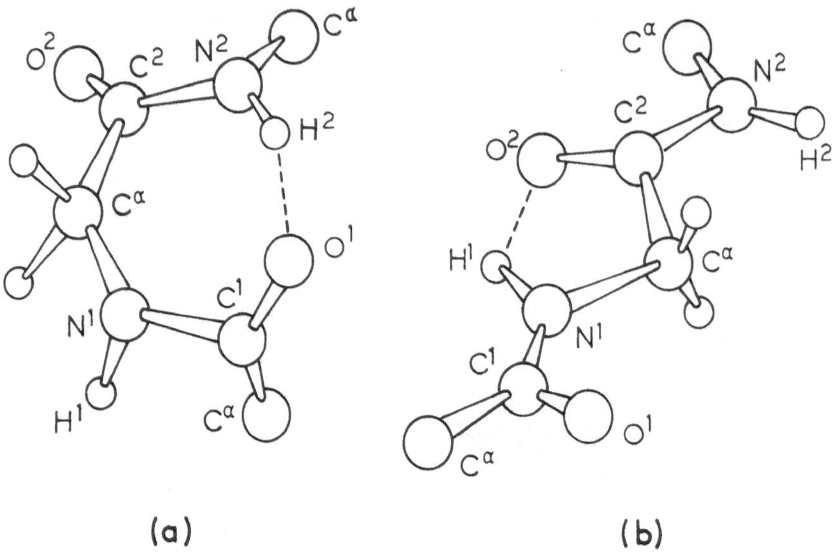

Fig. 6.   The preferred conformations of the GLY and ALA dipeptides.

Thus, the conformational energy maps of the amino acid residues constructed on the basis of the dipeptide model indicate correctly the overall conformational possibilities of the residues but should not be used for the prediction of their most probable conformations in proteins. In order to make this last prediction one needs to go beyond the dipeptide model so as to include at least medium-size interactions. Attempts in this direction have been made recently with promising results and are being developed actively in a number of laboratories, including ours.

On the other hand, I would like to stress that the theoretical determination of the preferred conformations of the model dipeptides has been one of the outstanding successes of PCILO calculations. Although nearly all theoretical computations whether 'empirical' or quantum-mechanical (carried out by other methods such as e.g. EHT, CNDO etc.) have utilized the dipeptide model, the PCILO computations are practically the only ones to have predicted the C_7 and C_5 forms as the preferred ones for these model compounds. A group of Russian authors, using empirical methods, have predicted the importance of the C_7 forms (Popov et al., 1968). Nowadays the predictions have been completely confirmed by NMR and infrared studies by Néel, Lascombe and coworkers (for a review see Néel 1973; Cung et al., 1973; Avignon and Lascombe 1973). The experiments by these authors, carried out in inert solvents (CCl_4), correspond to conditions not too far away from the isolated molecule used in the computations. The results indicate the existence of the model compounds of the GLY and ALA residues, in the very two preferred conformations predicted by the PCILO calculations: the seven-membered hydrogen-bonded ring (C_7) with angles $\Phi \approx 105°$, $\Psi \approx 130°$ and the fully extended form, involving a five-membered hydrogen-bonded ring (C_5), with angles $\Phi = \Psi = 0°$. A refinement of the experimental technique of quantitative analysis of the dipeptide leads to an approximate evaluation of the proportion of the two forms. The agreement with the theoretical predictions is again satisfactory, the C_7 form being the predominant one.

Similar results have by now been obtained for *all* the amino acid residues of proteins and they all lead to similar individual conclusions. They indicate also, however, some general features which enable a number of extensions of the application of the procedure. Thus, one of the basic conclusions which may be drawn from these results is that the general allowed conformational space, within a fixed limit above the individual deepest minima, is similar in all these residues and similar to the general contour obtained for the alanyl residue. This situation is due to the fact that atoms situated beyond C^β of the side chain have a much smaller effect on the conformational stability than does the introduction of C^β. Naturally, there are differences among the individual residues but they all more or less conform to the pattern obtained for alanyl. What essentially distinguishes these residues from alanyl (and among themselves) is the fine structure of the conformational space and the location of the different energy minima. Under these conditions it seems reasonable to assume that the conformational map of the alanyl residue may be considered as representing, at least at first approximation, the conformational map of all C^β containing residues, that is, of all residues with the exception of glycine. In order to check the validity of this proposition we may confront the theoretical predictions based on this assumption with the available experi-

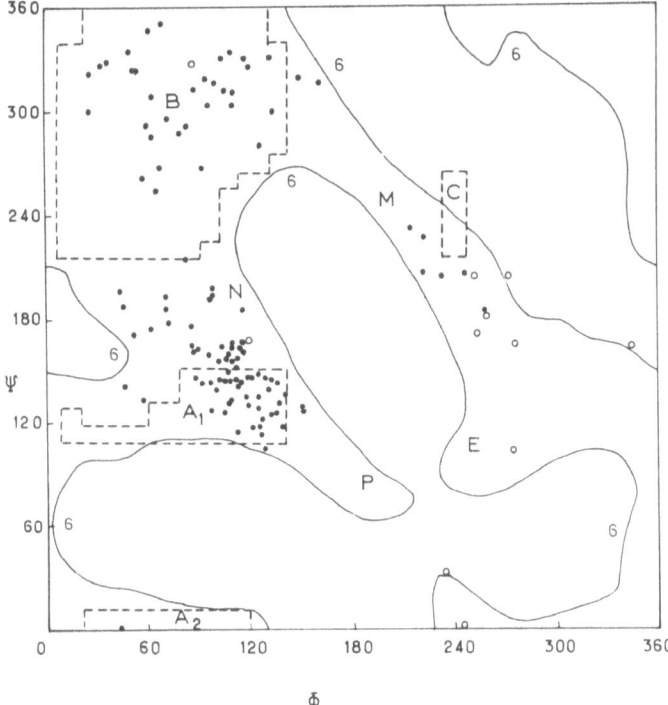

Fig. 7. The general significance of the conformational energy map for the alanyl residue. ● conformations of all the amino acid residues with C^β-carbons in lysozyme. ○ conformations of GLY residues in lysozyme.

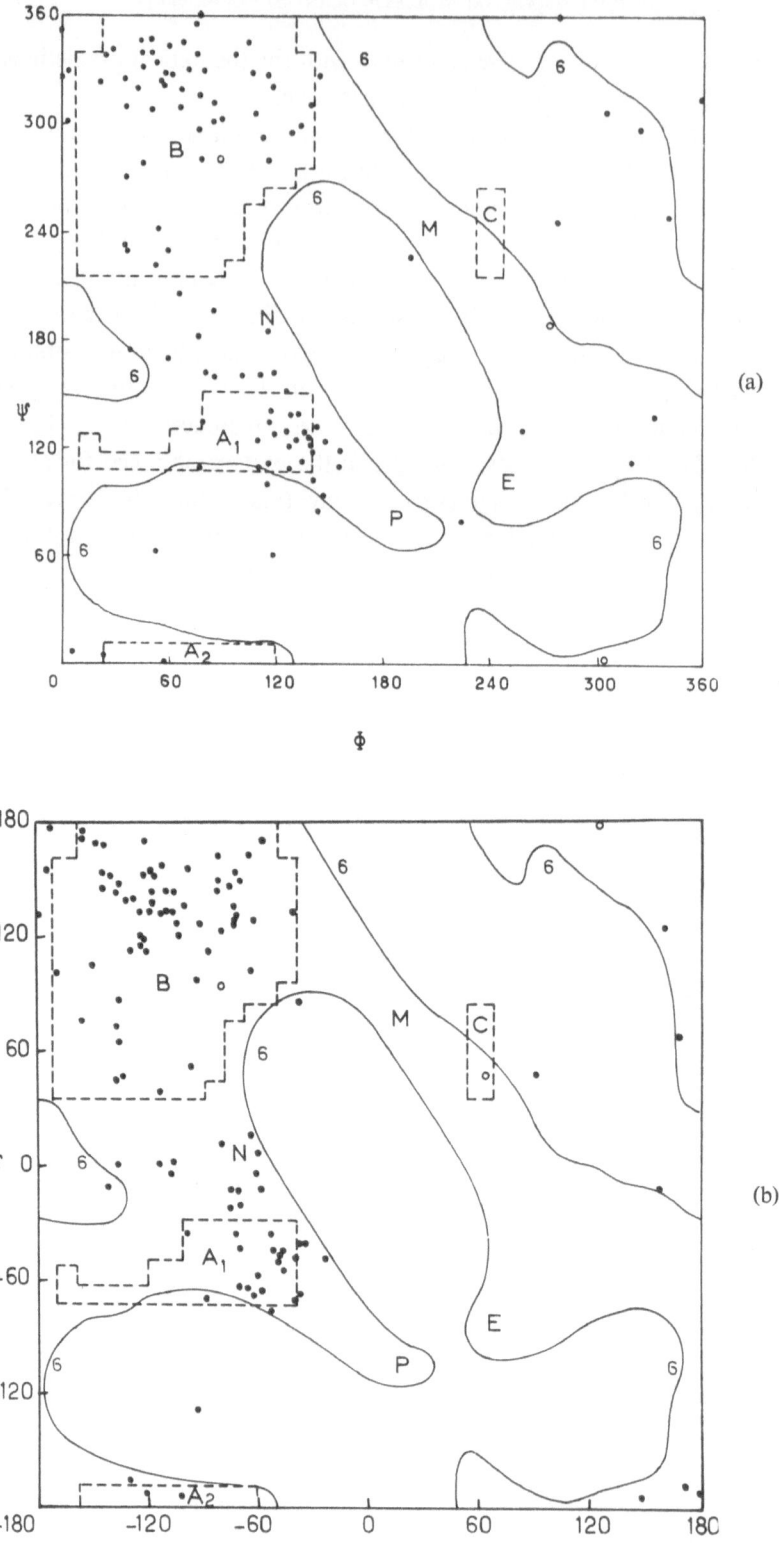

Fig. 8a–b. The general significance of the conformational energy map for the alanyl residue. ● conformations of all the amino acid residues with C^β-carbons in ribunoclease. ○ conformations of GLY residues in ribonuclease. (a) primitive X-ray data; (b) refined X-ray data.

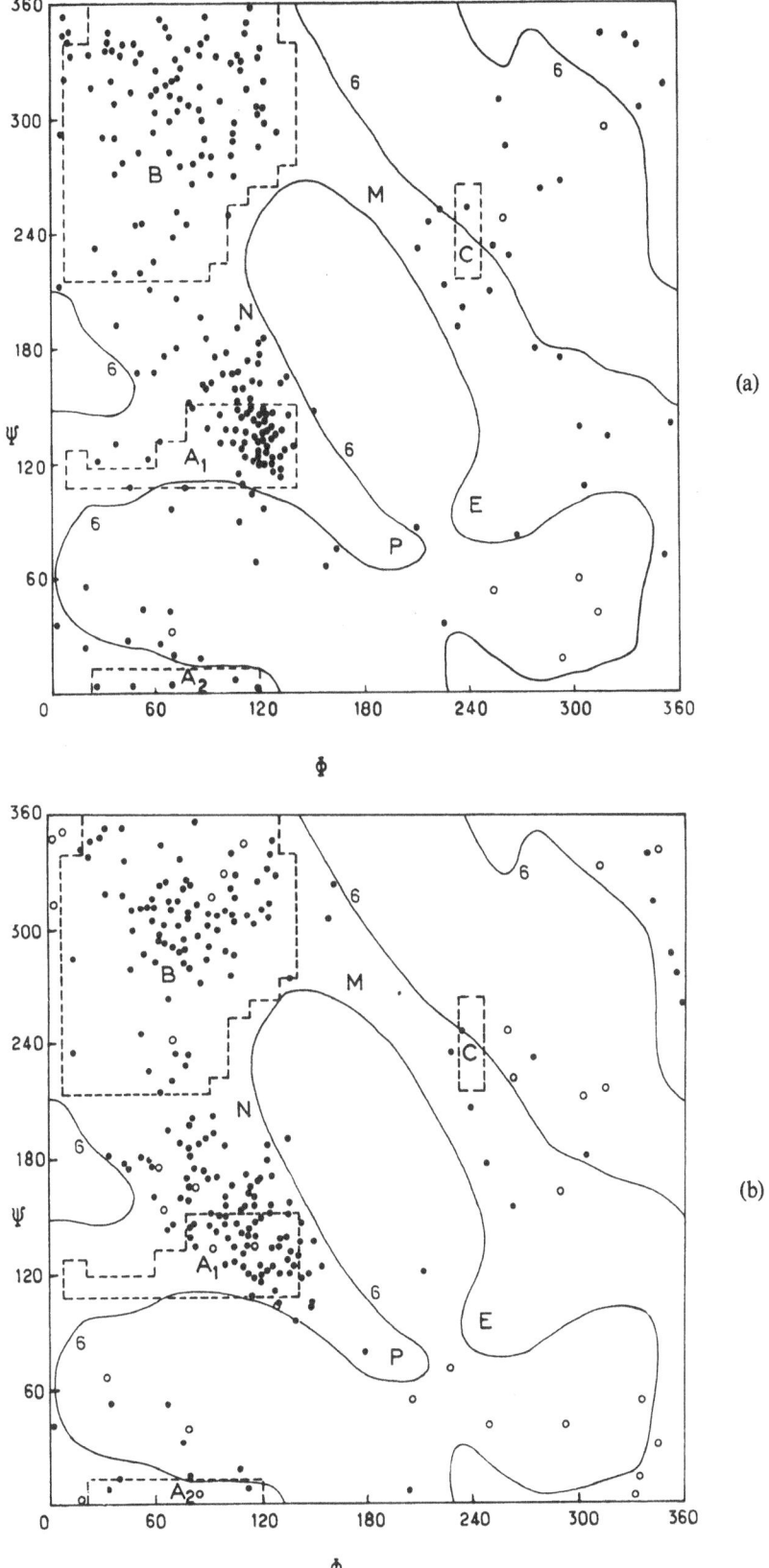

Fig. 9a–b. The general significance of the conformational energy map for the alanyl residue. ● conformations of all the amino acid residues with C^{β}-carbons in subtilisin-BPN'. ○ conformations of GLY residues in subtilisin-BPN'. (a) primitive X-ray data; (b) refined X-ray data.

mental data on all the aminoacid residues in any of the experimentally determined proteins. Such a confrontation is shown in Figure 7 for the illustrative case of lysozyme. The figure is self-explanatory and demonstrates the obvious success of the proposition. (Crystallographic data from Imoto *et al.*, 1972.)

In fact, the value of the theoretical calculations is such that they may be used to estimate the precision of the experimental data and sometimes to suggest their revisions or corrections. It must be borne in mind that the Φ and Ψ values are determined experimentally with the precision of about 10°–20°. In most cases, the data are published in two steps: preliminary data and refined data. What is particularly striking is that the refinement of the experimental data generally brings them closer to the theoretical predictions. This is illustrated in Figures 8a and 8b for ribonuclease (crystal data from F. M. Richards, private communication), and in Figures 9a and 9b for subtilisin-BPN′ (crystal data from Alden *et al.*, 1971). In both cases, figure a corresponds to primitive X-ray results, figure b to refined results. In both cases the refinement brings the experiment closer to theory.

3. Nucleic Acids and Their Constituents

The case of the nucleic acids presents us with a different challenge. The conformational properties of polynucleotide chains depend on a series of torsion angles (Figure 10) which may be divided into three groups.

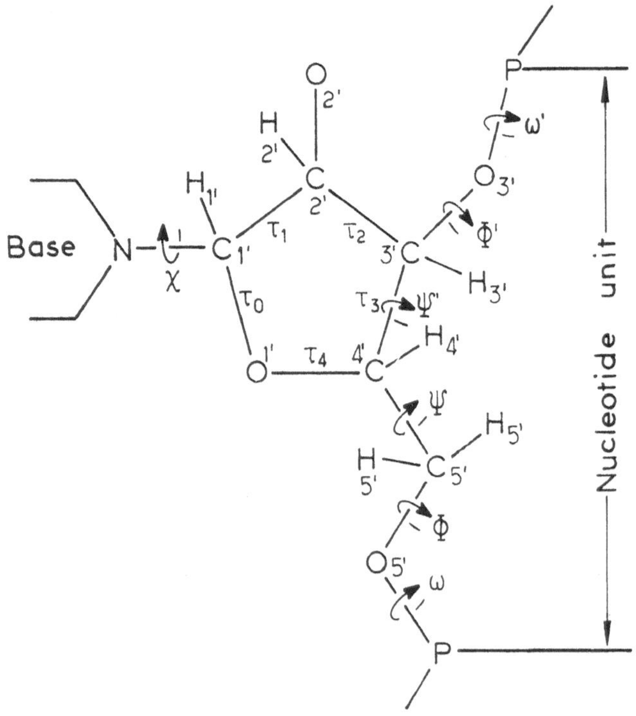

Fig. 10. The principal torsion angles in polynucleotides. Notations following Sundaralingam (1969, 1972).

(1) The glycosidic torsion angle χ, defining the relative orientation of the purine and pyrimidine bases with respect to the sugar.

(2) The torsion angles of the backbone Φ', ω', ω, Φ, Ψ, Ψ'.

(3) The torsion angles about the bonds of the sugars: τ_0–τ_4, defining the pucker of this constituent.

In the study of nucleosides and nucleotides special attention is frequently devoted to the orientation of the exocyclic CH_2OH or CH_2–O–PO_3H group of the sugar.

This represents a large number of degrees of freedom. Even if we put aside the torsions of the bonds of the sugars and adopt the usual empirical division of the pucke-

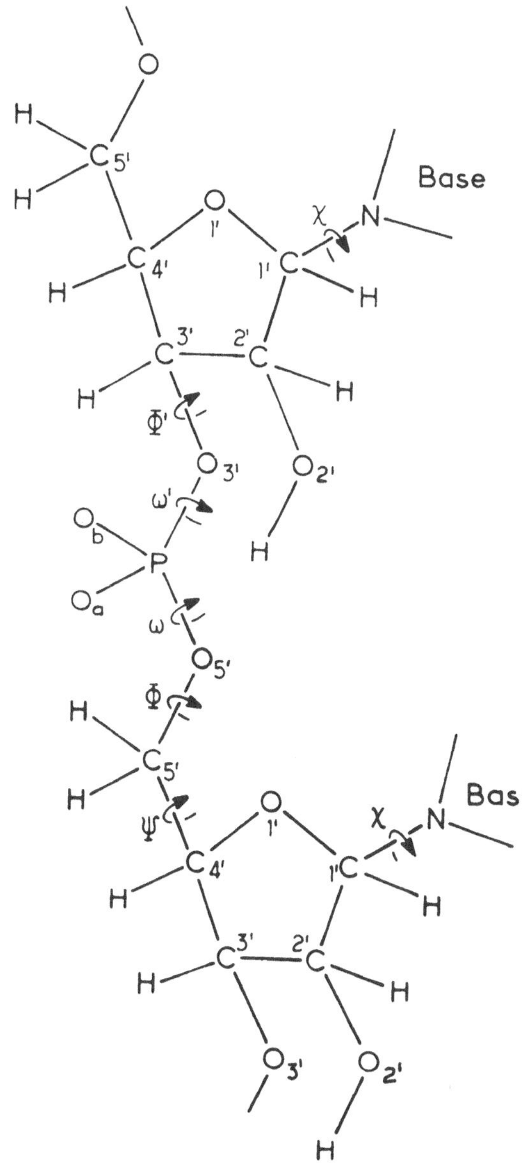

Fig. 11. The backbone of a dinucleoside monophosphate.

ring of the sugar into the classical four principal types: C(3')-endo, C(2')-endo, C(3')-exo and C(2')-exo, we are still left with six essential torsion angles in the nuc-cleotide unit. If we admit *a priori* that each of these torsions can adopt three preferred values (a reasonable *a priori* estimation), the number of possible combination is $3^6 =$ $= 729$. Taking into account the four principal puckerings of the sugars, a dinucleoside monophosphate (Figure 11) may have about 3000 acceptable conformations. One of the goals of the theoretical work is to operate a selection of the most stable and most probable among these possible conformations so as to reduce this number.

The work, although a hard one, has been highly successful. It has been and conti-nues to be carried out in steps.

At first sight the simplest to treat seems to be the glycosidic torsion angle χ between the base and the sugar because of its somewhat isolated position. It corresponds, in fact, to one of the most complex but also most important conformational problems in the field of nucleic acids and their constituents, which has attracted an enormous amount of theoretical and experimental investigations (Pullman and Berthod, 1973). The complexity of the problem (especially in nucleosides and nucleotides) stems from the dependence of this torsion angle upon the pucker of the sugar, the nature of the base and the orientations of the OH groups and of the exocyclic CH_2OH or CH_2O- $-PO_3H$ groups at the sugar (Figure 12). Different results are obtained for different

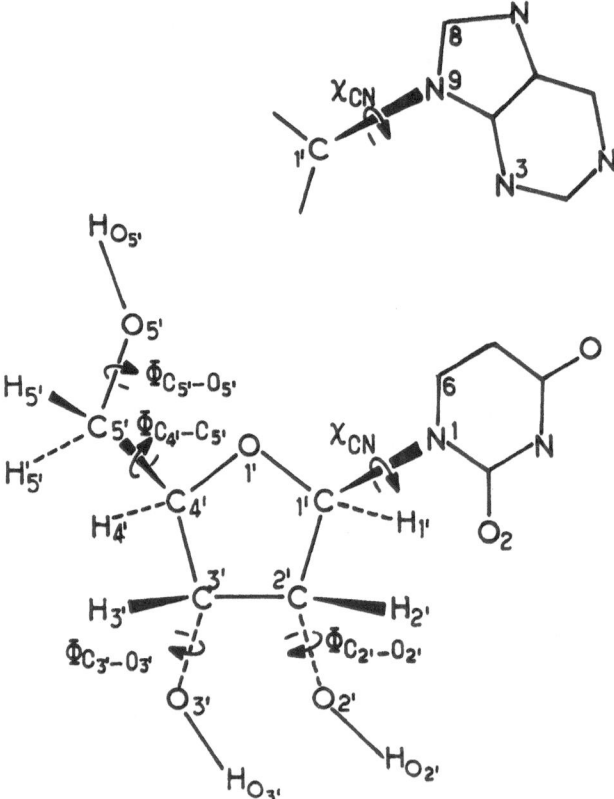

Fig. 12.   Torsion angles involved in the search for the *syn* and *anti* conformers.

TABLE I

The *syn-anti* equilibrium in nucleosides

Interaction of the base with	Theory		Experimental		
	Type of nucleosides	Major components	Type of compounds	Supposed conformation	Results
O(5')–H	2'-endo I, A, U, C	Syn	*i*-Nucleosides	3'-endo	I, A, G, U: *syn*
	3'-endo A, C	Syn	2'-Deoxy nucleosides	3'-endo⇌2'-endo	U: *anti*
	3'-endo G	Syn ≈ anti	Some crystals	2'-endo	Pu: syn
	3'-endo U	Anti	2', 3'-Cyclic nucleotides	°E, oE	C: syn, U: anti
O(2')–H	2'-endo I, A, C	Anti			cUMP, cIMP, cAMP: anti
	2'-endo U	Syn ≈ anti	3', 5'-Cyclic nucleotides	3'-endo ($^3T_4$)	cGMP: syn
	3'-endo A, G, U, C	Anti			
O(5')–H and O(2')–H	7'-endo A, I	Syn ≈ anti		Py: 3'-endo⇌2'-endo	Py, halo U: anti
	2'-endo U	Syn	Free nucleosides		U: anti⇌syn
	2'-endo C	Anti		Pu: 2'-endo predomin	Pu: anti ≈ syn
	3'-endo A, G	Syn ≈ anti			
	3'-endo C, U	Anti			
No interaction	3'-endo Pu, Py	Anti			
	2'-endo Pu, C	Anti	Majority of crystals		Anti
	2'-endo U	Syn ≈ anti			
	3'-exo, 2'-exo	Anti			

possible combinations of these factors. We have explored all of them and believe we are the only ones to have done so. The results are summarized, very schematically, in Table I, in the simplified terms of a competition between the so-called *syn* and *anti* conformations (Figure 13). Contrary to the usual opinion following which the *anti*

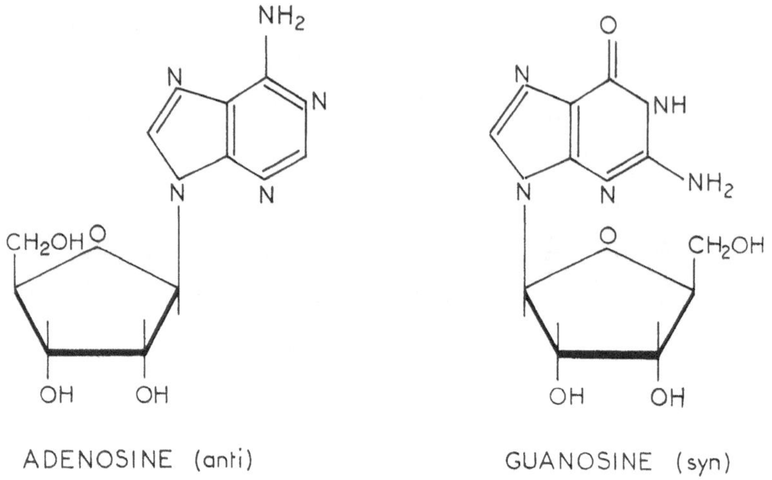

ADENOSINE (anti)   GUANOSINE (syn)

Fig. 13.   Schematic representation of *syn* and *anti* conformers.

conformations are the predominant ones (opinion based on X-ray crystal studies of nucleosides and nucleotides) the table shows a much more complex situation, and a more detailed examination of experimental results in *solution* confirms, that in a number of situations the *syn* forms are abundant and important.

    The essential problem appears, however, to be that of the backbone torsion angles (Pullman *et al.*, 1972; Perahia *et al.*, 1973). These are too numerous to be easily treated *en bloc*, although nowadays we could do it. The most economical treatment appears a stepwise one. Thus, it is reasonable *a priori* that the strongest interdependence will exist between *adjacent* torsion angles. This means that our main goal should be to construct the conformational energy maps corresponding to the four possible combinations of two consecutive such angles: $(\omega'-\omega)$, $(\Phi'-\omega')$, $(\omega-\Phi)$, $(\Phi-\Psi)$. To do so, preselected values have to be adopted for the torsions of the angles not involved in the particular map under consideration. These are obtained by a combination of general stereochemical arguments, experimental indications and indications from available calculations by simpler empirical or quantum-mechanical methods. They turn out to be: 300° (preferred), 60°, and 180° for ω and ω', 60° (preferred), 180° and 300° for Ψ, 180° for Φ and 240° for Φ'. The conformational energy maps have been established for each pair of adjacent rotation angles with the appropriate combinations of these different pre-selected values for the remaining torsion angles. The procedure may be repeated a number of times till self-consistency is being attained.

    What do the results look like?

    Let us illustrate them on the example of the $(\omega'-\omega)$ conformational energy map, which concerns what finally appears to be the two fundamental flexibility axis in polynucleotides. A representative one is given in Figure 14 (Perahia *et al.*, 1973). It corresponds to a (3′–5′) linked diribose monophosphate with both sugars in the C(2′)-endo pucker but represents in fact the whole family of both 3′–5′ and 2′–5′ linked diribose monophosphates independently of the puckers of the sugars. Its outstanding features are the existence of two preferred stable regions corresponding to the right-handed (GG) and left-handed (gg) helices at $\omega = 270°$, $\omega' = 300°$ and $\omega = \omega' = 90°$ respectively. In Figure 14, there is an energy difference between the two energy minima of 2.1 kcal/mole in favor of the GG conformation, which on the other hand is 5.5 kcal/mole more stable that the extended form ($\omega = \omega' = 180°$). The available experimental results, in crystals and in solution, confirm these predictions satisfactorily.

We may compare our results with those obtained by the empirical computations and in this case also by the Extended Hückel Theory. EHT results (Saran and Govil, 1973) predict the existence of *seven equivalent* regions of energy minima located around $(\omega', \omega) = (60°, 60°)$, $(60°, 180°)$ $(180°, 60°)$, $(180°, 180°)$, $(180°, 300°)$, $(300°, 180°)$ and $(300°, 300°)$. This means that although this procedure indicates minima in the actually stable regions it is unselective with respect to these minima and places at the same level regions which are of high energy in the PCILO computations and which, in fact, are unoccupied by experimental conformations. Obviously, the EHT procedure is less precise than PCILO.

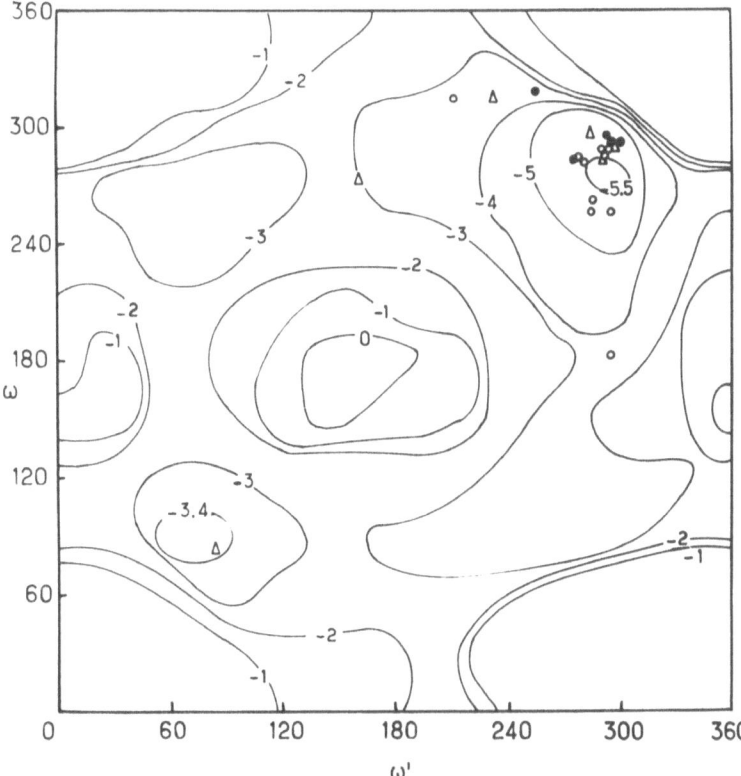

Fig. 14.   A typical (ω–ω') conformational energy map for dinucleoside monophosphates.

The results of the empirical computations (Sasisekharan and Lakshminarayanan, (1969) are very similar to those of the EHT procedure and indicate seven energy minima located in the same regions as those of EHT. Following the details of the computational procedure, some or others of these regions represent global minima: thus, when only nonbonded and torsional interactions are considered, the global minima are located at $(\omega', \omega) = (60°, 60°)$ and $(300°, 300°)$; when electrostatic interactions are included in the computations with the dielectric constant $\varepsilon = 4$, the global minimum shifts to $(180°, 180°)$; when $\varepsilon = 10$, however, the global minima move again to $(60°, 60°)$ and $(300°, 300°)$. These results suffer thus from the existence of too great a number of local minima, some of which in regions of apparently little significance and from the difficulty in choosing unambiguously the preferred ones. Very recently Olson and Flory (1972) devoted a large study to an attempt aimed at refining their empirical computations through the evaluation of the *statistical weights* of the low energy domains of the conformational energy maps. Their effort leads to what can only be considered as very unsatisfactory results. Truly in cases in which the older empirical computations indicated a number of equivalent local energy minima, the calculations of Olson and Flory select the most probably ones. This leads, however, in general to a clear disagreement with the available experimental data. In particular, one may note the occurrence of the highest statistical weights on the ω–ω' (Ψ'–Ψ'' in their notations) map around the 180°–180° values of the corresponding torsion angles (*tt* conformation) for which the conformations are practically never observed.

Very similar results are obtained for the other couples of the torsion angles. Altogether the calculations indicate that the most stable arrangements of two consecutive torsion angles consist generally of the combination of the two individually preferred values with, sometimes, a small deformation. Thus the most stable combinations of (ω, ω') correspond to 300° with 300° and 90° with 90° (large stable zones); of (Φ', ω) to 180°–240° with 270° (large stable zone) followed by 180°–210° with 90°; of (ω, Φ) to 270°–300° with 180°, followed by 60° with 180°; of $(\Phi$–$\Psi)$ to 180° with 60° followed by 180° with 300° or with 180°. The problem of hundreds of possible combinations has been reduced to a limited number of most probable ones. In distinction to the situation observed in proteins the experimental conformations of the backbone of polynucleotides generally correspond closely to the computed energy minima. This is due to the fact that the models used are more directly representative of the polymeric situation than was the dipeptide model in the case of proteins and involve less specific effects.

PCILO computations have by now been completed for the remaining degrees of freedom appearing in mono- or polynucleotide units. Detailed studies have been devoted, in particular, to the orientation of the exocyclic CH_2OH group of the sugar in nucleosides (the problem of the *gg*, *gt* or *tg* conformations) (Saran et al., 1972), to a comparative study of the nucleosides and nucleotides, in connection with the problem of the possible greater conformational rigidity of the latter in comparison with the former (Berthod and Pullman, 1973) and to the puckering of the ribose or deoxyribose rings (Saran et al., 1973).

This last study, carried out within the concept of pseudorotation represents a useful progress beyond the classical division of the furanose ring conformations. The results are summed up in Figure 15. They point to the existence of two stable conformational zones centeres around the C(3')-endo and C(2')-endo conformations which in the isolated furanose ring are separated by barriers of the order of 4 kcal/mole. In nucleosides one of the barriers (the one running through the O(1')-exo–C(2')-exo path) becomes very high. A more schematic way of presenting the results is given in Figure 16 and may be compared to a similar representation deduced by Hruska *et al.* (1973) from studies of vicinal proton–proton couplings in the sugar rings of nucleosides. These authors deduce from their studies that the furanose rings exist in dynamic equilibria involving well-defined states corresponding to the two stable conformational zones, centered around the C(3')-endo and C(2')-endo conformations, which they denote by the symbols τ_+ and τ_-, respectively, separated by two barrier zones, corresponding to the two pathways C(1')-exo –C(4')-exo (which they denote by the symbol

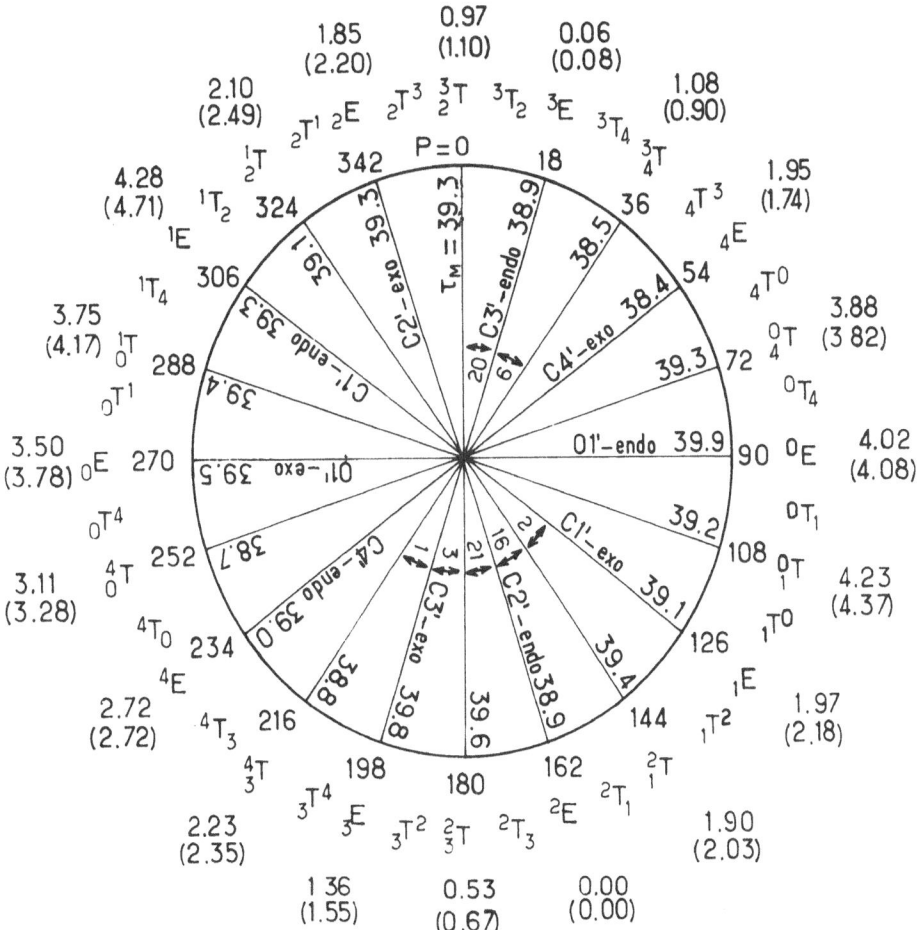

Fig. 15.   The conformational wheel, in pseudorotational representation, for D-ribose and deoxyribose (see Saran *et al.*, 1973, for details).

^{0}B) and C(3')-exo–C(2')-exo (which they denote by the symbol B$_0$). These authors give, however, a symmetrical representation of the overall situation (Figure 16a). An unsymmetrial representation such as given in Figure 16b seems to describe more adequately the real situation in nucleosides.

4. The Lessons of Acetylcholine

Among compounds of pharmacological interest none seems to have been as abun-.dantly investigated from the viewpoint of its conformational properties as acetylcholine (I) and its derivatives, with the understanding that these properties may play

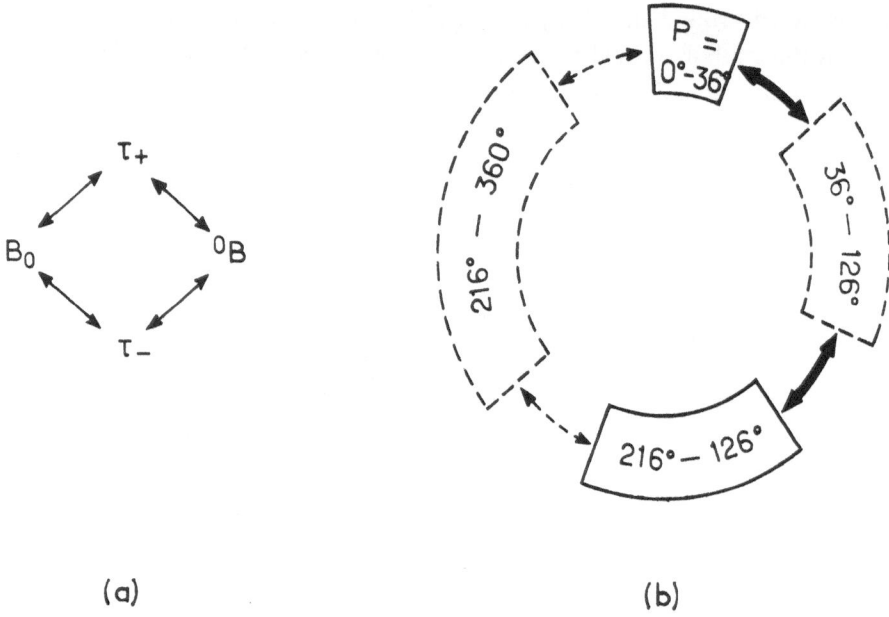

(a)   (b)

Fig. 16.   Classes of states of the pseudorotational itinerary of the sugar ring in nucleosides. (a) following Hruska *et al.* (1973); (b) following PCILO computations.

an essential role in the biological activity of these molecules. These investigations included both theoretical and experimental studies. Theoretically, the conformational problems of acetylcholine itself present the record figure of having been examined by four empirical (i.e. based on partitioned potential functions) computations and nine quantum-mechanical ones: three carried out by the Extended Hückel Theory, one by the PCILO method, one by the INDO method, two by the CNDO/2 method and two by *ab initio* procedures. (For references, general presentation and discussion see Pullman *et al.*, 1971; Pullman and Courrière, 1972, 1973a, b; Port and Pullman, 1973.) To these we may add a dozen theoretical studies on acetylcholine derivatives. Experimentally, the conformation of about 30 acetylcholine derivatives and analogs has been studied by X-ray crystallography and almost as extensive studies have also been made on the conformations of these molecules in solution.

We cannot discuss here the fascinating but too broad problem of the conformational

I. ACETYLCHOLINE

$$\tau_1 = \tau(C_6-O_1-C_5-C_4)$$

$$\tau_2 = \tau(O_1-C_5-C_4-N^+)$$

mysteries and subtleties of acetylcholines in relation to their parasympathomimetic activity. We simply wish to draw attention to a few observations of general interest in conformational studies in biochemistry or pharmacology.

The first observation relates to the importance of geometrical input data used as starting point in computations. Differences appeared between some quantum-mechanical treatments of acetylcholine, which were of importance for the theory of action of this neurotransmitter and it was originally thought that they were due to the differences in the method used in the computations until it was realized – when two CNDO/2 calculations carried out by two different groups of authors gave different results – that they were actually due to differences in the input geometries. Thus, Figure 17 presents the PCILO conformational energy map of acetylcholine, corresponding to the torsion about the two essential flexible bonds τ_1 and τ_2, constructed with the geometry of acetylcholine chloride as input data. The essential feature of this map is the positioning of the global minium at $\tau_1 = 180°$, $\tau_2 = 60°$ which represents a *gauche* conformation of the molecule with respect to the mutual arrangement of the N^+ and esteric O atoms, and corresponds closely to the crystal conformation of that molecule and of a large number of acetylcholine derivatives. The fully extended form ($\tau_1 = \tau_2 = 180°$) represents a local energy minimum 3 kcal/mole above the global one.

Figure 18 represents also a PCILO conformational map of acetylcholine constructed, however, with the geometry of its bromide as input data. This is an unusual geometry for acetylcholine compound, considered by the crystallographers as containing strain and deformation, leading to an unusual X-ray conformation with $\tau_1 = \tau_2 \approx 80°$. It can be seen that the general aspect of this map, especially the contours of stability within the same limit of 6 kcal/mole above the global minimum is extremely similar to that of Figure 17. There are, however, important differences in details, in particular the global energy minimum is displaced and occurs now at $\tau_1 = -90°$,

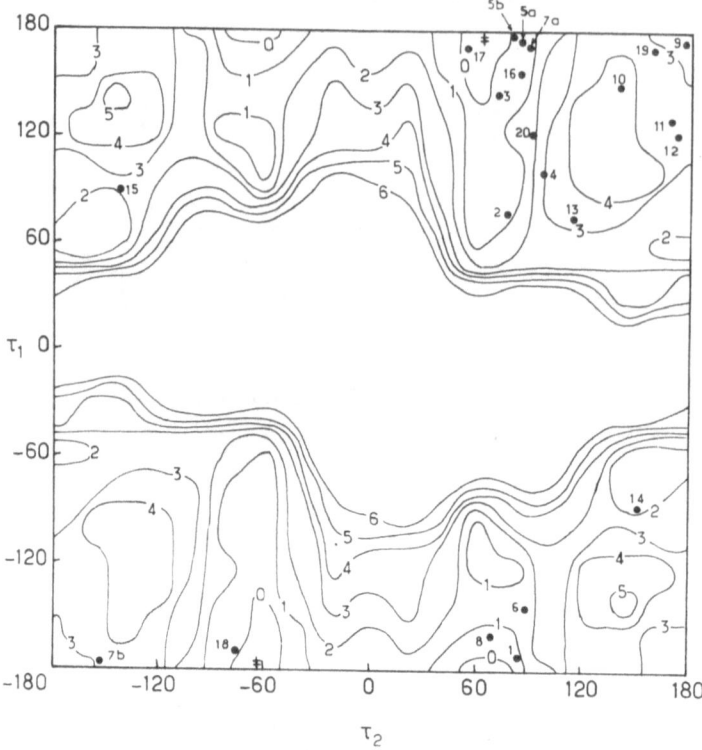

Fig. 17.   Conformational energy map of cetylcholine. Isoenergy curves in kcal/mole with respect to the global energy minimum (≠) taken as energy zero. X-ray geometry of acetylcholine chloride used as input data. ● crystal conformations of acetylcholine derivatives.

$\tau_2 = 60°$ in a region apparently devoid of any significance for acetylcholine derivatives. It does not even correspond to the crystallographic conformation of this compound. Therefore, it is obvious that the geometry of acetylcholine chloride is a better representative of acetylcholine derivatives than that of acetylcholine bromide and is thus susceptible to lead to more significant results.

The second observation concerns the individuality of the compounds studied. We have indicated in Figure 17 the experimental results for a large number of acetylcholine derivatives. All of them are located in the allowed low energy zone, many cluster even around the global energy minimum. At a closer look some of them raise, however, interesting questions. Thus, acetylthiocholine, an isolog of acetylcholine in which the esteric oxygen has been replaced by S (compound 11 in Figure 17), adopts in fact a *trans* conformation in its crystal and is, in Figure 17, in a region of relatively high conformational energy (4 kcal/mole above the global minimum). Now, because the replacement of O by S represents an important structural change, the inclusion of acetylthiocholine in Figure 17 may be questioned. We have therefore constructed a separate conformational energy map for this molecule. The results (Figure 19) are striking. First, the allowed conformational space (within the same limit of 3 kcal/mole

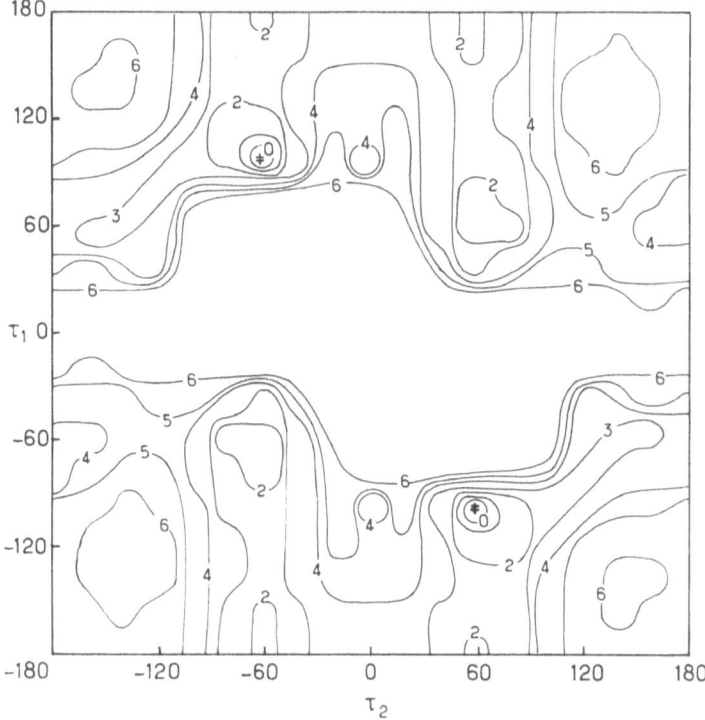

Fig. 18.   Conformational energy map of acetylcholine based on the X-ray geometry of acetylcholine bromide as input data. ╪ global energy minimum.

above the global minimum) has decreased considerably. Second, the region of the energy minimum corresponding to a *gauche* conformation disappeared. Finally, a new global minimum has appeared at $\tau_1 = 60\text{--}80°$ and $\tau_2 = 180°$, corresponding to a *trans* conformation, close to the observed one. The construction of the individual map represented thus an extremely useful development.

Another useful aspect of individualization may be illustrated in the following example. Both acetylcholine and acetylthiocholine conserve their preferred conformations, *gauche* or *trans*, in solution. While it is obvious from Figure 19 that the conformation of acetylthiocholine is expected to be relatively frozen and should not depart from the *trans* one, a larger degree of conformational freedom is conceivable for acetylcholine, in principle, on the basis of Figure 17. Our calculations indicate, however, that the transitions from the *gauche* to the *trans* conformation would involve a barrier height of about 4 kcal/mole. Apparently this barrier is sufficient to prevent the transition.

On the other hand, carbamoylcholine, a derivative of acetylcholine which differs from its parent compound only by the replacement of the methyl group of the acetyl fragment by an amino group, exists in the *trans* conformation in the crystalline state (although some of its derivatives substituted by more complex groups at the quaternary nitrogen exist in *gauche* forms in their crystals) but in the *gauche* conformation, similar

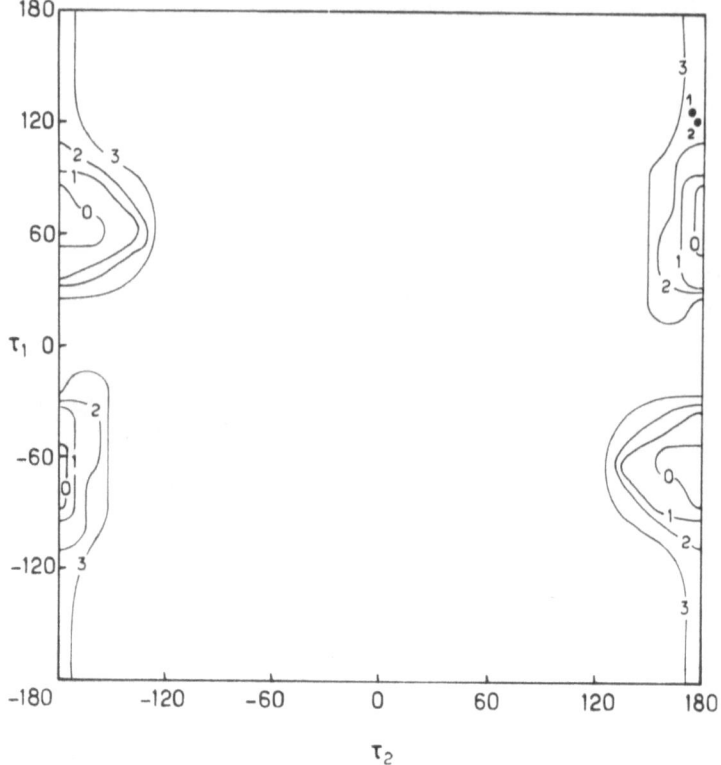

Fig. 19. Conformational energy map of acetylthiocholine.

to that of acetylcholine, in solution. The examination of the PCILO conformational energy map of this compound (the empirical computations failed to indicate any significant difference with respect to the map of acetylcholine) shows rewarding features (Figure 20). Although similar in its general aspect to that of acetylcholine, it shows distinct differences of particular significance for the present discussion. Thus, it contains three nearly equivalent global energy minima of which two represent *trans* forms (at $\tau_1 = 180°$, $\tau_2 = 180°$ and $\tau_1 = 80°$, $\tau_2 = -140°$–$180°$) and one the classical *gauche* form ($\tau_1 = 160°$–$180°$, $\tau_2 = 80°$). Carbamoylcholine itself occupies one of the *trans* minima, that associated with the most extended form. Three other carbamoylcholine derivatives are in conformations which are obviously related to the two remaining energy minima and thus substantiate their significance. Moreover, while the barrier between the *gauche* and *trans* forms was of 4 kcal/mole in acetylcholine it is only of 1 kcal/mole in carbamoylcholine a situation which may obviously be related to the prevention of a *trans-gauche* transition in solution in the former and its occurrence in the latter.

   It is thus apparent that a real understanding of the conformational problems of acetylcholine derivatives needs the construction of the conformational energy map for at least a large number of them.

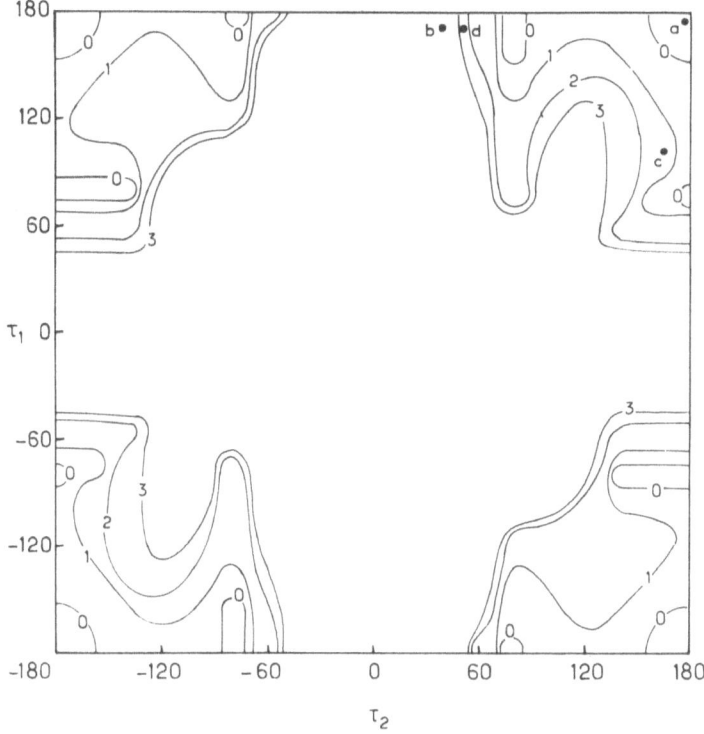

Fig. 20. Conformational energy map of carbamoylcholine.

This last example brings me to the last problem I would like to mention here. It is the problem of the relation of calculations such as those described here and which refer to isolated molecules to conformations observed experimentally, it means, in general, in crystals or in solution and to conformations possibly involved in biological phenomena. This is, of course, a general problem in quantum chemistry and biochemistry, not specific to conformations, the great majority of quantum-mechanical computations being carried out for isolated molecules. It will be discussed in our 5th symposium, and I cannot go into details here. I wish, however, to communicate to you the conclusion reached as a result of a large practice, one could say 'experimentation', in the field. It represents a practical, empirical conclusion which can be substantiated by a large number of explicit examples. I have presented it recently on the occasion of a large discussion on pharmacological molecules (Pullman and Courrière, 1973) and will quote myself from there:

... Although, as indicated at the beginning of this lecture, there was no *a priori* certitude that calculations carried out for isolated molecules should or could be representative or even related to their conformations in crystals or in solution, the results presented here indicate a frequent, if not constant close relationship between the two. If we put aside the case of cationic serotonin and histamine, which we intend to investigate further, the observed conformations of all the molecules reported here fall within the low energy, stable regions (3 kcal/mole above the global minimum) predicted by the theoretical (PCILO) computations. Moreover, each time that a definite and relatively distinct, stable global energy minimum is present the observed crystalline conformations either correspond exclusively to this minimum (nicotinamides, barbiturates, riboflavins, iproniazids, phenothiazines) or at least cluster in their majority around that minimum (acetylcholine derivatives). In some cases other energetically somewhat higher local minima are also populated (nicotine, acetylcholine). In cases in which there exist a few practically equivalent global energy minima, the crystal structure may indicate a preference for one of them (phenethylamines). In solution, however, the different forms co-exist, substantiating thus the significance of the multiple energy minima calculated ....

... Referring to a variety of different molecules, the regularities which these data show may reasonably be considered as having a significance. They point to a relation between the conformational possibilities and preferences computed for the isolated molecules and the structures observed in crystals or solutions. Broadly speaking it looks as if the role of the environmental forces consisted rather of stabilizing some of the computed stable conformations than of creating new ones. Frequently it is the originally most stable conformation(s) which wins the game, but it must be acknowledged that this is not always the case. Further work is therefore needed in order to be able to describe the situation with more precision ...

I would like to add that such work is progressing in our own laboratory, as well as elsewhere, of course.

The practical significance of the calculations, as carried out presently is, however, sufficiently established that, when we come across a strong disagreement between the results of computations and, say, the conformations observed in the crystal, our first thought is not necessarily about the possible role of environmental effects but about possible errors in computation or in the input data. In a number of cases it is these possibilities including the last one which appear to be true.

You have already seen in Figures 8 and 9 how the improvement in the analysis of the X-ray data on proteins improves the agreement between theory and experiment. Recently in our laboratory we came across a significant other example. We were trying to study the conformation of penicillins, in particular the conformation of the side chain of penicillin G (benzyl penicillin), II, with respect to the β-lactam-thiazoli-

II. PENICILLIN G

dine ring system. The conformational energy map constructed with the geometry available in the literature as input data (Growfoot et al., 1949) is given in Figure 21. We found it strangely unsatisfactory, the observed conformation lying outside the predicted stability zone, in a relatively high energy region. We have then inquired with Mrs Dorothy Hodgkin whether these data, which were relatively old, were satisfactory. In her answer she drew our attention to a refinement of these data (Pitt, 1952), adding, however, that even these refined data are still not completely satisfactory. The conformational energy map built with the refined input data is shown in Figure 22. It is significantly different from the preceding one and, at the same time, shows a much more satisfactory agreement between theory and experiment.

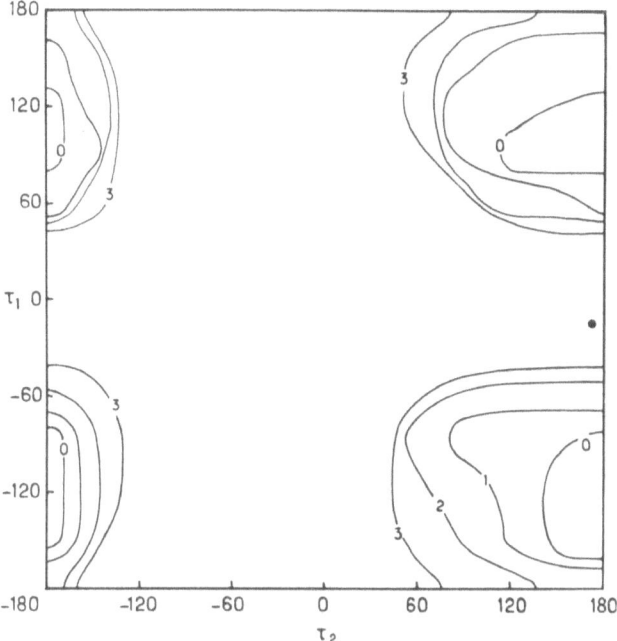

Fig. 21.   Conformational energy map of penicillin G (primitive geometry input data).

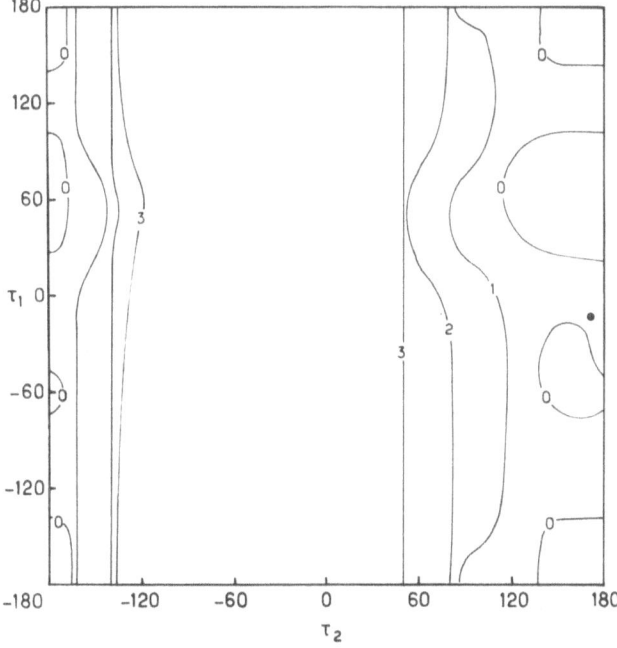

Fig. 22.   Conformational energy map of penicillin G (refined geometry input data).

In concluding, I feel thus justified to say that we seem to have today in quantum bio-chemistry and chemistry powerful methods for the study of conformational problems, which may be considered as quite competitive with, or if you prefer complementary to, experimental techniques, at least to those which work in solution and which frequently do not yield unambiguous results easily. This concerns both the preferred conforma-tions and the barriers between them. Although the possible effect of the environmental forces has always to be kept in mind and, still better, to be investigated, the results obtained by the theoretical methods seem of definite practical significance, indicating thereby that the molecular skeleton embodies generally information about its confor-mational destiny.

Acknowledgements

A number of workers from our laboratory contributed to the studies described here. The PCILO method is due to a joint effort by Diner, Malrieu, Daudey and Claverie with a contribution by Gilbert, Jordan and Pincelli. The studies on the conformation of amino acid residues of proteins have been carried out in collaboration with Maigret and Perahia, those on the conformation of polynucleotides and their constituents in collaboration with Mme Berthod, Perahia and Saran. To these may be added studies on the conformational properties of the peptide, ester and hydrogen bonds carried out by Mme Pullman in collaboration with Mme Berthod, Dreyfus and Perricaudet. The conformational properties of pharmacological molecules have been studied in collaboration with Courrière and Coubeils and more recently by Mme Pullman and Port. Finally, a series of studies on the theoretical analysis of RMN data in relation to conformational problems in biochemistry was carried out in collaboration with Mme Giessner-Prettre. We wish to thank all of them for their contribution.

This work was sponsored by R.C.P. No. 173 and A.T.P. No. A655-2303 of the C.N.R.S.

References

Alden, R. A., Birktoft, J. J., Kraut, J., Robertus, J. D., and Wright, C. J.: 1971, *Biochem. Biophys. Res. Comm.* **45**, 337.

Avignon, M. and Lascombe, J.: 1973, in *Conformation of Biological Molecules and Polymers, Proc. Fifth Jerusalem Symp.* (ed. by E. D. Bergmann and B. Pullman), Academic Press, New York, p. 97.

Berthod, H. and Pullman, B.: 1973, *FEBS Letters* **30**, 23.

Crowfoot, D., Dunn, L. W., Rogers-Low, B. W., and Turner-Jones, A.: 1949, *The Chemistry of Penicillin*, Princeton University Press.

Cung, M. T., Marraud, M., and Neel, J.: 1973, in *Conformations of Biological Molecules and Poly-mers, Proc. Fifth Jerusalem Symp.* (ed. by E. D. Bergmann and B. Pullman), Academic Press, New York, p. 69.

Diner, S., Malrieu, J. P., Jordan, F., and Gilbert, M.: 1969, *Theor. Chim. Acta* **15**, 100.

Edsall, J. T., Flory, P. J., Kendrew, J. C., Liquori, A. M., Nemethy, G., Ramachandran, G. N., and Scheraga, H. A.: 1966, *Biopolymers* **4**, 121.

Hruska, F. E., Wood, D. J., and Dalton, J. G.: 1973, *J. Am. Chem. Soc.*, in press.

Imoto, T., Johnson, L. N., North, A. C. T., Phillips, D. C., and Rupley, J. A.: 1972, in *The Enzymes* (ed. by P. D. Boyer), 3rd ed., Academic Press, New York, p. 665.

Maigret, B., Pullman, B., and Dreyfus, M.: 1970, *J. Theor. Biol.* **26**, 321.

Néel, J.: 1972, *Pure Appl. Chem.* **31**, 201.

Olson, W. R. and Flory, P. J.: 1972, *Biopolymers* **11**, 25.
Perahia, D., Saran, A., and Pullman, B.: 1973, in *Conformation of Biological Molecules and Polymers, Proc. Fifth Jerusalem Symp.* (ed. by E. D. Bergmann and B. Pullman), Academic Press, New York, p. 225.
Pitt, G. J.: 1952, *Acta Crystallog.* **5**, 770.
Pople, J. A. and Segal, G. A.: 1965, *J. Chem. Phys.* **43**, S136.
Pople, J. A. and Segal, G. A.: 1966, *J. Chem. Phys.* **44**, 3289.
Popov, E. M., Lipkind, G. M., Arkhipova, S. F., and Dashevskii, U. G.: 1968, *Mol. Biol.* **2**, 498.
Port, J. and Pullman, A.: 1973, *J. Am. Chem. Soc.* **95**, 4059.
Pullman, B.: 1971, in *Aspects de la Chimie Quantique Contemporaine* (ed. by R. Daudel and A. Pullman), CNRS, Paris, p. 261.
Pullman, B. and Berthod, H.: 1973, in *Conformation of Biological Molecules and Polymers, Proc. Fifth Jerusalem Symp.* (ed. by E. D. Bergmann and A. Pullman), Academic Press, New York, p. 209.
Pullman, B. and Courrière, Ph.: 1972, *Mol. Pharmacol.* **8**, 612.
Pullman, B. and Courrière, Ph.: 1973a, *Theor. Chim. Acta* **31**, 19.
Pullman and Courrière, Ph.: 1973b, in *Conformation of Biological Polymers and Molecules, Proc. Fifth Jerusalem Symp.* (ed. by E. D. Bergmann and B. Pullman), Academic Press, New York, p. 547.
Pullman, B. and Maigret, B.: 1973, in *Conformation of Biological Molecules and Polymers, Proc. Fifth Jerusalem Symp.* (ed. by E. D. Bergmann and B. Pullman), Academic Press, New York, p. 13.
Pullman, B. and Pullman, A.: 1963, *Quantum Biochemistry*, Wiley-Interscience, New York.
Pullman, B. and Pullman, A.: 1973, *Adv. Protein Res.* (in press).
Pullman, B., Courrière, Ph., and Coubeils, J. L.: 1971, *Mol. Pharmacol.* **7**, 391.
Pullman, B., Perahia, D., and Saran, A.: 1972, *Biochim. Biophys. Acta* **269**, 1.
Ramachandran, G. N. and Sasisekharan, V.: 1969, *Adv. Protein Chem.* **23**, 283.
Saran, A. and Govil, G.: 1971, *J. Theor. Biol.* **33**, 407.
Saran, A., Perahia, D., and Pullman, B.: 1973, *Theor. Chim. Acta* **30**, 31.
Saran, A., Pullman, B., and Perahia, D., 1972, *Biochim. Biophys. Acta* **287**, 211.
Sasisekharan, V. and Lakshminarayanan, A. V.: 1969, *Biopolymers* **8**, 505.
Scheraga, H. A.: 1968, *Adv. Phys. Org. Chem.* **6**, 103.
Sundaralingam, M.: 1969, *Biopolymers* **7**, 821.
Sundaralingam, M.: 1973, in *Conformation of Biological Molecules and Polymers, Proc. Fifth Jerusalem Symp.* (ed. by E. D. Bergmann and B. Pullman), Academic Press, New York, p. 417.

PART III

GENERAL THEORY OF CHEMICAL REACTIVITY

Chairman: Raymond Daudel

During the opening ceremony, from left to right, in the first row: Recteur Mayer, Deputy Mayor of Menton, Professor Raymond Daudel, Senator Francis Palméro, Mayor of Menton, Professor Bernard Pullman, Mme. Alberte Pullman and Professor Robert Mulliken.

INTRODUCTION TO SYMPOSIUM III:

GENERAL THEORY OF CHEMICAL REACTIVITY

RAYMOND DAUDEL

Sorbonne et Centre de Mécanique Ondulatoire Appliquée du C.N.R.S., Paris

1. Fundamental Problems

Quantum theory of chemical reactivity is too wide a field to be discussed during one morning only. It will be necessary to focus the discussion on certain points.

Our two distinguished lectures, Martin Karplus and Kenichi Fukui will draw attention to some interesting problems. The purpose of my talk is also to mention briefly some topics which could be starting points for lively discussions.

One of the fundamental problems which seems interesting to me is related to the Eyring transition state theory. This problem is based on the *assumption* that, during collision, molecules form transition states $M^{\neq}$ which would have a sufficiently long life to be in thermodynamical equilibrium with the intial molecules:

$$A + B \rightleftharpoons M^{\neq} \rightarrow C + D.$$

From such an hypothesis it is possible to derive a formula like:

$$k = \frac{\chi T}{h} \frac{f_{M^{\neq}}}{f_A f_B} e^{-\Delta \varepsilon^{\neq}/\chi^T}$$

giving an expression of the rate constant k as a function of the *potential barrier* $\Delta \varepsilon^{\neq}$ and of the various partition functions associated with the distribution of the transition state $M^{\neq}$ and the molecules A and B on the various energy levels, for the temperature T.

Thanks to this formula it has been possible to rationalize many experimental facts, mainly when relative rates are concerned for a homogeneous family of reactions. In that case it turns out very often that the ratio of the partition functions remains approximately a constant. *This is why the potential barrier is 'staring' in quantum chemistry.*

Recently Karplus *et al.* [1] have carefully discussed that hypothesis. They studied the exchange reaction:

$$H_2 + D \rightleftharpoons HD + H$$

using a quasiclassical procedure. The electronic energy of the intermediate complex HDH is calculated in the Born–Oppenheimer approximation as a function of the internuclear distances R_1, R_2 and R_3 (Figure 1) by solving approximately the corresponding Schrödinger equation. The resulting function is introduced in a classical Hamiltonian and, by solving the corresponding Hamiltonian equation, it is possible to obtain the nuclear trajectories for various initial conditions. A Monte Carlo procedure is used to calculate the probability of reaction as a function of initial conditions

R. Daudel and B. Pullman (eds.), The World of Quantum Chemistry, 93–100. All Rights Reserved

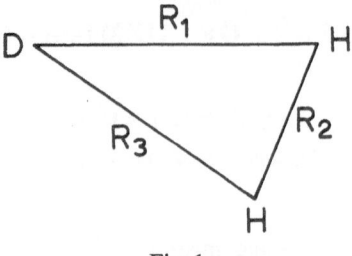

Fig. 1.

and therefore the total cross section from which it is easy to derive the rate constant. The result obtained (which is completely *ab initio*) is in perfect agreement with the molecular beams experiment.

Figure 2 gives an idea of typical trajectories for initial conditions leading to reaction. *It is seen that there is no evidence for a long-lived three-particule system.* During the first part of the process the molecule H_2 vibrates and the atom D is approaching. Suddenly the atom D forms a DH molecule and a hydrogen atom moves freely. The reaction time is of the order of 10^{-14} s. It appears that for this reaction no transition state is formed which could be in thermodynamical equilibrium with the initial molecules: the phenomenon is too short. The *Eyring theory* and the concept of potential barrier tumble down.

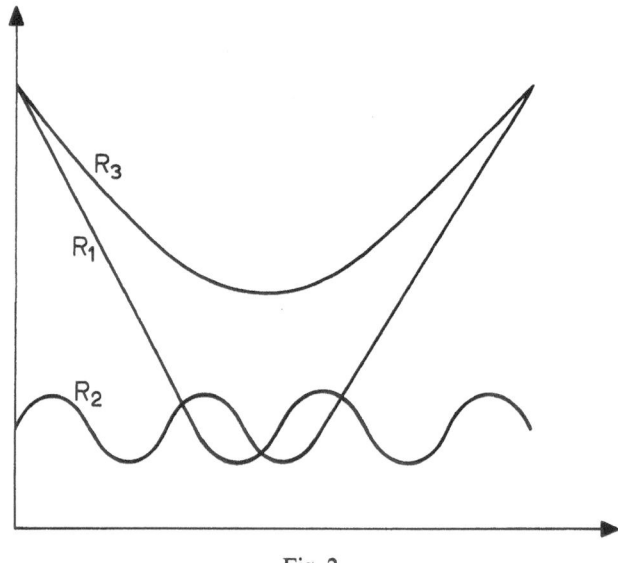

Fig. 2.

But we know that the Eyring theory is the only basis of the quantum theory of chemical reactivity for large molecules, as for that kind of molecules collision theory leads to untractable equations. Therefore, a serious theoretical difficulty arises which could be the origin of interesting discussions.

Let us add that a way seems to be open to avoid the difficulty: to build a theory, as handy as Eyring theory but in which the existence of a long-lived collision complex is not required.

That possiblity has been investigated recently by Christov [2] He started from another quantum-mechanical formulation of Eyring *et al.* [3]. In the framework of this problematic statistical mechanics is used to calculate the distribution of the reactants among the various energy levels and for each of them it is possible to calculate the probability of a reaction through or over the potential barrier. Finally the rate constant is written as:

$$k = \frac{\chi T}{h} K_t \eta \frac{\tilde{f}_{A,B}}{f_A f_B} e^{-\Delta \varepsilon^{\neq}/\chi T},\tag{1}$$

where K_t refers to a tunnel effect, $\hat{f}_{AB,B}$ is a partition function describing the distribution of the reactants among the various degrees of freedom *except for* those which are related with the reaction coordinate. Furthermore, $\Delta \varepsilon^{\neq}$ is the height of the saddle point. Therefore, this approach would be a *rehabilitation of the potential barrier.*

2. Potential Energy Surfaces; Reaction Paths

Finally two different approaches to quantum theory of chemical reactivity appear: collision theory and formula (1). Whatever procedure is used the knowledge of the electron energy as a function of internuclear distances seems to be needed. The collision theory starts from that function (called potential energy surface) to determine the nuclear trajectories. Equation (1) requires the calculation of the potential barrier, that is to say the position of a saddle point. Furthermore, the tunnel effect factor K_t directly depends on the shape of the potential surface.

They are three main procedures of computing potential energy surfaces [4]:

(a) large configuration interaction;

(b) both Hartree–Fock calculation and correlation energy approximated as a sum of Bethe–Goltstone pair energies;

(c) suitably chosen multiconfiguration calculation.

The calculation of potential energy surface is highly computer-time consuming. Surfaces have been computed only for very small reactants. For larger molecules it is impossible, but we can restrict that calculation to the bottom of potential valleys marking out the reaction path. Such a procedure, conveniently used, provides information about the structure of the intermediate complex at the saddle point, and

therefore permits the calculation of the potential barrier. Dedieu and Veillard [5] have recently calculated the energy of the collision complex which is responsible for the reaction

$$F_{(2)}^- + CH_3F_{(1)} \rightarrow CH_3F_{(2)} + F_{(1)}^-$$

by starting from an SCF LCAO MO wave function built on a Gaussian basis. Figure 3 shows the nuclear conformations for which the calculations have been performed.

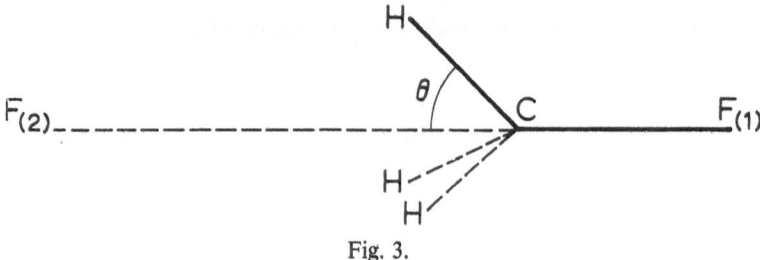

Fig. 3.

Table I contains the energy values for various conformations and for a certain basis of Gaussian functions denoted by IV in the paper referred to. These conformations correspond to the most stable ones for a given value of the angle θ, the CH internuclear distance being fixed to the normal CC bond length.

TABLE I

$R_{CF_{(2)}}$ (Å)	∞	3.70	2.11	1.88	1.5	1.44	1.42
$R_{CF_{(1)}}$ (Å)	1.42	1.44	1.5	1.88	2.11	3.70	∞
θ	107°	108°	104°	90°	76°	72°	63°
Energy (AU)	−238.508	−238.515	−238.507	−238.491	−238.507	−238.515	−238.508

It is seen that the top of the potential barrier (or saddle point) corresponds to a conformation possessing a symmetry plane ($\theta=90°$) as assumed by many chemists. The height of the potential barrier is about 6 kcal/mole ($−238.491-(−238.508)$) A.U.

Dedieu and Veillard have shown that the calculated value for that potential barrier depends on the size of the Gaussian basis: only large bases are convenient. Veillard [6] has found that a large CI calculation does not change significantly the value of the potential barrier.

In other cases the role of the configuration interaction can be very important. Peyerimhoff [7] has recently studied the potential energy surface for the cyclization of butadiene. She observed that all cis-butadiene is correctly predicted by CI to be more stable than cyclobutene by 0.34 eV (exp. 0.40 eV) in contrast to the SCF calculation alone which finds butadiene less stable by 0.2 eV.

3. Dynamic Indices

Potential barriers can be considered to be a dynamical index of chemical reactivity as it depends on the reaction path that is to say to something which is related to the dynamics of the reaction. The estimation of dynamic indices remains a central problem in the quantum theory of chemical reactivity. In principle the knowledge of the reaction path is needed as we must know the energy of the saddle point.

But in fact any procedure is convenient which permits the structure of the corresponding state to be obtained. We shall continue to call it transition state even if it is not a long-lived state in thermodynamical equilibrium with the initial molecules.

Various authors starting from chemical intuition suggested possible structures for the transition states of some particular reactions. The Wheland model (I) for substitution reactions on conjugated molecules is well known. An alternative model is the model with extension of the delocalized system (II).

(I)   (II)

The contributions of the delocalized systems to the potential barrier are respectively called localization and delocalization energies. They are dynamic indices. The most practical procedure to test the models is to compare the theoretical relative rates which correspond to them with the experimental values. It appeared that model (II) is better than model (I) specially because it permits us to predict the dependence of the site of reaction on the nature of the reagent and also because contrary to model (I) it leads to good results for many photochemical reactions [8–11].

Some principles can help to predict the structure of transition states Eyrenson [12] recently investigated the use of the least motion principle: "A reaction path is such to produce the less change in nuclear positions and in electronic configurations".

The calculation of the energy of various possible models can also help to select the most convenient one.

A discussion about simple procedures giving useful information about transition states would be welcome.

4. Static Indices

Establishing relationships between electronic structures of molecules and chemical reactivity is always a central problem in quantum chemistry. A priori, this purpose seems to be difficult to achieve because during the collision of two molecules there is a complete reorganization of the electronic system and it is not obvious that the height of the potential barrier is simply related to the organization of the electrons in the isolated molecules. However, for special families of compounds and particular reactions, it has been possible to establish such a relationship. The oldest one is perhaps the decreasing relations observed by Daudel et al. [13] between the free valence number of a carbon atom belonging to an alternant hydrocarbon and the Wheland localization energy for a substitution reaction.

Many others have been discovered and it became customary to call static indices quantities as atomic charges, free valence, bond orders, polarizabilities, Fukui frontier electronic charges and superdelocalizabilities which are associated with a particular region of a molecule and related to the potential barrier of a reaction taking place on that region.

These quantities are very interesting, but not completely satisfactory. The regions of the molecule with which they are associated are not well known. For example, the frontiers of an atom in a molecule are not usually precised, furthermore atomic orbitals overlap, therefore we do not know where lies an atomic charge. The classical static indices are only defined for certain kinds of approximate wave functions. The classical static indices do not take account of the nature of the reagent.

An attempt has recently been made [14] to propose static indices:

(a) which are factors in the expression of the interaction energy of the colliding molecules;

(b) defined for any kind of wave function;

(c) associated with very precise regions of each molecule;

(d) depending on both reacting molecules.

The starting point to define such new indices is the fact that any expectation value associated with a bielectronic operator for a given molecule can be expressed as a sum of contributions of each loge and of contributions of each loge pair. These indices allow us to understand what parts of a molecule play the most important role in a given reaction and what are for that reaction the most significant electronic events in these regions. From this viewpoint it appears that two identical electronic charges lying in analogous loges may possess complete different behaviours during the course of a chemical reaction. This would be the case, for example, of two electronic charges equal to say one electron: the first corresponding to a probability unity of finding one electron and one only in the loge (one electronic event), the second corresponding to a probability of one third of finding one electron and one only in the loge and to a probability of one third of finding in it a pair of electrons of opposite spins (two electronic events).

Other kind of static indices have been proposed recently by using a functional partitioning [15–17].

Static indices generated by a space partitioning are unhappily rather difficult to compute. Static indices generated by a functional partitioning do not possess a clear physical meaning. Research is necessary to make more tractable the calculation of static indices generated by space partitionings and to give physical meaning to those generated by functional partitionings.

Finally a new kind of static index has been introduced by Scrocco *et al.* [17]. They simply draw the electric potential maps created in the neighbourhood of a molecule by its nuclei and electronic cloud. When a molecule possesses lone pairs the potential map shows electrostatic potential minima in their neighbourhood. A relationship has been found between these minima and the corresponding SCF values for the protonation energy. A. Pullman [18] has shown that the origin of that relation lies probably in the fact that the polarization effect acts so as to reinforce the tendency set by the electrostatic potential and that charge transfer and geometrical deformation are generally not strong enough to reverse it. This is why these minima can be considered to be static indices but contrary to the previous ones they are not really created by a particular region of the molecule, they are generated by the all electronic density. This is perhaps the reason why they are more able to predict the acido-basic strength of heterocycles than localized charges as shown by Bonaccorsi *et al.* [19].

The isopotential curves take account of the repulsive role of the peripheral hydrogen atoms attached to the molecule during the approach of a proton. They make it possible to compare protonation sites of different kinds (oxygen with nitrogen, pyridine-like nitrogen with amino nitrogens, etc.). Let us add that in the case of protonation the static indices based on the loge theory permits us to calculate the contribution of each loge to the electrostatic potential minima.

A general discussion on static indices could be a great help to proceed farther.

5. Solvent Effect

It would be difficult to conclude this brief introduction without mentioning the solvent effect.

Let us recall that Dedieu and Veillard found a potential barrier of 6 kcal/mole for the exchange reaction:

$$F_{(1)}^- + CH_3F_{(2)} \rightarrow CH_3F_{(1)} + F_{(2)}^-.$$

The activation energy in acetone solution is 16 kcal/mole for:

$$Br_{(1)}^- + CH_3Br_{(2)} \rightarrow CH_3Br_{(1)} + Br_{(2)}^-,$$

and 20 kcal/mole for

$$Cl_{(1)}^- + CH_3Cl_{(2)} \rightarrow CH_3Cl_{(1)} + Cl_{(2)}^-.$$

It is unknown in the case of the fluorine compounds but by extrapolating it could be estimated at 25 kcal/mole.

It is well known that potential barrier and activation energy are never identical, but they cannot be as different as 6 and 25. The solvent is certainly responsible for the greater part of that difference.

The solvent effect has recently been calculated for the protonation of the FH molecule. It has been observed that the potential barrier is exclusively produced by the solvation effect [20].

The importance of the solvent effect on the potential barrier suggests that when a potential surface shows two possible intermediate states the preferred pathway of the reaction, and therefore its mechanism, can depend on the nature of the solvent.

A discussion on this important point would be interesting, but perhaps it is better to keep that topic for the symposium devoted to the environment effect.

References

[1] Karplus, M., Porter, R. N., and Sharma, R. D.: 1965, *J. Chem. Phys.* **43**, 3259.
[2] Christov, S. G.: 1972, *Berichte Bunsen-Gesellschaft für Physikalische Chemie* **76**, 507.
[3] Eyring, H., Walter, J., and Kimbal, G. E.: 1946, *Quantum Chemistry*, Wiley.
[4] For details see 'Potential Energy Surface in Chemistry', IBM Research Laboratory (1970).
[5] Dedieu, A. and Veillard, A.: *J. Am. Chem. Soc.*, in press.
[6] Veillard, A.: 1973, in Sixth Jerusalem Symposium on Chemical and Biochemical Reactivity.
[7] Peyerimhoff, S.: 1973, in Sixth Jerusalem Symposium on Chemical and Biochemical Reactivity.
[8] Chalvet, O., Daudel, R., Ponce, C., and Rigaudy, J.: 1968, *Int. J. Quant. Chem.* **II**, 521.
[9] Bertran, J., Chalvet, O., Daudel, R., McKillop, T. F. W., and Schmid, G. H.: 1970, *Tetrahedron* **26**, 339.
[10] Chalvet, O., Daudel, R., and McKillop, T. F. W.: 1970, *Tetrahedron* **26**, 349.
[11] Chalvet, O., Daudel, R., and Schmid, G. H.: 1970, *Tetrahedron* **26**, 365.
[12] Eyrenson, S.: 1973, in Sixth Jerusalem Symposium on Chemical and Biochemical Reactivity.
[13] Daudel, R., Sandorfy, C., Vroelant, C., Yvan, P., and Chalvet, O.: 1950, *Bull. Soc. Chim. France* **17**, 16.
[14] Daudel, R., Chalvet, O., Constanciel, R., and Esnault, L.: 1973, in Sixth Jerusalem Symposium on Chemical and Biochemical Reactivity.
[15] Coulson, C. A., and MacLagan, R. G. A. R.: 1971, *Adv. Chem. Phys.* **21**, 303.
[16] Constanciel, R.: 1972, *Chem. Phys. Letters* **16**, 432.
[17] Scrocco, E., Bonaccorsi, R., and Tomasi, J.: 1970, *J. Chem. Phys.* **52**, 5270.
[18] Pullman, A.: 1973, in Sixth Jerusalem Symposium on Chemical and Biochemical Reactivity.
[19] Bonaccorsi, R., Pullman, A., Scrocco, E., and Tomasi, J.: 1972, *Theor. Chim. Acta* **24**, 51.
[20] Leibovivi, C.: *Int. J. Quant. Chem.*, in press.

QUANTUM MECHANICS OF SIMPLE CHEMICAL REACTIONS

MARTIN KARPLUS

Department of Chemistry, Harvard University, Cambridge, Mass., U.S.A.

The exact calculation of cross-sections for the collisions leading to chemical reactions requires, in general, the solution of the Schrödinger equation for all of the particles, electrons and nuclei, involved. Fortunately, this impossibly difficult problem can be simplified for most reactions at the energies of interest for chemistry by means of the Born–Oppenheimer approximation. Thus, the scattering calculation can, just as in the more commonly studied bound-state case, be divided into two parts – the first, that of determining the potential surface or surfaces involved in the reaction and the second, that of the motion on the surface of the atoms treated as point particles (with spin, if necessary).

In my lecture, I shall concentrate on the second aspect of the problem, namely that of the motion of the atoms, given the potential surface. Although the determination of the surface is clearly a quantum-mechanical problem – that of solving the electronic Schrödinger equation for fixed positions of the nuclei – the motion of the atoms on the surface is at the borderline of quantum mechanics; that is, the de Broglie wavelengths $(\lambda = h/p)$ are sufficiently short and the variation in the potential with distance is sufficiently slow that it reasonable to expect a classical calculation to provide a good approximation to the quantum-mechanical motion. The question that needs an answer is "How good an approximation?" and that is the question I shall consider in this lecture. The reason it is an important question is that classical calculations of reaction cross sections by trajectory techniques are now relatively easy for simple systems [1], while quantum calculations are still extremely difficult. For a reaction involving molecules with a total of six or so atoms, classical studies can be done with no approximations other than those of the potential surface used to represent the interactions between the atoms. Thus, if the classical limit were adequate, the problem of understanding chemical reactions would be back again at the door of the quantum chemist. Once he provided an adequate surface, people doing trajectory calculations would take care of the rest. By an 'adequate surface', I mean one that is sufficiently accurate and also sufficiently simple so that the forces on the atoms can be calculated without using excessive computer time. Of course, the required accuracy depends on the problem under consideration. If one is interested in distinguishing processes with large differences in the energy barriers (e.g., as in an examination of the Woodword–Hoffman rules) a rather crude surface can be adequate; on the other hand, for the accurate evaluation of rate constants in the low energy region, an error of ± 0.5 kcal or better may be needed. An additional point worth emphasizing, particularly since this is a quantum chemistry congress, is that there is still a practical difficulty in going from a surface known at a series of points to a formulation which permits one to calculate the forces (i.e., the derivatives of the energy) as needed for following a classical trajectory.

R. Daudel and B. Pullman (eds.), The World of Quantum Chemistry, 101–111. All Rights Reserved
Copyright © 1974 by D. Reidel Publishing Company, Dordrecht-Holland

For three atoms it is possible to find an analytic function to fit the calculated points and to use the function for evaluating the derivatives. However, if there are more than three atoms, it becomes very difficult to make the required multidimensional analytic fit. Studies on how best to do this might well be made as a supplementary investigation by those in the audience concerned with calculating potential surfaces.

As an alternative to such fits of potential surfaces, we have been looking at the possibility of calculating directly the forces on the atoms along the trajectories. Dr. Iris Wang has recently completed a study [2] of the insertion reaction

$$H_2 + (singlet) CH_2 \rightarrow CH_4 .$$

For this investigation the CNDO approximation was used because it provides a way of rapidly evaluating the forces at each step; that is, one calculates the energy for a given configuration, differentiates it analytically to obtain the forces, follows the trajectory to the next point, and so on. In this procedure, there is no need to fit a series of calculated points and the trajectories can be calculated without any constraints on the particles in the full 15-dimensional space. A related approach has been used by Dr. Arieh Warshel in looking at *cis-trans* photoisomerization in π-electron systems (e.g., 2-butene) [3]. Here again the surface has a semiempirical form that permits rapid evaluation of the energy derivatives. Of course, there is the very interesting additional problem that surface crossing can occur and this is treated by a semiclassical procedure involving an extension of the Landau–Zener approximation. In Professor Salem's group, Chapuisat and Jean [4] have been using a somewhat different approach to calculate trajectories on an *ab initio* surface. For the thermal isomerization of cyclopropane, they fit the surface locally by the use of spline functions. It appears as if this procedure will permit the calculation of accurate trajectories, though constraints on the motion have had to be introduced to somewhat simplify the difficult multidimensional problem posed by cyclopropane.

Returning to the question of whether it is necessary to introduce quantum corrections to the classical results (i.e., how large they are for different reactions and for what properties of the reaction they are more or less important), we can imagine proceeding in two ways – by comparing classical results with experiments or by comparing classical results with quantum calculations. There have been a variety of experimental comparisons in recent years – primary with crossed molecular beam scattering measurements – that suggest that classical calculations provide an adequate semiquantitative picture of reaction attributes such as the angular distribution of products and the reaction energy partitioning among the degrees of freedom of the products. However, quantitative tests of the rate constants obtained by classical trajectory techniques are much more difficult because they require accurate potential surfaces for reactions with accurately known rate constants. Only if these conditions are met is it possible to assign deviations between the two sets of results – classical trajectory and experimental – to quantum corrections. At the present time the only candidate for such a comparison is the $H + H_2$ exchange reaction. However, even for this simple system the best complete potential surface yields a barrier that is signifi-

cantly too high [5]. A more recent surface calculation [6], which may be of the required accuracy, is restricted to the linear H–H–H geometry and, therefore, not sufficient for a three-dimensional trajectory calculation that can be compared with experiment. As to measurements for $H + H_2$, there are now good results for the gas phase
reaction over a reasonable temperature range, Also for isotopic analogues, such as
$D + H_2$, there are data available. The latter is a reaction that has been studied by a
number of workers [7] who suggested that the curvature in the standard $\log K$ vs $1/T$
plot corresponded to a significant quantum correction ('tunneling') at low temperature.
However, a recent redetermination [8] of the low temperature rate constants demonstrates that the curvature is much less than it appeared to be in the earlier studies. This
is not to say that quantum corrections are never important – at a sufficiently low temperature the rate constant for a reaction with a barrier as in $H + H_2$ must be dominantly due to tunneling – only that it has been difficult to conclusively demonstrate directly from experimental results that quantum corrections to the reaction rate are important in the region where it is measurable with high accuracy. Isotope effect comparisons for series of reactions do suggest that quantum corrections are important in
$H + H_2$ itself and perhaps in some other reactions.

In comparing classical and quantum results obtained from calculations for the
same potential surface, most work has been done on the $H + H_2$ exchange reaction.
This choice was made for two reasons – first, although the known surface for the
reaction is not quite accurate enough for direct experimental comparison, it is of such
quality that meaningful results are obtained when two calculations made on the same
surface are compared; second, because H atoms are the lightest atoms, quantum
corrections might be more important for this reaction than for those involving only
heavier atoms. In fact, the $H + H_2$ reaction has served as the testing ground for
theoretical studies since the beginning of 'modern' chemical kinetics more than forty
years ago; it was the first system to which transition state theory was applied [9].

To carry out an accurate quantum-mechanical scattering calculation for a realistic
reactive system is, as I have already mentioned, much more difficult than the corresponding classical treatment. In fact, there does not exist, at present, a single calculation of satisfactory accuracy for any exchange reaction with the collision taking place
in three-dimensional space. To introduce the approximations that have been used, let
me outline schematically the quantum-mechanical formulation. We consider the
reaction $A + BC \rightarrow AB + C$, where A, B and C are structureless particles (except for
nuclear spin) that interact through a known potential surface $\mathscr{V}$. The differential
scattering cross-section for rearrangement from the reactant channel i to the product
channel f can be written in the centre-of-mass coordinate system [10]

$$\sigma_{fi}(\hat{k}_f) = \frac{\mu_i \mu_f}{(2\pi\hbar^2)^2} \frac{k_f}{k_i} |T_{fi}|^2. \tag{1}$$

Here i corresponds to the quantum numbers for a particular rotation-vibration state of
molecule BC and to momentum k_i of A relative to BC, μ_i is the reduced mass (A, BC),
the index f has the same connotation for the final channel. The quantity T_{fi} is the

transition matrix (T matrix) defined by the two equivalent expressions

$$T_{fi} = \langle \Phi_f | \mathscr{V}_f | \psi_i^{(+)} \rangle = \langle \psi_f^{(-)} | \mathscr{V}_i | \Phi_i \rangle, \tag{2}$$

where $\mathscr{V}_i$, $\mathscr{V}_f$ are initial- and final-state interaction potentials; i.e. $\mathscr{V}_i$ is the part of $\mathscr{V}$ that goes to zero as the initial atom-molecule relative coordinate goes to infinity, and $\mathscr{V}_f$ is correspondingly defined for the final channel. In terms of these, the total system Hamiltonian H is written

$$H = K + \mathscr{V} = H_i + \mathscr{V}_i = H_f + \mathscr{V}_f, \tag{3}$$

where K is the total (center-of-mass system) kinetic-energy, and H_i and H_f are the noninteracting initial- and final-state Hamiltonians, which have solutions Φ_i and Φ_f (normalized to unit density):

$$\Phi_i = \exp(i\mathbf{k}_i \cdot \mathbf{R}) \eta_{BC}(\mathbf{r})_i, \qquad \Phi_f = \exp(i\mathbf{k}_f \cdot \mathbf{S}) \eta_{AB}(\mathbf{s})_f. \tag{4}$$

Here $\mathbf{R}$ is the relative co-ordinate, $\mathbf{r}$ the molecular coordinate, and $\eta_{BC}(\mathbf{r})_i$ the molecular (rotation-vibration) wave function for the initial channel; the quantities $\mathbf{S}$, $\mathbf{s}$ and $\eta_{AB}(\mathbf{S})_f$ are defined correspondingly for the final channel. The functions $\psi_i^{(\pm)}$, $\psi_f^{(\pm)}$ are, respectively, initial- and final-channel eigenfunctions of the entire Hamiltonian with energy E and outgoing $(+)$ or incoming $(-)$ sperical-wave boundary conditions; e.g. the $\psi_i^{(\pm)}$ satisfy the Lippmann–Schwinger equations,

$$\psi_i^{(\pm)} = \Phi_i + \frac{1}{E - H \pm i\varepsilon} \mathscr{V}_i \Phi_i \tag{5}$$

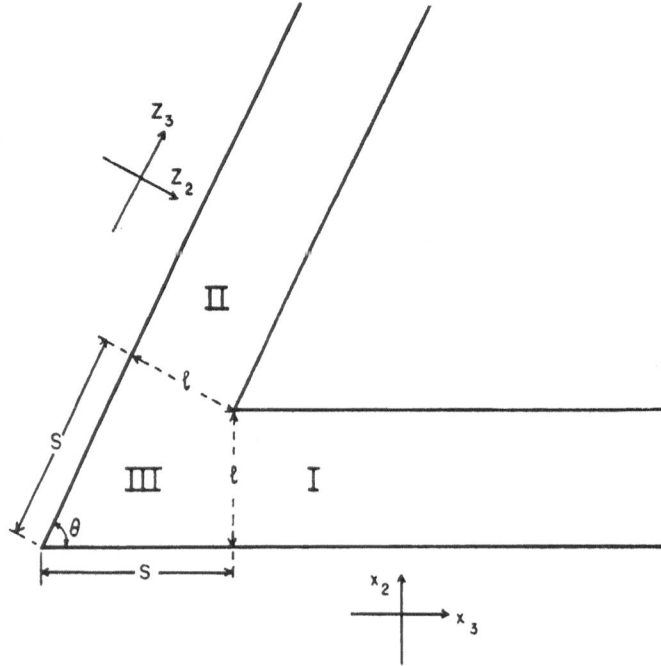

Fig. 1.   Simplified potential surface for collinear reaction; for description see text.

with the positive infinitesimal ε introducing the appropriate asymptotic behavior. The functions Φ_i, Φ_f, $\mathscr{V}_i$ and $\mathscr{V}_f$ being known, the difficulty in obtaining T_{fi}, and from it the differential cross-section, $\sigma_{fi}(\hat{k}_f)$, occurs in the determination of $\psi_i^{(+)}$ or $\psi_f^{(-)}$. Since their exact evaluation, which is equivalent to solving the three-body Schrödinger equation with appropriate boundary conditions, is not feasible at present, approximations must be introduced at this point.

There are two types of approximations that have been used – the first is to reduce the dimensionality of the problem so that it can be solved exactly and the second is to approximate $\psi_i^{(+)}$ or $\psi_f^{(-)}$. The first type of approach has been applied to the reactive scattering problem by many people in recent years. The essential idea is to restrict the collision to line. With this constraint, the reactive scattering calculation can be done 'exactly' by a variety of numerical and/or expansion techniques with present-day digital computers. The pioneering study in this area, as in many others involving serious calculations in chemical kinetics, was made by Hulburt and Hirschfelder [11]. Figure 1 shows the case considered – a collinear collision in a simplified potential with X_2, Z_2 the initial channel coordinates and X_3, Z_3 the final channel coordinates:

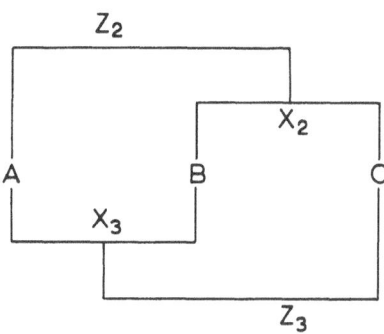

The reactant and product molecule potential (V_{I} and V_{II}) are 'square wells' while the interaction region V_{III} is represented by a constant potential which can be positive (barrier) or negative (complex or well) with respect to V_{I} and V_{II}. Hulburt and Hirschfelder considered $V_{\mathrm{III}}=0=V_{\mathrm{I}}=V_{\mathrm{II}}$ (i.e., no barrier or well). Because of some mathematical difficulties in the solution of the problem they did not get quite the correct result. It was only in 1968 that the correct solution was obtained [12]. A number of different potentials and masses were considered, including the case with a barrier and three atoms of equal masses; i.e., a simple model for $H+H_2$. The results obtained for reaction probabilities as a function of total energy are shown in Figure 2. The quantity T_1^* is the quantum-mechanical transmission coefficient for reaction from ground-state reactant to ground-state product, while T_2^* corresponds to going from ground-state reactant to first excited state product. For comparison, the classical calculation for the reaction probability from ground-state reactant to a sum over all states of the product, is also shown. It can be seen that the classical result is very similar to the quantum-mechanical reaction probability, both rising rapidly to a sharp maximum and than dropping off rapidly to zero, with small peaks at higher energies. However, it is clear that the quantum probability, $|T_1^*|^2$, has a significantly lower threshold than the classical calculation. This would produce a nonnegligible tunneling correction for this case. Figure 3 presents corresponding results obtained in a collinear study [13] with a more realistic semiempirical potential [14] for $H + H_2$. Again the classical and quantum calculations show similar behavior, but the quantum threshold falls below the classical value.

Semiclassical studies have also been made for this collinear systems with methods

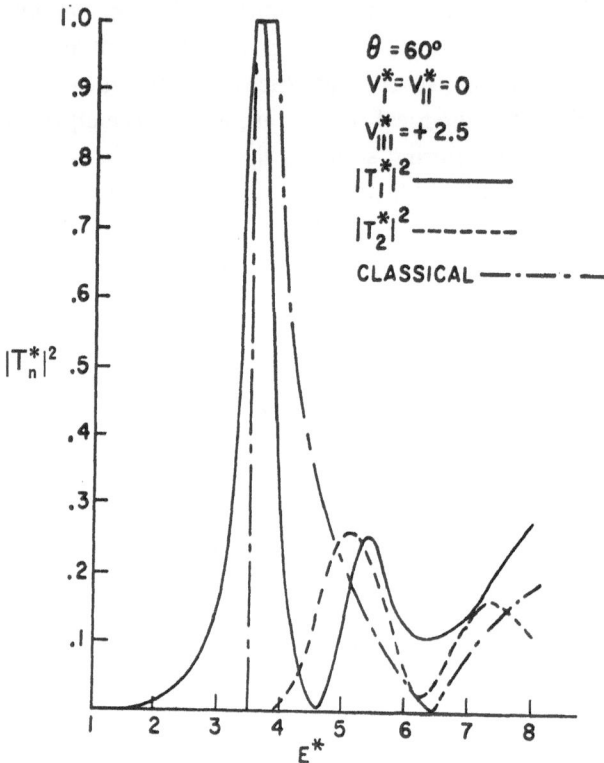

Fig. 2. Comparison of classical and quantum-mechanical reaction probabilities for $H + H_2$ on surface shown in Figure 1.

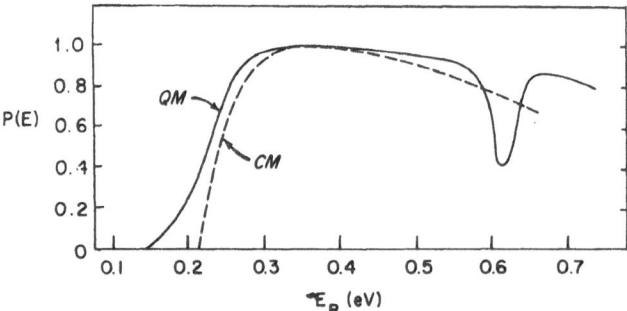

Fig. 3. Reaction probability for collinear $H + H_2$ versus relative translational energy: (QM) quantum mechanical and (CM) classical calculation.

developed by Marcus and by Miller [15]. In these treatments, classical paths chosen to connect with the initial and final quantum states are followed in the classically allowed regions, and extension to the classically forbidden regions is made by the appropriate generalization of the WKB approximation. For $H + H_2$ at very low energies (much below the classical threshold), excellent agreement with the exact quantum results have been obtained by semiclassical calculations [15]. However, so far the correct semiclassical treatment has not been applied in the threshold region where the tunneling effects are expected to be most interesting; approximate semicassical calculations for this region have not yielded satisfactory results [16].

Another comparison of interest for the collinear case can be made with transition state theory [13]. Figure 4 shows the probability of reaction as a function of the total

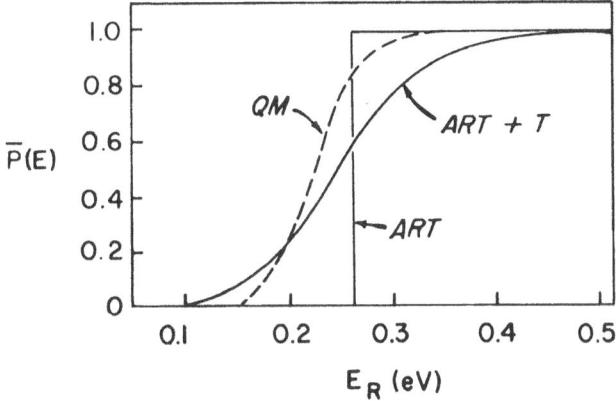

Fig. 4. Comparison of weighted average quantum-mechanical reaction probability: (QM) exact quantum mechanical, (ART) transition state theory, and (ART + T) transition state theory with Eckart tunnelling correction.

system energy obtained from the exact quantum calculation (same as Figure 3) and that determined from the quantum-mechanical form of transition state theory (without tunneling) by a method described previously [17]. Also included in the figure is the transition-state-theory probability with an approximate (Eckart-type) one-dimensional tunneling correction. The latter brings the transition state theory result closer to the quantum calculation but is not sufficiently good for obtaining accurate reaction rate constants. Since, in the purely classical limit for $H + H_2$ at low energies, transition state theory gives excellent results [18], it would be very valuable to have available a simple procedure for introducing accurate quantum corrections; no such procedure has been developed.

Turning now from the exact collinear to the approximate three-dimensional treatments, we outline some of the approximations that have been used:

(a) *Born approximation.* The simplest treatment replaces $\psi_i^{(+)}$ or $\psi_f^{(-)}$ in the transition matrix T_{fi} by Φ_f or Φ_i respectively; i.e.,

$$T_{fi} \Rightarrow T_{fi}(\mathbf{B}) = \langle \Phi_f | \mathscr{V}_f | \Phi_i \rangle = \langle \Phi_f | \mathscr{V}_i | \Phi_i \rangle, \tag{6}$$

which is the well-known Born approximation. Although it may be useful for some atomic rearrangement problems (e.g., high-energy, low-activation barrier) $T_{fi}(B)$ has been shown to yield incorrect results for the $H + H_2$ reaction in the thermal region.

(b) *Distorted-wave Born approximation.* To account in part for the strong repulsive interaction (high-energy barrier) between the approaching or receding atom and the molecule, it is appropriate to separate $\mathscr{V}_i$ and $\mathscr{V}_f$ into two parts:

$$\mathscr{V}_i = \mathscr{V}_i^0 + \mathscr{V}_i', \qquad \mathscr{V}_f = \mathscr{V}_f^0 + \mathscr{V}_f', \tag{7}$$

where $\mathscr{V}_i^0$ and $\mathscr{V}_f^0$ are distortion potentials that cannot produce rearrangement. They are chosen to account for the interaction as completely as possible, subject to the condition that the Hamiltonians $H_i + \mathscr{V}_i^0$ and $H_f + \mathscr{V}_f^0$ have solutions $\chi_i^{(\pm)}$ and $\chi_f^{(\pm)}$, respectively, which can be evaluated exactly or, at least, to a high degree of approximation. Here the $\chi_f^{(\pm)}$ satisfy the equation

$$\chi_f^{(\pm)} = \Phi_f + \frac{1}{E - (H_f + \mathscr{V}_f^0) \pm i\varepsilon} \mathscr{V}_f^0 \Phi_f, \tag{8}$$

and a corresponding equation exists for $\chi_i^{(\pm)}$. Introduction of $\chi_f^{(-)}$ and use of Equation (7) and (8) in the first expression for T_{fi} in Equation (2) yields

$$T_{fi} \Rightarrow \langle \chi_f^{(-)} | \mathscr{V}_f' | \psi_i^{(+)} \rangle. \tag{9}$$

If now $\psi_i^{(+)}$ is approximated by $\chi_i^{(+)}$, the so-called distorted-wave Born approximation (DWB) is obtained; i.e.

$$T_{fi} \Rightarrow T_{fi}(\text{DWB}) = \langle \chi_f^{(-)} | \mathscr{V}_f' | \chi_i^{(+)} \rangle. \tag{10}$$

As compared with $T_{fi}(B)$ (Equation (6)), Equation (10) should be considerably better for $H + H_2$ because it includes distorsion of the relative motion wave functions in both the initial and final channels. However, the replacement of $\psi_i^{(+)}$ by $\chi_i^{(+)}$ in Equation (9) to obtain (10) is generally a serious approximation, the accuracy of the resulting scattering matrix $T_{fi}(\text{DWB})$ depending both on the nature of the problem and on the judicious choice of $\mathscr{V}_i^0$ and $\mathscr{V}_f^0$.

(c) *Coupled-equation approach.* The next step beyond the distorted-wave Born approximation is to introduce some type of expansion for the functions $\psi_i^{(\pm)}$ and $\psi_f^{(\pm)}$. This type of approach has been applied to a variety of scattering problems (e.g., rotational excitation in atom-molecule collisions, electron scattering including rearrangement) and is applicable to chemical reactions as well. We write the wave function ψ (where we drop channel subscripts and boundary-condition superscripts for simplicity) as

$$\psi = \sum_n \gamma_n(r, R) + \sum_m \lambda_m(s, S). \tag{11}$$

The functions γ_n, λ_m can be determined by a variation principle or related methods corresponding to those used in bound state problems, though the fact that the varia-

tion principle is frequently not a maximum or minimum principle, introduces difficulties. To simplify the expansion given in Equation (11) one can introduce a finite
number of terms of the form $\gamma_n(\mathbf{r}, \mathbf{R}) = \eta_{BC}(\mathbf{r})_n f_n(\mathbf{R})$, $\lambda_m(\mathbf{s}, \mathbf{S}) = \eta_{AB}(\mathbf{s})_m g_m(\mathbf{S})$, where
the $\eta_{BC}(\mathbf{r})_n$, $\eta_{AB}(\mathbf{s})_m$ represent the bound states of the molecules BC and AB. Solution
for the relative motion function f_n, g_m gives rise to a coupled set of integrodifferential
equations [19].

All of these approximate methods have been applied to the $H + H_2$ reaction in
three dimensions with the same semiempirical potential [14] as used for the collinear
case. As already mentioned the Born approximation is very poor for the $H + H_2$
reaction because the repulsive potential corresponding to the energy barrier has an
important effect on the wave functions. Once the distortion due to the repulsive potential is introduced, as in the distorted-wave Born approximation, reasonable results are
obtained for many reaction attributes [10]; e.g., the angular dependence of the differential cross-section appears to be given correclty and in near agreement with the
classical results for the same surface. However, to obtain the correct energy dependence of the total cross section, the coupled equation technique must be used. Recently,
George Wolken has completed a very difficult calculation with a basis set consisting of
ground and excited molecular rotational states for the reactant and product molecules
[19]. Figure 5 shows the quantum results for the reactive $0 \rightarrow 1$ molecular rotational
excitation cross section and compares it with the classical cross section for reaction
from the ground rotational state to all final rotational states. As found already in the
linear case, there is a significantly lower quantum threshold. To indicate the importance of this for the reaction rate constant, we make a comparison of the rate con-

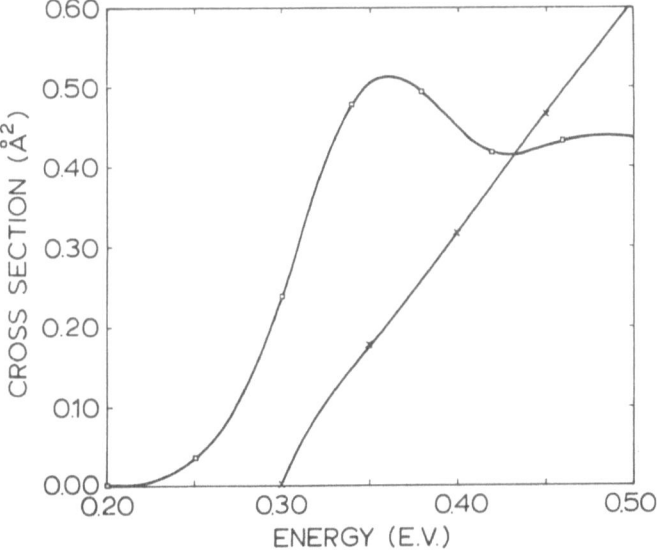

Fig. 5.   Comparison of classical and quantum mechanical reaction cross section for $H + H_2$ exchange
reaction on the surface given in [14]; (—□—□—) quantum-mechanical result, (—×—×—) classical
result.

MARTIN KARPLUS

stants calculated values obtained with the two cross sections of Figure 5. The results at a series of temperatures are shown in Table I. At 300 K, the quantum result is ten times the classical value, while at 1000 K the difference is a factor of 2. Although the quantum calculations are still only approximate, they suggest that there is an important tunneling correction for this reaction. This manifests itself also by the significantly lower effective activation energy obtained in fitting the quantum rate constants by an Arrhenius equation as compared with the classical value.

TABLE I
Quantum and quasiclassical $H + H_2$ rate constants

Temp (K)	$K_{class} \times 10^{-11}$ (cm^3 $mole^{-1}$ sec^{-1})	$K_{quant} \times 10^{-11}$ (cm^3 $mole^{-1}$ sec^{-1})
300	0.00253	0.029
400	0.0512	0.31
500	0.321	1.54
700	2.76	7.0
900	9.56	18
1000	14.9	25

In this lecture I have considered the importance of quantum corrections to classical trajectory calculations of reaction attributes. By looking at the threshold region for the $H + H_2$ exchange reaction, I have focused on the situation for which quantum effects are expected to play the largest role. It is clear that they are significant here if one is concerned with the accurate evaluation of low temperature rate constants. However, for most other reactions, and even for other properties of the $H + H_2$ reaction itself, quantum corrections are expected to be much less important and it can be hoped that the much simpler classical calculations will prove adequate.

Acknowledgement

This work was supported in part by grants from the National Science Foundation.

References

[1] M. Karplus, R. N. Porter, and R. Sharma, *J. Chem. Phys.* **43** (1965), 3259.
[2] I. Wang and M. Karplus, *J. Am. Chem. Soc.* **96** (1973).
[3] A. Warshel and M. Karplus, to be published.
[4] X. Chapuisat, Y. Jean, and L. Salem, private communication.
[5] I. Shavitt, R. M. Stevens, F. Minn, and M. Karplus, *J. Chem. Phys.* **48** (1968), 2700.
[6] B. Liu, *J. Chem. Phys.* **58** (1973), 1925.
[7] For a review, see F. S. Rowland, *Proc. International School of Physics*, 'Enrico Fermi' LXIV (ed. by Ch. Schlier), Academic Press, New York, 1970.
[8] D. N. Mitchell and D. J. Leroy, *J. Chem. Phys.* **58** (1973), 3449.
[9] H. Pelzer and E. Wigner, *Z. Physik Chem.* **B15** (1933), 445.
[10] M. Karplus and K. T. Tang, *Dis. Faraday Soc.* **44** (1967), 1.

[11] H. Hulburt and J. O. Hirschfelder, *J. Chem. Phys.* **11** (1943), 273.

[12] K. T. Tang, B. Kleinman and M. Karplus, *J. Chem. Phys.* **50** (1969), 1119.

[13] D. J. Diestler and M. Karplus, *J. Chem. Phys.* **55** (1971), 5832.

[14] R. N. Porter and M. Karplus, *J. Chem. Phys.* **40** (1964), 1105.

[15] See, for example, W. H. Miller, *Adv. Chem. Research* **4** (1971), 161; T. F. George and W. H. Miller, *J. Chem. Phys.* **56** (1972), 5722; J. Stine and R. A. Marcus, *Chem. Phys. Letters* **15** (1972), 536.

[16] J. M. Bowman and A. Kuppermann, *Chem. Phys. Letters* **19** (1973), 160.

[17] K. Morokuma, B. C. Eu, and M. Karplus, *J. Chem. Phys.* **51** (1969), 5193.

[18] K. Morokuma and M. Karplus, *J. Chem. Phys.* **35** (1971), 63.

[19] G. Wolken and M. Karplus, *J. Chem. Phys.* **60** (1974).

THE CHARGE AND SPIN TRANSFERS IN
CHEMICAL REACTION PATHS

KENICHI FUKUI

Dept. of Hydrocarbon Chemistry, Kyoto University, Kyoto, Japan

Contents:

1. Introduction

The role of the interaction of particular molecular orbitals (MO) in chemical reactions in general has long been recognized as an essential factor determining the orientation (Fukui *et al.*, 1952, 1954a; Fukui, 1964, 1965a, 1970a, b, 1971b; Klopman and Hudson, 1967; Klopman, 1968; Salem, 1968, 1969a, b). The interaction of particular MO's also relates to the stereoselection rule and is believed to control the stereochemical path (Fukui, 1964, 1965b, 1971a; Woodward and Hoffmann, 1965, 1969a, b; Fukui and Fujimoto, 1969a; Anh, 1970), and the mechanism of such particular MO controls has qualitatively been discussed in relation to the nature of chemical reactions (Fukui *et al.*, 1954b; Fukui and Fujimoto, 1968, 1969b; Fujimoto *et al.*, 1971b; Fujimoto and Fukui, 1972).

The particular MO's which play such an essential part in chemical reactions are the highest occupied MO (or sometimes high-lying occupied MO's) (HOMO), and the lowest unoccupied MO (or sometimes low-lying unoccupied MO's) (LUMO). In donor molecules, the behavior of HOMO is more important than that of LUMO, while in acceptor molecules this is reversed. In open-shell molecules, i.e. in radicals and excited-state molecules, the above-stated circumstances are valid in the unrestricted Hartree–Fock sense. In the restricted Hartree–Fock sense, the singly occupied MO's (SOMO) play this part. The overlapping interaction between HOMO of one specie and LUMO of another one, hereafter simply referred to as HOMO–LUMO interaction, has an essential significance in chemical reactions between two closed-shell molecules. Similarly, HOMO–SOMO, SOMO–LUMO, and SOMO–SOMO interactions, in the restricted Hartree–Fock sense, are the other types of particular orbital interaction.

These circumstances are essentially general, irrespective of whether the molecule is saturated or unsaturated, aliphatic or aromatic, and whether the intermolecular reaction is unicentric or multicentric. The criterion of particular orbital interaction can be applied also to intramolecular reactions in which two parts in the molecule essential to the reaction are regarded as if they were two molecules.

R. Daudel and B. Pullman (eds.), The World of Quantum Chemistry, 113–141. All Rights Reserved
Copyright © 1974 by D. Reidel Publishing Company, Dordrecht-Holland

The purpose of the present paper is to give a more detailed, quantitative description for the mode of particular orbital interactions in chemical processes. Stress is laid on the role and significance of charge and spin transfers in the molecular deformation and the bond interchange occurring in mutually reacting chemical species. It is also expected that such a treatment serves to give an illuminating interpretation for the selection of favorable paths in actual chemical reactions.

2. Description of Mutually Interacting Systems

Consider most generally a system composed of N nuclei and a definite number of electrons. We are able to calculate the adiabatic potential of the system with *fixed* nuclei in the frame of Born–Oppenheimer approximation. The location of a nucleus α in usual three-dimensional Cartesian space is denoted by X_α, Y_α. Z_α, and the distance of two nuclei α and β by $R_{\alpha\beta}$. The adiabatic potential W is given by a function of $(3N-6)$ independent $R_{\alpha\beta}$'s. As these $R_{\alpha\beta}$'s, we may choose for instance R_{12}, R_{13}, R_{23}, $R_{i\mu}$ $(i=1, 2, 3; \mu=4, 5, ... N)$ by specifying any three nuclei which are not collinear. We write these $(3N\text{-}6)$ particular pairs of nuclei as $(\alpha\beta)$.

Since we consider here a nonequilibrium system with fixed nuclei in general, each nucleus is subject to a force in the direction of grad W. If we permit an *infinitely slow* motion from the initial fixed nuclear configuration, the displacement of each nucleus will be given by

$$\cdots = \frac{M_\alpha dX_\alpha}{\partial W/\partial X_\alpha} = \frac{M_\alpha dY_\alpha}{\partial W/\partial Y_\alpha} = \frac{M_\alpha dZ_\alpha}{\partial W/\partial Z_\alpha} = \cdots, \quad (\alpha = 1, 2, ... N) \tag{1}$$

in which M_α is the mass of nucleus α, and dX_α, dY_α, dZ_α are the components of displacement vector of nucleus α, $d\mathbf{R}_\alpha$. This displacement obviously satisfies the conditions

$$\sum_\alpha M_\alpha dX_\alpha = \sum_\alpha M_\alpha dY_\alpha = \sum_\alpha M_\alpha dZ_\alpha = 0$$

$$\sum_\alpha M_\alpha (Y_\alpha dZ_\alpha - Z_\alpha dY_\alpha) = \sum_\alpha M_\alpha (Z_\alpha dX_\alpha - X_\alpha dZ_\alpha) =$$

$$= \sum_\alpha M_\alpha (X_\alpha dY_\alpha - Y_\alpha dX_\alpha) = 0, \tag{2}$$

since no force acts to translate the center of gravity nor to rotate the system around it.

The locus of Equation (1) is given by solving this equation regarded as a differential equation. If we consider such a locus passing through a transition state, this becomes an '*intrinsic*' reaction coordinate (Fukui, 1970c). An *equilibrium point* of function W may be defined as satisfying the condition

$$\partial W/\partial R_{\alpha\beta} = 0 \quad \text{(with respect to all } (\alpha\beta)). \tag{3}$$

Accordingly, a reaction intermediate, an initial state or final state of a reacting system, and a transition state correspond to equilibrium points. The initial and final states of a reacting system which is composed of two independent species are represented by those points of function W which correspond to the infinite mutual separation of these species. At an equilibrium point Equation (1) is modified to be

$$\cdots = \frac{M_\alpha dX_\alpha}{\sum\limits_\beta \left(\text{grad}_\beta \dfrac{\partial W}{\partial X_\alpha} \cdot d\mathbf{R}_\beta\right)} = \frac{M_\alpha dY_\alpha}{\sum\limits_\beta \left(\text{grad}_\beta \dfrac{\partial W}{\partial Y_\alpha} \cdot d\mathbf{R}_\beta\right)} = \frac{M_\alpha dZ_\alpha}{\sum\limits_\beta \left(\text{grad}_\beta \dfrac{\partial W}{\partial Z_\alpha} \cdot d\mathbf{R}_\beta\right)} = \cdots, \tag{4}$$

in which grad_β signifies the vector $(\partial/\partial X_\beta, \partial/\partial Y_\beta, \partial/\partial Z_\beta)$. Equation (4) leads to a usual secular equation of $3N$ dimensions to give $(3N-6)$ directions of displacement which are nothing but the normal coordinates. The other six zero-roots of the secular equation correspond to the freedom of translation and rotation as a whole.

In order to treat Equation (1) further we have to know the change of adiabatic potential with nuclear displacement. By the use of the Hellmann–Feynman theorem we have

$$\frac{\partial W}{\partial X_\alpha} = \frac{\partial V_{nn}}{\partial X_\alpha} + \int \varrho(1) \frac{\partial V_n(1)}{\partial X_\alpha} \, dv(1), \quad \text{etc.,} \tag{5}$$

where V_{nn} is the nuclear–nuclear repulsion, $V_n(1)$ is the nuclear attraction of electron 1, $\int dv(1)$ means the integration over the space coordinates of electron 1, and $\varrho(1)$ is the first-order density given by

$$\varrho(1) = n_e \int d\sigma(1) \int \Psi^*(1, 2, 3, \ldots) \, \Psi(1, 2, 3, \ldots) \, d\tau(2) \, d\tau(3) \ldots, \tag{6}$$

where n_e is the total number of electrons, Ψ is the electronic wave function of the total system, $\int d\sigma(1)$ implies the integration with respect to the spin coordinate of electron 1, and $\int d\tau(2)$ signifies the integration over space and spin coordinates of electron 2.

At an equilibrium point the displacement of nuclei can be described in terms of normal coordinates, $Q^{(1)}, Q^{(2)}, \ldots Q^{(3N-6)}$. The adiabatic potential W is expressed in the form

$$W = W_0 + \tfrac{1}{2} \sum_{i=1}^{3N-6} \kappa^{(i)} Q^{(i)2} \tag{7}$$

in the vicinity of the equilibrium point. The coefficient $\kappa^{(i)}$ represents $\partial^2 W/\partial Q^{(i)2}$ and is generally positive with respect to $(3N-6)$ normal coordinates in stable equilibrium points. At a transition state one of $\kappa^{(i)}$'s is negative, and the corresponding direction is that of reaction coordinate. At initial and final states of intermolecular reactions in which two species are infinitely separate from each other, the six $\kappa^{(i)}$'s corresponding to one mutual approach, and five mutual rotations including torsion with respect to the axis connecting the two centers of gravity of each species are all infinitesimal. The others are those of normal vibrations localizing at each species. In such systems the intrinsic reaction coordinate at infinite separation is obviously the normal coordinate of mutual approach.

The $\kappa^{(i)}$ can be a measure of the favorableness of normal coordinate $Q^{(i)}$ as a reaction coordinate. The reaction starts from a given initial state most favorably along the normal coordinate of minimum κ. In intermolecular reactions the infinite-simal κ's never serve as the measure of favorableness, and the comparison of stability of the combined system at finite separation of the components becomes significant.

The second-order perturbation treatment of the total system at an equilibrium point in which the nuclear displacement is regarded as a perturbation provides a convenient expression of κ (Bader, 1962), i.e.

$$\kappa = \frac{\partial^2 V_{nn}}{\partial Q^2} + \int \varrho_{00}(1) \frac{\partial^2 V_n(1)}{\partial Q^2} \, dv(1) + 2 \sum_k{}' \frac{\left| \int \varrho_{0k}(1) \frac{\partial V_n(1)}{\partial Q} \, dv(1) \right|^2}{W_0 - W_k} .$$

(8)

In this expression, the symbols are the same as employed in Equation (5) except that subscript 0 and k signify the ground and kth excited electronic states at the equilibrium point. The transition density $\varrho_{0k}(1)$ is defined by

$$\varrho_{0k}(1) = n_e \int d\sigma(1) \int \Phi_0^*(1, 2, 3\ldots) \, \Phi_k(1, 2, 3\ldots) \, d\tau(2) \, d\tau(3)\ldots .$$

(9)

The usual symbol $\sum_k'$ implies the summation over the excited states k(excepting 0). The electron distribution for the perturbed system, i.e. for a point on the normal coordinate Q corresponding to a nuclear configuration near the equilibrium point is given by

$$\varrho(1) = \varrho_{00}(1) + 2Q \sum_k{}' \varrho_{0k}(1) \frac{\int \varrho_{k0}(2) \frac{\partial V_n(2)}{\partial Q} \, dv(2)}{W_0 - W_k},$$

(10)

where Q is the nuclear displacement from the equilibrium point along the normal coordinate Q (Bader, 1962). This expression plays a significant role in the discussion of favorableness of the reaction coordinate in relation to actual occurrence (see Section 7).

3. Wave Function Suitable for Configuration and Population Analyses

We present the electronic wave function, Ψ, for the combined system consisting of two reacting species, A and B, with a nuclear configuration fixed at one point on the reaction coordinate. Each of the species A and B in a mutually interacting condition may have in general a geometry which is different from that of the equilibrium state, i.e. the most stable one possessed in the free condition. The total system is supposed to be composed of such a nonequilibrium species of generally *deformed* geometry. The condition that the total system lies on a reaction coordinate is never an essential one, but we can treat a more general case similarly. The procedure of constructing the wave function for the total system is chosen so as to be suitable for the present purpose to elucidate the mechanism of chemical reaction in terms of particular orbital interaction. Two methods are introduced here.

A. CI WAVE FUNCTION

In this procedure the wave function Ψ is represented by a linear combination of many configuration functions, Φ_M's, i.e. a configuration interaction (CI) technique is adopted. Each Φ_M is built from the MO's calculated by an appropriate Fock–Roothaan process with respect to each of the species A and B which exists separately with the *same* geometry as in the combined system (Fukui and Fujimoto, 1968). We call hereafter these species as 'isolated' species A and B. Never confound this with a species in the free, equilibrium condition. Each normalized configuration function Φ_M is antisymmetrized in a determinantal form and is constructed so as to represent a spin eigenstate. In this way, we have

$$\Psi = \sum_M C_M \Phi_M . \tag{11}$$

We designate the configuration function corresponding to the electron configuration which is the same as in each of the isolated species A and B by Φ_0 especially. The coefficient C_M in Equation (1) is determined so as to minimize the total energy

$$W = \int \Psi^* H \Psi \, d\tau \Big/ \int \Psi^* \Psi \, d\tau \tag{12}$$

by solving the secular equation

$$\begin{vmatrix} H_{00} - W & H_{01} - S_{01} W \dots \\ H_{10} - S_{10} W & H_{11} - W & \dots \\ \dots\dots\dots\dots\dots\dots\dots \end{vmatrix} = 0, \tag{13}$$

in which H is the Hamiltonian operator of the combined system and

$$H_{MN} = \int \Phi_M^* H \Phi_N \, d\tau$$
$$\tag{14}$$
$$S_{MN} = \int \Phi_M^* \Phi_N \, d\tau .$$

The wave function, energy, and the coefficients may be specified by the electronic state to which they belong in such a way as $\Psi^{(n)}$, $W^{(n)}$, and $C_M^{(n)}$.

In order to explain how a configuration function, Φ_M, is constructed, we take as an example the case of two closed-shell species. We denote an occupied MO and an unoccupied MO of isolated species A by a_i and a_j, respectively, and those of species B by b_k and b_l, respectively. 'Unoccupied' MO's are defined here as corresponding to the nonrealistic solutions of the Fock equations for the ground state of an isolated species.

Consider an electron configuration in which one electron originally occupying the MO a_i jumps into the MO a_j. The function Φ_M corresponding to this configuration is denoted by $\Phi_{i \to j}$. In a similar manner $\Phi_{i \to l}$ stands for the configuration in which one electron in a_i of species A is transferred to one of the unoccupied MO's of species

B, b_l. The configuration $\Phi_{i \to j}$ may be referred to as a *monoexcited* one, and $\Phi_{i \to l}$ a *monotransferred* one. In the same way we can consider a *diexcited* configuration, say $\Phi_{i' \to j'}^{i \to j}$, a *ditransferred* one, say $\Phi_{i' \to l'}^{i \to l}$, and a *monoexcited monotransferred* one, say $\Phi_{i' \to j'}^{i \to j}$, and so forth. It is a matter of course that the accuracy of the wave function Ψ increases with more configurations taken into account. The function Φ_M may be specified by indicating the spin eigenvalue as $^1\Phi_M$ (singlet function), $^3\Phi_M$ (triplet function), and so on. The original configuration, Φ_0, may be called *zero*-configuration.

In this way, for instance $^1\Phi_{i \to l}$ is easily written as

$$^1\Phi_{i \to l} = 2^{-1/2}\{\mathcal{N}\mathscr{A}[a_1\bar{a}_1 a_2 \bar{a}_2 \dots a_{i-1}\bar{a}_{i-1} a_i a_{i+1}\bar{a}_{i+1} \dots a_n\bar{a}_n b_l\bar{b}_l \dots b_m\bar{b}_m\bar{b}_l]$$
$$- \mathcal{N}\mathscr{A}[a_1\bar{a}_1 a_2 \bar{a}_2 \dots a_{i-1}\bar{a}_{i-1} \bar{a}_i a_{i+1}\bar{a}_{i+1} \dots a_n\bar{a}_n b_l\bar{b}_l \dots b_m\bar{b}_m b_l]\}, \quad (15)$$

in which m and n are the numbers of doubly occupied MO's of isolated species A an B, respectively, a_i and $\bar{a}_i$ represent the spin orbitals with spin function of α and β, respectively, and $\mathcal{N}\mathscr{A}$ implies the normalization-antisymmetrization operator.

The coefficient C_M in Equation (1) represents the degree of participation of electron configuration Φ_M in the wave function for the mutually interacting system. The difference in electron distribution between the isolated species and the interacting system is caused by the contribution of these configurations which represent local excitations or mutual transfers of electrons, since each Φ_M has an electron distribution generally different from Φ_0.

The total energy of the combined system is interpreted in terms of the MO's of isolated species A and B. The picture adopted here concerns the rearrangement of electrons in the deformed nonequilibrium species on the occasion of mutual interaction. This energy of rearrangement may be referred to as 'interaction energy', which can be attributed to the mixing of Φ_M and the accompanying electron 'migration' in and between the species A and B. The merit of such a treatment is to enable an analysis of the origin of interaction energy in terms of MO's of each reactant molecule.

We assume here the normal case of chemical reactions between two ground-state species in which H_{00} is lower than any of H_{MM}'s.

The treatment mentioned above leads to the following form of the ground-state wave function:

$$\Psi = C_0\Phi_0 + \left\{ \overset{\text{monoex.}}{\underset{M}{\sum}} + \overset{\text{monotr.}}{\underset{M}{\sum}} + \overset{\text{diex.}}{\underset{M}{\sum}} + \overset{\text{ditr.}}{\underset{M}{\sum}} + \overset{\substack{\text{monoex.}\\\text{monotr.}}}{\underset{M}{\sum}} + \dots \right\} C_M\Phi_M \dots, \tag{16}$$

in which -ex. and -tr. stand for -excited and -transferred. The interaction energy is defined by

$$\Delta W = W - (W_{A0} + W_{B0}), \tag{17}$$

where W is the lowest energy of the system composed of mutually interacting species A and B, W_{A0} and W_{B0} being the lowest energy of the isolated species A and B. In order to catch the essential feature of the interaction, it is often convenient to adopt an adequate procedure of approximation. For instance, the energy constituents can be recognized by means of the following approximate expression:

$$\Delta W \sim \{H_{00} - (W_{A0} + W_{B0})\} - \left(\overset{monoex.}{\underset{M}{\sum}} + \overset{monotr.}{\underset{M}{\sum}} \right) \frac{|H_{0M} - S_{0M}H_{00}|^2}{H_{MM} - H_{00}},$$

(18)

in which multiexcited and multitransferred terms and higher perturbation terms are neglected. The contribution of each electron configuration is separately estimated by this equation. The term of $\{H_{00} - (W_{A0} + W_{B0})\}$ contains the effect of interaction without change in electron configuration and is divided into usual *Coulomb* and *exchange* terms. The second term refers to the influence of mixing of different configurations, and the terms corresponding to the excited configurations may be named *polarization* term and those corresponding to the transferred configurations can be called *delocalization* term.

In conformity with this approximation, the wave function of Equation (16) can be approximated as

$$\Psi \sim C_0 \Phi_0 + \left(\overset{monoex.}{\underset{M}{\sum}} + \overset{monotr.}{\underset{M}{\sum}} \right) C_M \Phi_M,$$

(19)

where

$$C_0 \sim 1 - \left(\overset{monoex.}{\underset{M}{\sum}} + \overset{monotr.}{\underset{M}{\sum}} \right) \frac{S_{M0}(H_{0M} - S_{0M}H_{00})}{H_{00} - H_{MM}},$$

$$C_M \sim \frac{H_{M0} - S_{M0}H_{00}}{H_{00} - H_{MM}}.$$

(20)

Since the magnitude of the numerator of C_M parallels the overlap integral S_{M0}, the mixing of a configuration into the original one is proportional to the overlapping between these and inversely proportional to the level separation. This relation is important with regard to the mechanism of chemical interaction.

B. LCMO WAVE FUNCTION

A similar object is accomplished by another method of constructing the wave function. We calculate the MO's of the combined system $A + B$, ϕ_p's, as well as the MO's of each isolated species A and B. We proceed with the closed-shell case in which the concept of double occupation is employed, for simplicity. Each MO, ϕ_p, is expressed in the following form:

$$\phi_p = \overset{occ}{\underset{i}{\sum}} C_i^{(p)} a_i + \overset{uno}{\underset{j}{\sum}} C_j^{(p)} a_j + \overset{occ}{\underset{k}{\sum}} C_k^{(p)} b_k + \overset{uno}{\underset{l}{\sum}} C_l^{(p)} b_l, \quad (p = 1, 2, \ldots v)$$

(21)

where $\overset{occ}{\sum}$ and $\overset{uno}{\sum}$ respectively denote summation over all the occupied and the unoccupied MO's (Fujimoto *et al.*, 1974), and v is the total number of occupied and unoccupied MO's of A and B. It is supposed here that all MO's are composed of linear combinations of independent atom-centered functions, usually atomic orbitals (AO). Since the number of AO's adopted agrees with that of MO's in each of systems A, B, and $A + B$, the coefficients in the right-hand side of Equation (21) can easily be determined by comparing the AO coefficients of both sides of the equation. The ground-state wave function using the MO's of Equation (21) leads to the following density function:

$$\varrho(1) = \sum_{p}^{occ} \phi_p^*(1)\,\phi_p(1),$$
$$= \varrho^{(I)}(1) + \varrho^{(II)}(1) + \varrho^{(III)}(1) + \varrho^{(IV)}(1), \qquad (22)$$

in which $\varrho^{(I)}$ contains the terms of the type $a_i^*(1)a_i(1)$, $a_j^*(1)a_j(1)$, $b_k^*(1)b_k(1)$, and $b_l^*(1)b_l(1)$ and reflects the intramolecular change in bonding situation; the term $\varrho^{(II)}$ consists of the terms $a_i^*(1)a_{i'}(1)$, $a_i^*(1)a_j(1)$, etc.; $\varrho^{(III)}$ is a similar term concerning the MO's of B; $\varrho^{(II)}$ and $\varrho^{(III)}$ are the origin of intramolecular rearrangement of electrons in each of species A and B and vanish upon integration with respect to the space coordinates; and the last term $\varrho^{(IV)}$ is composed of the products of an MO of A and an MO of B, and is principally accountable for the formation of new bonds between A and B. The overlapping of the occupied MO's of A and the occupied MO's of B gives an antibonding effect for the intermolecular bond formation through exchange repulsion. The electron density accumulating in the intermolecular region necessary for the formation of new bonds chiefly results from the occupied-unoccupied orbital overlapping between A and B. Of course, in contrast with the remarkable contributions of these orbital overlappings, the unoccupied–unoccupied orbital over-lapping between two species is much less significant.

As a result of interaction, each of the initially doubly occupied MO's of isolated species A and B turns to have an occupation number less than 2 in general, and each of the initially unoccupied MO's comes to have a finite occupation number. These changes can be discussed in terms of each MO of the isolated species A and B.

The wave function of the combined system composed of MO's of Equation (21) is decomposed into the products of MO's of constituent species A and B. Comparison with the result of similar decomposition of the right-hand side of Equation (16) easily gives the coefficients C_0 and C_M's of Equation (16) in terms of $C_i^{(p)}$, etc. of Equation (21). This is the procedure of configuration analysis (Baba *et al.*, 1969), and we can thus discuss the degree of contribution of each electron configuration.

The essential feature of the methods, described above under A and B, is that these methods enable configuration and population analyses in terms of electrostatic, exchange, polarization, and delocalization effects separately. Each effect can further be divided into the contributions of each MO, if necessary. The role of particular orbital interactions can be elucidated in this way. Also the chemically obscure concept of 'donation and back-donation' can easily be made quantitative.

4. Transfer of Charge and Spin Through Orbital Interactions

A. SEMIEMPIRICAL CALCULATION BY CI APPROACH

It is worthwhile to show how the methods mentioned in the previous section work in elucidating the mechanism of chemical interaction on molecular, or even 'orbital' basis. Several examples in which a semiempirical all-valence-electron SCF (self-consistent-field) MO calculation is employed have already been reported (Fukui, 1971b; Fukui *et al.*, 1972; Fujimoto *et al.*, 1972c). The reactions treated there are as follows:

$$Cl^- + CH_3Cl \rightarrow ClCH_3 + Cl^-$$

(Fukui, 1971b; Fukui *et al.*, 1972; Fujimoto *et al.*, 1971a);

$$H^- + CH_2-CH_2 \rightarrow CH_3 - CH_2 - O^-$$
$$\diagdown \diagup$$
$$O$$

(Fukui, 1971b; Fujimoto *et al.*, 1972c);

$$HCl + C_2H_4 \rightarrow CH_2 = CH_2$$
$$\vdots$$
$$H$$
$$|$$
$$Cl$$

(Fukui, 1971b).

Also the dimerization of methylenes to ethylene (Fujimoto *et al.*, 1972a), the addition of singlet methylene to ethylene to form cyclopropane (Fujimoto *et al.*, 1972b), the hydrogen abstraction from methane by methyl radical (Fujimoto *et al.*, 1972d), and the addition of methyl radical to ethylene (Fuljimoto *et al.*, 1972d) have been dealt with by similar calculations. Calculations of these examples are carried out by Method A (CI wave function) mentioned in the previous section. A brief summary is given here.

(1) *Bimolecular Nucleophilic Substitution of Methyl Chloride by Chloride Ion*

The preference of inversion path to retention path is shown to result from a difference in stabilization energy. The stability of inversion path is attributed mainly to the large delocalization energy, which is understood to originate from the effective overlapping of the lone-pair orbital of Cl^- with LUMO of CH_3Cl extending along the C—Cl bond on the back of the C atom. This consists of a favorable HOMO–LUMO inter-action. Population analysis along this reaction path can show that a considerable amount of electrons is transferred from Cl^- to the antibonding LUMO localizing on the C—Cl bond of CH_3Cl, resulting in the weakening of original C—Cl bond and an accumulation of electrons in the intermolecular region to form a new C—Cl bond.

(2) *Ring Opening of Ethylene Oxide by Hydride Anion*

The two lowest lying unoccupied MO's of ethylene oxide have large amplitude on the back of C atoms. The unsymmetrical approach of H^- along the path of *trans*-opening is shown to be more favorable than *cis*-opening. This way of approach is explained by the shape of two LUMO's, the overlapping of which with the hydride anion orbital brings about the delocalization stabilization. The deformation of C_2H_4O molecule due to the unsymmetrical approach of H^- causes a growth of LUMO amplitude near the reaction center which makes the charge transfer easier and easier.

(3) *Addition of Hydrogen Chloride to Ethylene*

The LUMO of hydrogen chloride has a large amplitude at hydrogen atom, extending to the direction opposite to the chlorine atom, and has a node between $Cl-H$ bond. The HOMO of ethylene is π bonding orbital. The symmetrical approach favors the HOMO–LUMO overlapping. The configuration Φ_0 in Equation (16), which represents the adiabatic interaction without mixing of transferred or excited configurations, produces a density distribution in which electrons are expelled from the intermolecular region by exchange repulsion. Calculation shows that the mixing of transferred configurations, chiefly of HOMO–LUMO interaction, gathers electrons in the region between two molecules, which contribute to form a new bond.

(4) *Dimerization of Singlet Methylenes to Ethylene*

Singlet methylene has a C_{2v} bent structure, HOMO being the lone-pair orbital extending in the direction of the molecular plane and LUMO the π-orbital. The HOMO–LUMO criterion is very useful for guessing the favorable path. A rough inference at once leads to the path indicated in Figure 1a, in which $HOMO^{(1)}$ of $CH_2^{(1)}$ overlaps with $LUMO^{(2)}$ of $CH_2^{(2)}$ along the line of approach to form a σ-bond of ethylene, while $HOMO^{(2)}$ parallels $LUMO^{(1)}$ to form a π-bond. However, a further detailed investigation suggests a little modification of the model (a), since $LUMO^{(1)}-HOMO^{(2)}$ bonding would more effectively favor the stabilization in σ-type one rather than π-type. A model expected as the result would be between (a) and (b) of Figure 1. An

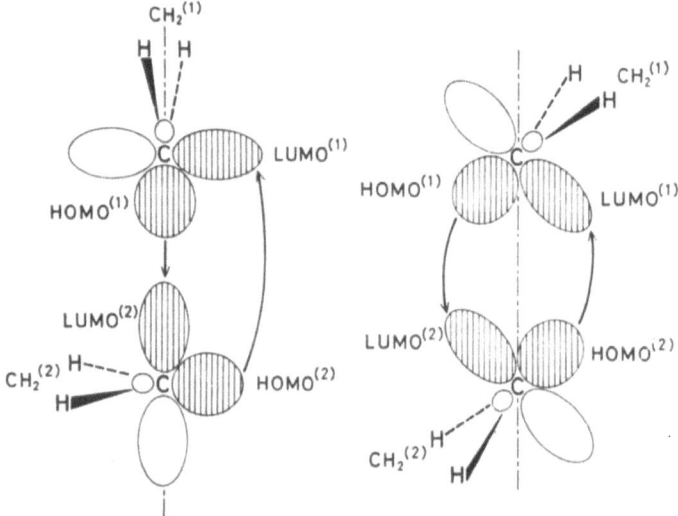

Fig. 1. Unsymmetrical (a) and symmetrical (b) approach of two methylenes favorable to double HOMO–LUMO overlappings.

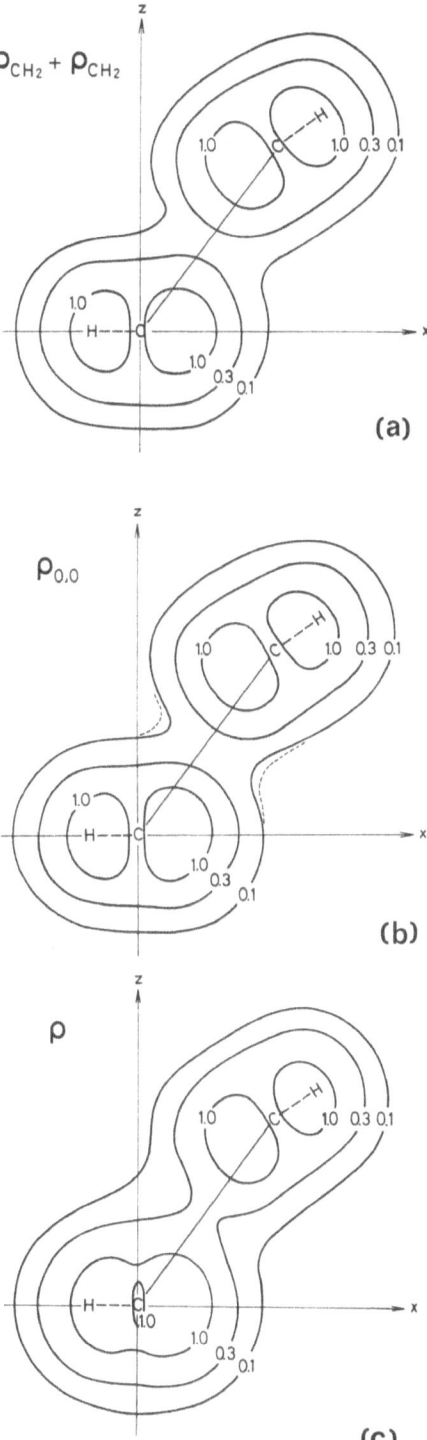

Fig. 2. Changes of electron density distribution in the dimerization of methylenes. (a) The sum of electron densities of two methylenes with no interaction. (b) The electron density of the system of two adiabatically interacting methylenes. (c) The electron density of the system of two interacting methylenes with the mixing of transferred configurations.

extended Hückel calculation (Hoffmann *et al.*, 1970) indicates that the deviation of each angle between CH_2-plane and the C—C axis from the model (a) to the direction of the model (b) is 6° at the C—C distance of 3.0 Å. This model shows that the contribution to the large delocalization energy mainly comes from the HOMO–LUMO interaction.

The changes in electron densities are shown in Figure 2. The simple sum of electron distributions of two methylenes with no interaction at all is (a), and (b) indicates the electron density of two methylenes which interact with each other adiabatically, i.e., without mixing of charge-transfer or excited configurations. It is seen in (b) that the exchange repulsion excludes electrons out of the intermolecular region. It is the effect of mixing of charge-transfer configurations that the electron density in that region grows to contibute to the bond formation.

(5) *Addition of Singlet Methylene to Ethylene*

Similar investigation is made with regard to the addition of singlet methylene to ethylene to produce cyclopropane. The HOMO–LUMO overlapping is effectuated by the mode of interaction shown in Figure 3, which is essentially the same as obtained by an extended Hückel calculation (Hoffmann, 1968). No explanation will be needed with regard to Figure 3. The calculation with respect to this model by the method described in Section 3 gives the following result. The two transferred configurations (from the HOMO of etylene to the LUMO of methylene, and from the HOMO of methylene to the LUMO of ethylene) have a dominant contribution to the wave function. The electron density maps of Figure 4 show the important role of the HOMO–LUMO interaction in the genesis of newly forming bonds. The exchange repulsion expels electrons from the intermolecular region and the mixing of charge-transfer states calls electrons back in this region. But almost the same maps of (c) and (d) imply that the contribution of the HOMO–LUMO interaction is decisive also in this case.

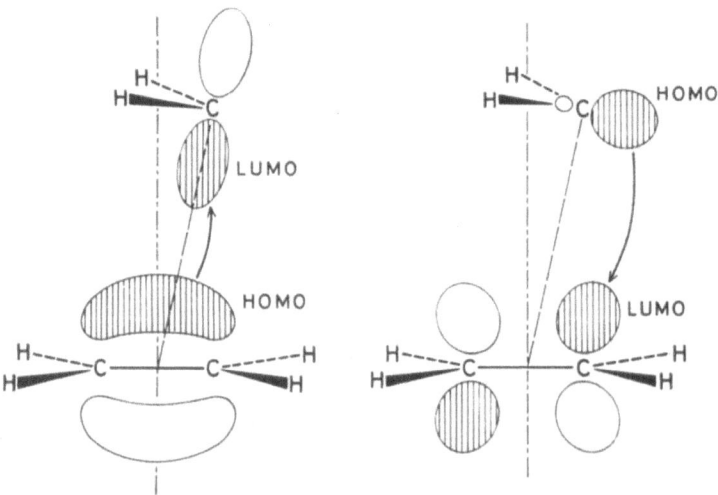

Fig. 3.   The mode of HOMO–LUMO overlapping in the interaction between methylene and ethylene.

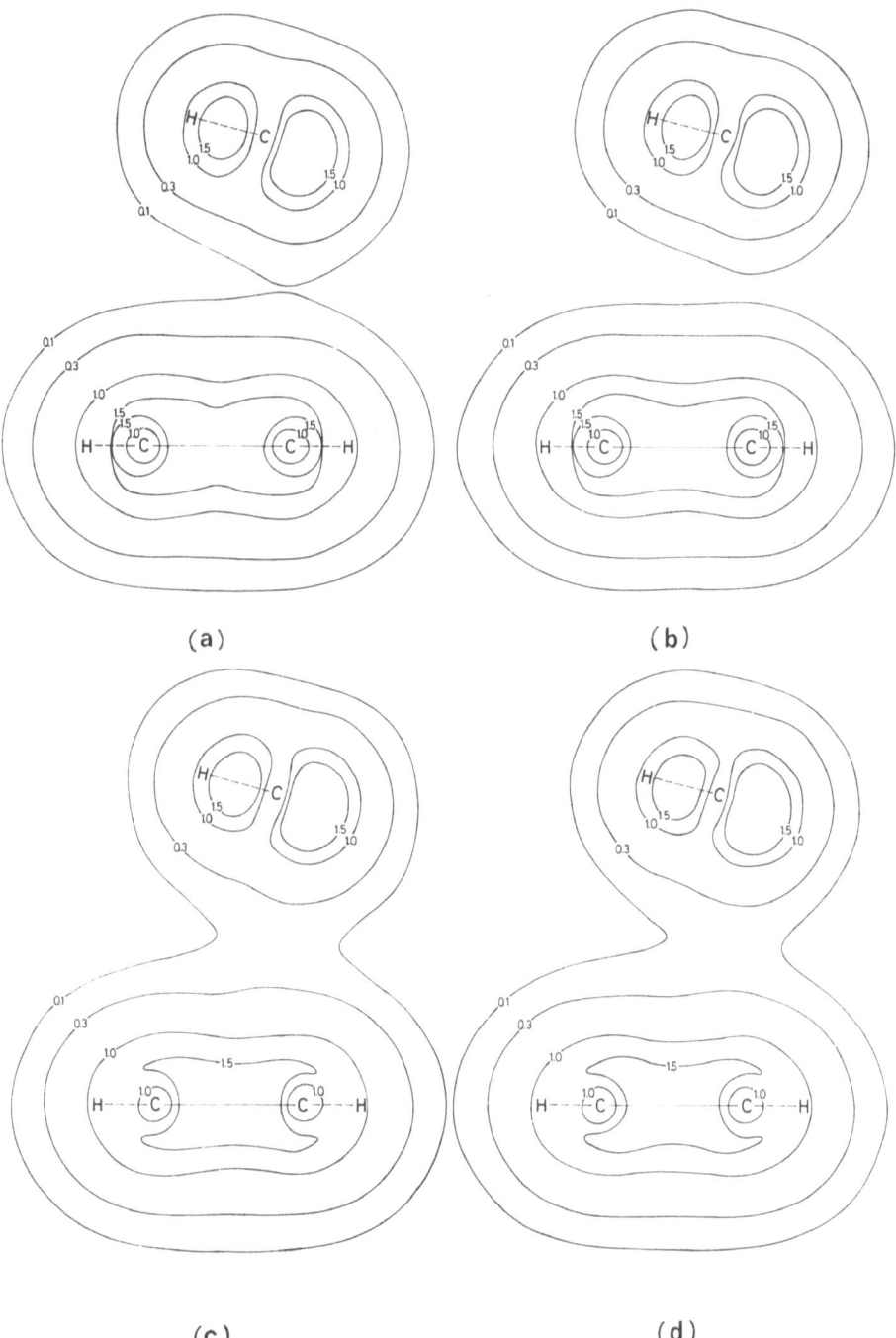

Fig. 4. Electron density maps for the system $(CH_2 + C_2H_4)$. (a) The simple sum of electron densities of CH_2 and C_2H_4. (b) Adiabatic interaction without mixing of charge-transfer configurations. (c) Interaction with mixing of charge-transfer configurations. (d) With the consideration of the HOMO–LUMO interaction only.

(6) *Hydrogen Abstraction of Methane by Methyl*

The model used for the treatment of this reaction is shown in Figure 5. The unpaired electron of methyl radical is assumed to have α spin without loss of generality. The unrestricted open-shell calculation is made to obtain electron density distribution. Figure 6 indicates the density maps with regard to spins α and β, respectively.

As in the case of interactions of two closed-shell molecules, the electron exchange causes expulsion of electrons from the intermolecular region also in the system composed of a radical and a closed-shell, as is seen by the comparison of Figures 6a and b. The difference of maps (b) and (c) indicates that the electron density in the intermolecular region is supplied by the charge-transfer interaction. The most important contribution is the β-spin transfer from CH_4-HOMO to the LU-β-MO of methyl radical, and that of the α-spin transfer from HO-α-MO of methyl to CH_4-LUMO is the next.

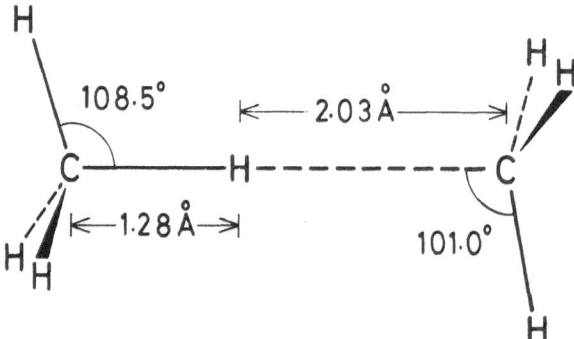

Fig. 5.   Model for the interaction of CH_4 and CH_3.

The increased amount of α-spin in the region between methane and methyl is almost cancelled by the exchange repulsion as is illustrated in (a-1) and (b-1), and the maps of (c-1) and (a-1) are eventually not much different. It is hence β-spin electrons in this case that contribute to the formation of a new C–H bond on account of the trend of pairing with α-spin electrons. In this fashion, the exchange mechanism of spin migration is really remarkable, but for the problem of chemical reactions the charge-transfer mechanism has a more decisive significance. The polarization mechanism is less important than these two.

The donor–acceptor relation in regard to α- and β-spins may depend on the polarity of the radical and on the combination of radical and molecule. Population analysis in Figure 7 shows that the methyl radical behaves as an electrophile and attracts β-spin electrons from methane molecule.

(7) *Addition of Methyl to Ethylene*

A similar calculation is applied to the addition of methyl radical to ethylene molecule. An asymmetrical model is used (Hoyland, 1971). The charge-transfer interaction is

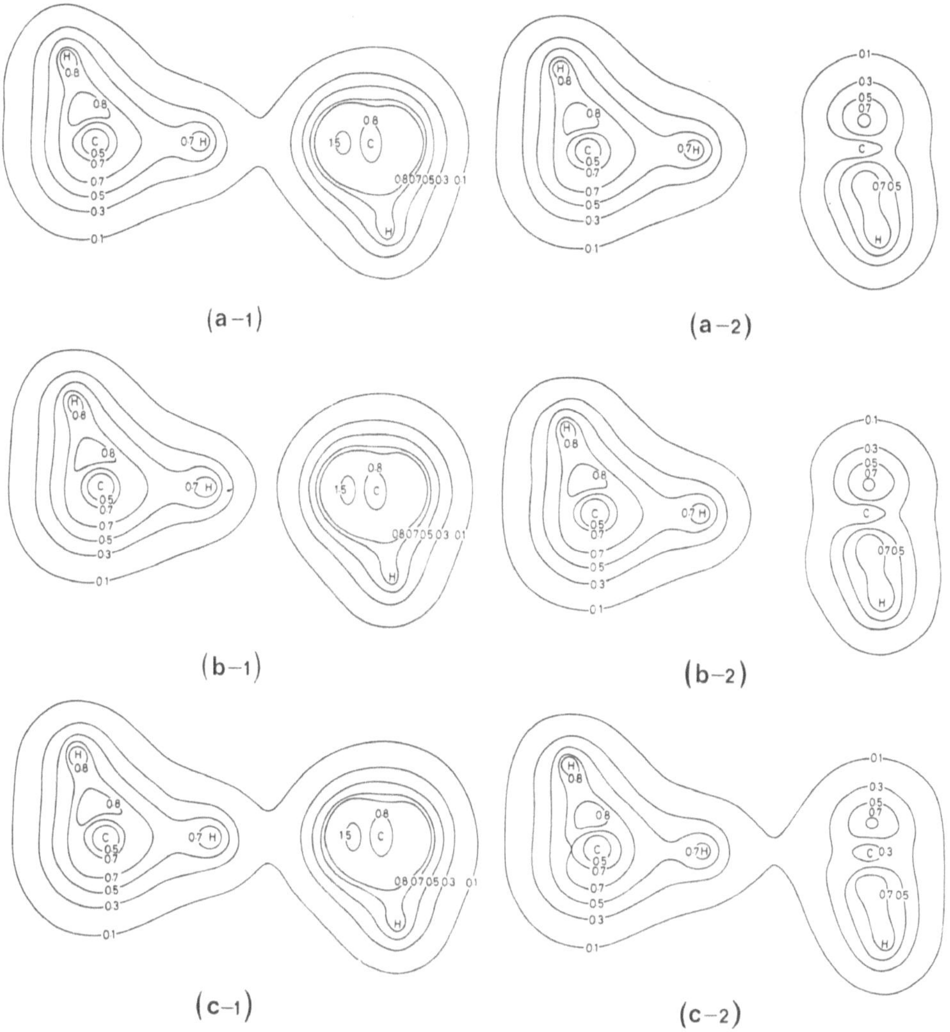

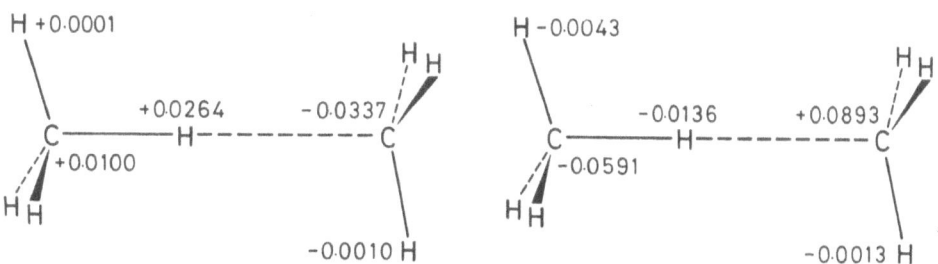

Fig. 6. The spin distributions of the system composed of methane and methyl. (a) The simple sum of electron densities of CH_4 and CH_3 without interaction. (b) The electron density of the system $(CH_4 + CH_3)$ with adiabatic interaction. (c) The electron density with the mixing of charge-transfer taken into account. -1 and -2 refer to α and β spins.

Fig. 7. The changes in atomic populations of methane and methyl due to interaction.

shown to be essential also in this case, and the α-spin transfer from the HO-α-MO of CH$_3$ to LUMO of ethylene exhibits the largest contribution and the β-spin transfer from HOMO of ethylene to the LU-β-MO of CH$_3$ is the next. The former causes an increase of electron density in the intermolecular region but this is wiped off by exchange interaction. The net accumulation of electrons between methyl carbon and one of ethylene carbons is effectuated by the latter β-spin transfer. The population analysis gives a net positive charge on methyl radical. This time methyl radical behaves as a nucleophile in the course of interaction.

B. AB INITIO CALCULATION BY LCMO APPROACH

In addition to the results of semiempirical calculation described above, an *ab initio* MO calculation has been applied to similar problems for the present purpose. The examples to be discussed here are diborane (B$_2$H$_6$) as a dimeric form of BH$_3$ (Yamabe *et al.*, 1973a), the NH$_3$—HF system (Yamabe *et al.*, 1973b), borazane as a complex of NH$_3$ and BH$_3$ (Fujimoto *et al.*, 1974), and borine carbonyl as a complex of BH$_3$ and CO (Kato *et al.*, 1974). All of these calculations have recourse to method B (LCMO wave function) described in the previous section. The minimal basis set of Slater-type orbitals, each of which is expanded in three Gaussian-type orbitals, is used for the *ab initio* calculation.

(1) B$_2$H$_6$ *Molecule as a Dimer of BH$_3$*

The electronic structure of diborane, supposed to be of bridged D_{2h} geometry, is a suitable example to be discussed conveniently by the present method. Two BH$_3$ molecules are deformed so as to make up the D_{2h} dimer. The nature of the bridge bonds in B$_2$H$_6$ is elucidated in terms of the MO's of such deformed BH$_3$. The HOMO of one of these deformed BH$_3$'s and the LUMO of the other are indicated in Figures 8a and 8b, respectively. The energy of HOMO is elevated and the LUMO is stabilized in comparison with undeformed BH$_3$, as well as the orbital lobes grow so as to make the HOMO–LUMO interaction easier (cf. Section 5). We can clearly see how the deformation of molecules favors the HOMO–LUMO interaction by overlapping. Reflecting this circumstance, the configuration analysis gives an outstanding contribution of the configurations corresponding to the (HOMO)$_I$ → (LUMO)$_{II}$ and (HOMO)$_{II}$ → → (LUMO)$_I$ transfers, and also of the ditransferred configuration corresponding to the *mutual* HOMO → LUMO transfers. The concept of donation and back-donation is thus made distinct.

   The electron density appearing in the intermolecular region results largely from the mixing of transferred configurations, among which the two HOMO–LUMO transfers are conspicuous. It is obvious that this *double* HOMO–LUMO overlapping becomes the main origin of bridge bonding structure of B$_2$H$_6$. Such *double* single-center interaction illustrated in Figure 9 should not be confounded with the two-center interaction which frequently causes the stereoselectivity in the concerted interaction (see Section 5). In this case, the exchange repulsion is also shown to expel electrons from intermolecular area to result in exerting an antibonding effect.

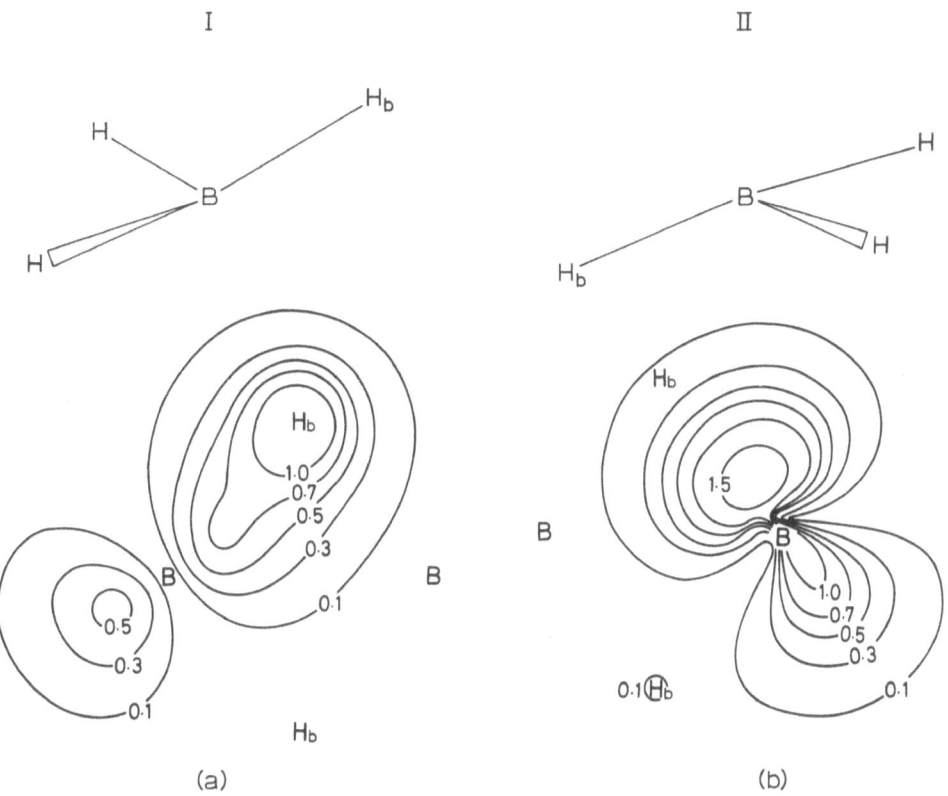

Fig. 8.   The density maps of HOMO and LUMO of deformed BH_3. (a). The HOMO of $(BH_3)_I$. (b). The LUMO of $(BH_3)_{II}$.

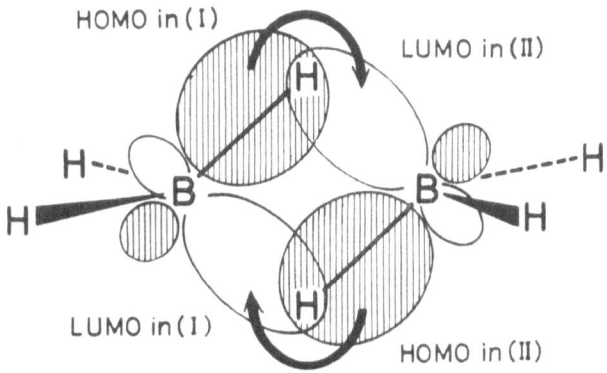

Fig. 9.   The double HOMO–LUMO overlapping in B_2H_6 ((I) and (II) imply $(BH_3)_I$ and $(BH_3)_{II}$ of Figure 8, respectively).

(2) NH_3-HF complex

The nature of the hydrogen bond in NH_3–HF complex is also a good example. A linear C_{3v} model is employed. The electron density in the intermolecular region is shown to originate chiefly from the electron transfers from occupied NH_3 MO's to unoccupied HF MO's, among which the transfer from HOMO of NH_3 to LUMO of HF contributes 96.31% of the total contribution of transferred configurations. The back donation from HF to NH_3 is small in this case.

(3) Borazane as a Complex of NH_3 and BH_3

The staggered C_{3v} geometry is assumed for the model of the NH_3-BH_3 complex. With the use of a geometry shown in Figure 10, the total amount of electrons transferred from NH_3 to BH_3 is calculated as 0.4370. The decrease in occupation number

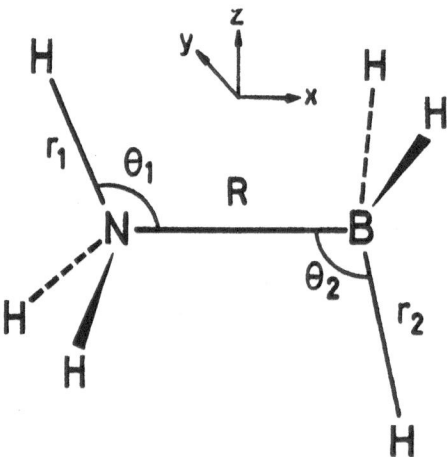

Fig. 10.   The model of the geometry of NH₃—BH₃ complex.
$R = 1.561$ Å; $r_1 = 1.011$ Å; $r_2 = 1.222$ Å; $\theta_1 = 116.0°$; $\theta_2 = 105.4°$.

of NH_3-HOMO $(3a_1)$ is 0.4394, while the increase of BH_3-LUMO $(3a_1)$ is 0.4759. The overlap population of the B—N bond comes largely from that between axially extending $2p$ orbitals and $2s$ orbitals of N and B, arising from the interaction between occupied MO's of NH_3 and unoccupied MO's of BH_3, reflecting the relation that the directions of extension of NH_3-HOMO and BH_3-LUMO are both axial and opposite to the bonds with hydrogen atoms.

The configurational expansion of the wave function of the combined system, represented by Equation (16), is as follows:

$$\Psi = 0.60755 \, \Phi_0 + 0.36786 \, \Phi_{HO \to LU} + 0.08742 \, \Phi_{HO \to LU}^{HO \to LU} + \cdots, \tag{23}$$

where HO and LU signify the NH_3-HOMO and the BH_3-LUMO, and, hence, $\Phi_{HO \to LU}^{HO \to LU}$ denotes the ditransferred configuration in which two electrons shift from the NH_3-HOMO to the BH_3-LUMO. The coefficient of this ditransferred configuration is

TABLE I

The constituents of atom bond population of N—B in NH₃BH₃

		ϱ	π
$(NH_3 \cdot BH_3$	$NH_3 \cdot BH_3$)	-0.0971	-0.0242
$(NH_3 \cdot BH_3$	$NH_3^+ \cdot BH_3^-)$	$+0.3723$	-0.0145
$(NH_3 \cdot BH_3$	$NH_3^- \cdot BH_3^+)$	$+0.0004$	$+0.0142$
$(NH_3^+ \cdot BH_3^-$	$NH_3^+ \cdot BH_3^-)$	$+0.0227$	-0.0089
$(NH_3^- \cdot BH_3^+$	$NH_3^- \cdot BH_3^+)$	-0.0009	-0.0002
$(NH_3 \cdot BH_3$	$NH_3^* \cdot BH_3$)	-0.0003	-0.0033
$(NH_3 \cdot BH_3$	$NH_3 \cdot BH_3^*)$	$+0.0530$	-0.0002
$(NH_3^* \cdot BH_3$	$NH_3^* \cdot BH_3$)	-0.0021	-0.0004
$(NH_3 \cdot BH_3^*$	$NH_3 \cdot BH_3^*)$	-0.0070	-0.0005

larger than any of monotransferred configurations except HO → LU and also any of the monoexcited configurations.

Table I shows the contributions of each 'state' to the atom bond population of N—B in NH_3BH_3. The notation $(NH_3 . BH_3 \mid NH_3^+ . BH_3^-)$ represents the contribution from the overlap densities

$$\varrho_{0K}(1) = \int \Phi_0^*(1, 2, 3, ...) \, \Phi_K(1, 2, 3, ...) \, d\tau(2) \, d\tau(3) ... ,$$

in which Φ_K represents one of the configurations corresponding to one electron transfer from NH_3 to BH_3. The others imply similar meaning, an asterisk denoting excited state. Figure 11 indicates the difference density, $\varrho_{NH_3BH_3}(1) - (\varrho_{NH_3}(1) + \varrho_{BH_3}(1))$, which clearly illustrates the effect of interference between two interacting species. All of the facts pointed out above represent the great disparity in donor–acceptor relationship between NH_3 and BH_3.

(4) *Borine Carbonyl as a Complex of* BH_3 *and CO*

The geometry of borine carbonyl is assumed here as a C_{3v} structure, i.e.

H
     \B···C=O.
H   /
 H

The total wave function is expanded into the following form:

$$\Psi(BH_3 - CO) = 0.5475 \; \Psi(BH_3 . CO) + 0.2977 \; \Psi(BH_3^- . CO^+) +$$
$$+ 0.1470 \; \Psi(BH_3^+ . CO^-) + 0.0560 \; \Psi(BH_3 . CO^*) +$$
$$+ 0.0705 \; \Psi(BH_3^* . CO) + 0.0602 \; \Psi(BH_3^{2-} . CO^{2+}) +$$
$$+ 0.0243 \; \Psi(BH_3^{2+} . CO^{2-}) + 0.0683 \; \Psi(BH_3^{\mp} . CO^{\pm}) +$$
$$+ 0.0074 \; \Psi(BH_3^* . CO^*) + \cdots . \tag{24}$$

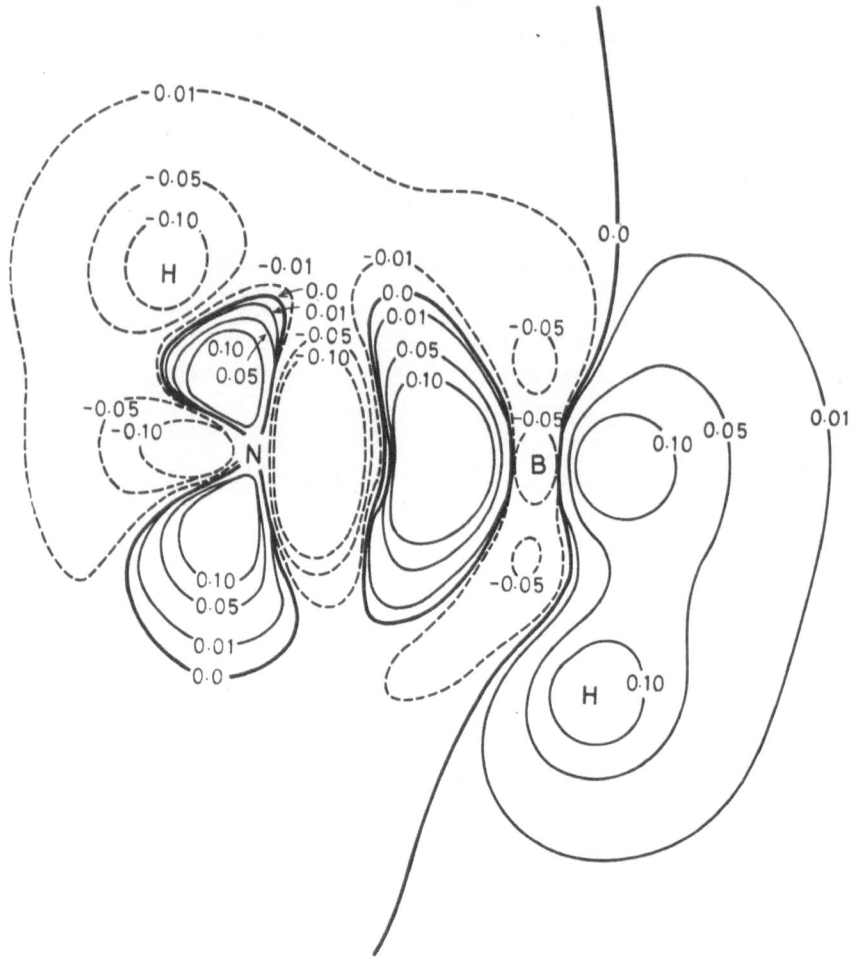

Fig. 11.   The difference density, $\varrho_{NH_3BH_3}(1) - (\varrho_{NH_3}(1) + \varrho_{BH_3}(1))$.

The square of coefficient of each 'state function' is the sum of the squares of coefficients of all possible configurations. The charge-transfer interaction plays a distinct role in the formation of this complex. The contribution of the HOMO–LUMO interaction is of discriminative importance also in this case. The constitution of the B–C 'bond' is analysed in terms of each state. The B–C atom bond population chiefly consists of the σ-component of $(BH_3^- . CO^+)$ state and, next, the π-component of $(BH_3^+ . CO^+)$ state. These components come from the overlapping of these respective states with the zero-configuration of Equation (16) in the expression of the electron density function; i.e., the density terms of such a type as

$$\sum_{M}^{\text{monotr.}} 2C_0 C_M \varrho_{0M}(1), \tag{25}$$

in which $\varrho_{0M}(1)$ implies the overlap density like the form of Equation (19) between

TABLE II

The constituents of atom bond population of B—C in BH_3CO

	ϱ	π
$(BH_3.CO \mid BH_3.CO)$	-0.0954	-0.0007
$(BH_3.CO \mid BH_3^-.CO^+)$	$+0.2521$	-0.0052
$(BH_3.CO \mid BH_3^+.CO^-)$	-0.0124	$+0.0314$
$(BH_3^-.CO^+ \mid BH_3^-.CO^+)$	$+0.0280$	-0.0023
$(BH_3^+.CO^- \mid BH_3^+.CO^-)$	-0.0069	-0.0003
$(BH_3.CO \mid BH_3.CO^*)$	-0.0021	-0.0007
$(BH_3.CO \mid BH_3^*.CO)$	$+0.0441$	-0.0006
$(BH_3.CO^* \mid BH_3.CO^*)$	-0.0007	-0.0001
$(BH_3^*.CO \mid BH_3^*.CO)$	-0.0060	-0.0001

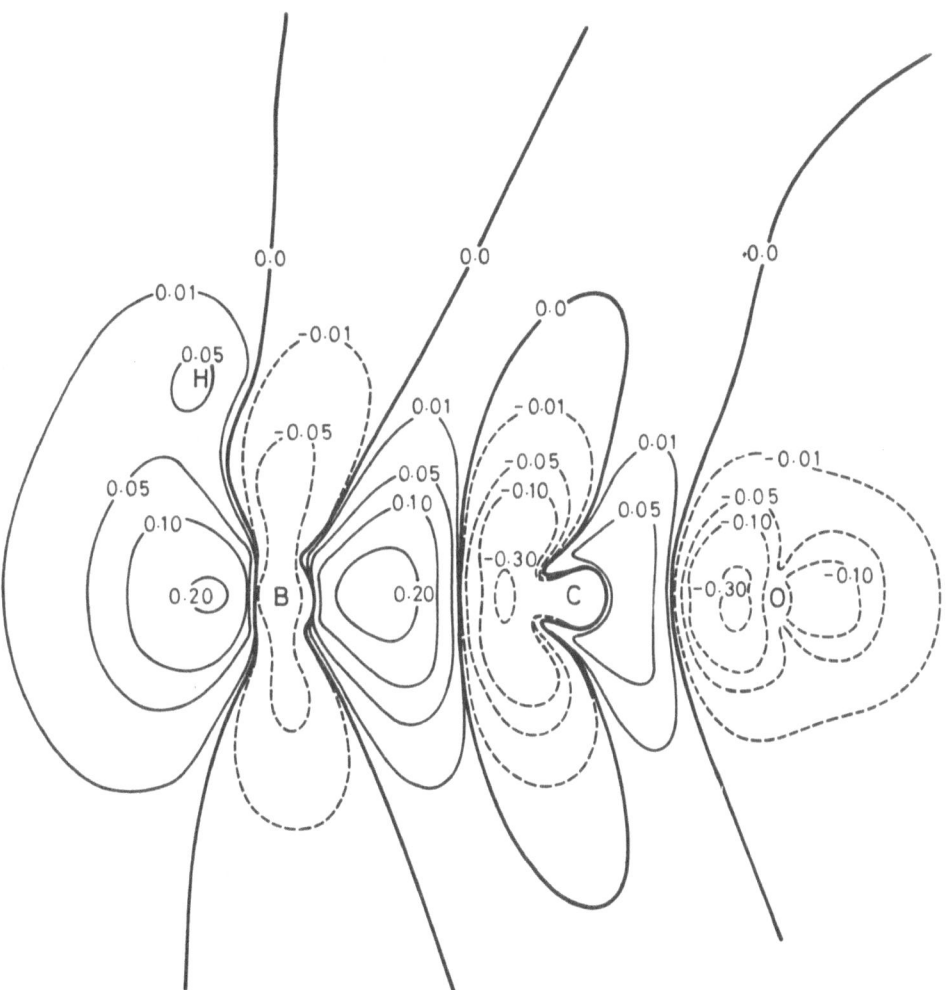

Fig. 12. The difference density, $\varrho_{BH_3CO}(1) - (\varrho_{BH_3}(1) + \varrho_{CO}(1))$.

configurations 0 and M. The σ- and π-components are distinguished by means of the type of overlapping of each pair of constituent atomic orbitals.

Table II indicates the constituents of the B—C bond in borine carbonyl, in which $(BH_3 \cdot CO \mid BH_3^- \cdot CO^+)$ signifies the contribution to the atom bond population from the overlap densities $\varrho_{KL}(1)$ where configurations K and L belong to the state functions $\Psi(BH_3 \cdot CO)$ and $\Psi(BH_3^- \cdot CO^+)$, respectively.

Figure 12 shows the difference density, $\varrho_{BH_3CO}(1) - (\varrho_{BH_3}(1) + \varrho_{CO}(1))$.

5. Favorableness of a Reaction Coordinate

The conspicuous role played by the particular orbital interaction in charge and spin transfers is obvious in view of the various examples mentioned in the last section, and should be closely related to the condition for actual occurrence of a chemical reaction and further, to the selection of favorable path of chemical reactions.

Consider a chemical reaction occurring between molecules A and B. The 'intrinsic' reaction coordinate mentioned in Section 2 begins from the infinite separation of A and B. But, practically, we are forced to discuss the most stable mutual arrangement of A and B at large, but finite, separation. This can be accomplished by using Equations (18) and (19). As the various results given in the previous section clearly show, the dominant contribution to the stabilization originates from monotransferred configurations, particularly from HOMO-to-LUMO transfers. The most contributing terms among each summation of the right-hand side of Equations (18) and (19) are

$$\sim -|H_{0,HO \to LU} - S_{0,HO \to LU}H_{00}|^2/(H_{HO \to LU, HO \to LU} - H_{00}) \tag{26}$$

and

$$C_{HO \to LU} \sim (H_{HO \to LU, 0} - S_{HO \to LU, 0}H_{00})/(H_{00} - H_{HO \to LU, HO \to LU}), \tag{27}$$

respectively. Roughly speaking, these terms parallel with the HOMO–LUMO overlapping and are inversely proportional to the HOMO–LUMO level separation between reactant molecules. Hence, the consideration of HOMO–LUMO overlapping can be a criterion of the stable spatial arrangement of A and B at large separation. At a long distance, the far-reaching effect of charge distribution is believed to dominate the orientation in charged or polar molecules. But even in such cases, the consequence of HOMO–LUMO criteria almost parallels with that of electrostatic consideration.

It happens occasionally that the species A and B form a loose intermediate and the reaction starts from such a complex. The formation of such a complex can be discussed in terms of HOMO–LUMO overlapping, similarly as mentioned before. When the intrinsic reaction coordinate begins from this equilibrium point, this will start along one of normal coordinates of this complex (cf. Section 2). The favorableness of this normal coordinate is characterized by κ of Equation (8). Evidently, the smaller the value of κ is, the more favorable is that normal coordinate. The reduction of the value of κ has recourse to the negative third term of the right-hand side of Equation (8) (Bader, 1962). Very roughly, the wave function of the ground state can be approximated by the zero-configuration function and that of an excited state k by each corresponding configuration function, say K: i.e.

$$\Psi_0 \sim \Phi_0 \quad \text{and} \quad \Psi_k \sim \Phi_K,$$

provided that the interaction is small, so that the transition density ϱ_{0k} is also approximated by

$$\varrho_{0k}(1) \sim \varrho_{0K}(1),$$

in which $\varrho_{0K}(1)$ is overlap density between configurations 0 and K. Obviously, the main contribution to the third term of the right-hand side of Equation (8) will come from the term in which Φ_K is $\Phi_{HO \to LU}$, since we are discussing the complex favored with effective HOMO–LUMO overlapping. The normal coordinate corresponding to the reaction coordinate must be related to the direction of closer mutual approach. But the overlap density $\varrho_{0, HO \to LU}$ has appreciable values in the intramolecular region, since

$$\varrho_{0, HO \to LU}(1) \sim c\{a_{HO}(1) b_{LU}(1) + 0(s_2)\}, \tag{28}$$

where c is a numerical constant, a_{HO} and b_{LU} is HOMO of A and LUMO of B, respectively, and $0(s_2)$ denotes small terms of the square of overlap integrals between the MO's of A and B. Accordingly, the nuclear motion to this intermolecular region will cause stabilization. Such a circumstance reduces the value of κ and contributes to the favorableness of this reaction coordinate, so that in such cases the overall reaction will be facilitated by this complex formation.

Once the reaction coordinate departs from an equilibrium point, the energy change, ΔW, due to the small nuclear displacement, is given by Equation (5), i.e. as

$$\Delta W = \Delta V_{nn} + \int \varrho(1) \, \Delta V_n(1) \, dv(1), \tag{29}$$

where ΔV_{nn} is the change of nuclear–nuclear repulsion due to this nuclear displacement and the $\Delta V_n(1)$ is the electron–nuclear attraction for electron 1. The favorableness of a reaction coordinate depends on the degree of the reduction of unstabilization due to ΔV_{nn} by the negative contribution of the second term of the right-hand side of Equation (29) in case of mutual approach of two reactant molecules along the reaction coordinate. If the electron density function, $\varrho(1)$, shows an accumulation in the intermolecular area in the ground state of the combined system, the approach of nuclei to this area will give a *negative* value of the integral. In this connection, it seems useful to consider a simple case in which only the zero-configuration and one HO $\to$ LU transfer configuration, denoted simply by subscript K, are mixed. Ground states and excited states are specified by superscripts (g) and (e), respectively. The approximate wave functions are represented by

$$\psi^{(g)} \simeq \left\{ 1 - \frac{S_{K0}(S_{0K}H_{00} - H_{0K})}{H_{KK} - H_{00}} \right\} \Phi_0 + \frac{S_{K0}H_{00} - H_{K0}}{H_{KK} - H_{00}} \Phi_K,$$

$$\Psi^{(e)} \simeq -\frac{S_{0K}H_{KK} - H_{0K}}{H_{KK} - H_{00}}\Phi_0 + \left\{1 - \frac{S_{0K}(S_{K0}H_{KK} - H_{K0})}{H_{KK} - H_{00}}\right\}\Phi_K, \qquad (30)$$

in the approximation of Equation (19). The density distribution is written as

$$\varrho(1) = F_{00}(1) + F_{KK}(1) + F_{0K}(1), \qquad (31)$$

in which F_{00}, etc. stand for the contribution from $\Phi_0\Phi_0$, etc., and it follows that

$$F_{0K}^{(g)}(1) \sim \frac{S_{0K}(S_{K0}H_{00} - H_{K0})}{H_{KK} - H_{00}} \cdot \frac{\varrho_{0K}(1)}{S_{0K}},$$

$$F_{0K}^{(e)}(1) \sim -\frac{S_{K0}(S_{0K}H_{KK} - H_{0K})}{H_{KK} - H_{00}} \cdot \frac{\varrho_{K0}(1)}{S_{K0}}. \qquad (32)$$

The function of the second factor of the right-hand side obviously has positive values in the intermolecular region, and the coefficient has positive definite sign in the ground-state and negative sign in the excited state in usual charge-transfer interactions. Hence, the ground-state charge-transfer interaction will contribute to stabilization in the nuclear motion of mutual approach of the reactants. If the circumstances favor the HOMO–LUMO interaction to make this effect strong enough for overcoming the effect of exchange repulsion which expels electrons out of the intermolecular region, the electron density will become positive in this region to aid the new bond formation. It has been shown that all of the various reactions cited in Section 4 satisfy this condition for actual occurrence.

As has been shown in Equation (10), the difference of electron density, $\Delta\varrho(1)$, and the difference of electron–nuclear attraction, ΔV_n, are connected by

$$\Delta\varrho(1) = 2\sum_k{}' \varrho_{0k}(1) \frac{\displaystyle\int \varrho_{k0}(2)\, \Delta V_n(2)\, dv(2)}{W_0 - W_k}. \qquad (33)$$

The notation of $\Delta V_n(2)$ signifies the quantity referred to the coordinates of electron 2. This equation is applicable to any small nuclear displacement. If it is permitted that only one important term in the summation is taken, this equation becomes

$$\Delta\varrho(1) \sim 2\varrho_{0, HO \to LU}(1) \frac{\displaystyle\int \varrho_{HO \to LU, 0}(2)\, \Delta V_n(2)\, dv(2)}{H_{00} - H_{HO \to LU, HO \to LU}}, \qquad (34)$$

where the energies W_0 and W_k are approximated by H_{00} and $H_{HO \to LU, HO \to LU}$, respectively. Since the overlap density $\varrho_{HO \to LU, 0}$ has appreciable values only in the intermolecular region as has been mentioned above, the nuclear displacement to this region gives a negative value of the numerator of the right-hand side of Equation (34) in the intermolecular region. Accordingly, the mutual approach of two molecules A and B, keeping the arrangement of HOMO–LUMO overlapping, causes an accumulation of electrons in the intermediary region between two molecules. This is the genesis of a chemical bond between two reactants.

The electron accumulation in the intermolecular region originates largely in the charge-transfer from HOMO of one molecule A to LUMO of the other B, or vice versa. But both the electron escape from a bonding MO and the electron acceptance in an antibonding MO will cause the weakening of bonds. In this fashion, the formation of the new bonds is accomplished at the sacrifice of the bond fission. This is the bond exchange in chemical reactions and the cause of molecular deformation. As the chemical interaction becomes stronger in favored cases, the following three principles work cooperatively, giving rise to an accelerating effect for the promotion of reaction (Fukui and Fujimoyo, 1969b; Fujimoto and Fukui, 1972):

(1) *Positional Parallelism between the Charge-Transfer and the Bond Interchange*

At an early stage, the interaction begins between the positions where HOMO of one reactant and LUMO of the other overlap effectively. But in a majority of organic compounds, saturated or unsaturated, the position of large amplitude of HOMO is at the same time the position where the bond order with the other part of molecule is large and, hence, the position liable to undergo loosening of the bonds with the other part of the molecule in case of electron runaway from HOMO. Similarly, the position of large LUMO extension is also easily subject to weakening of the neighboring bonds in case of electron acceptance in LUMO (Fukui and Fujimoto, 1969b). This relation holds also in deformed molecules at a point on the reaction coordinate, since such a parallel relation is somewhat of mathematical nature, being almost valid in any nuclear arrangement (Fukui and Fujimoto, 1969b).

(2) *Successive Narrowing of the HOMO-LUMO Level Separation*

Since a free molecule takes a nuclear arrangement of the least energy, the deformation due to bond exchange gives rise to unstabilization, which is caused by the elevation of levels of occupied MO's. The MO which dominantly contributes to this unstabilization must be HOMO, since in this MO the actual bond loosening takes place at the bonds where the partial bond order is large, resulting from the positional parallelism between the reaction and the bond weakening. In consequence, the HOMO level rises conspicuously. Similarly, unoccupied orbitals fall according to the molecular deformation in general, among which LUMO goes down most remarkably. Many instances are available (Fukui and Fujimoto, 1969b). As the reaction proceeds, the HOMO–LUMO separation between the two reactants becomes smaller. The charge-transfer interaction takes place more easily when the HOMO–LUMO level difference becomes smaller. This is clearly understood by Equation (20).

(3) *Gradual Increase of the Localizing Character of HOMO and LUMO at the Reaction Center*

The molecular deformation accompanied by the proceeding of reaction gives rise to a growth of HOMO or LUMO amplitude near the reaction center. This relation applies in a wide range of compounds, and is understood by the following consideration. The conjugation between the reaction site and the remainder decreases successively with proceeding of the reaction. At the limit of vanishing conjugation, an MO localizing at the reaction center appears. In this way, the MO is confined within a narrow region in the vicinity of the reaction site. Such a tendency is valid even not in this extreme case. The HOMO and the LUMO amplitudes increase at the reaction center, so that the HOMO–LUMO transfer interaction grows, since the numerator of C_M in Equation (20) is roughly proportional to the product of HOMO and LUMO amplitudes of both reactants.

As has been mentioned, the charge-transfer interaction takes place between the positions of large HOMO and LUMO extension. Such positions are at the same time those which are liable to loose the neighboring bonds. The bond interchange causes molecular deformation, giving rise to the narrowing of HOMO–LUMO separation and the growth of HOMO and LUMO extension at the reaction center. Consequently, the charge-transfer and the bond exchange are promoted. In this way, initially very weak interaction is magnified to be a strong one, so that the reaction is accelerated. This is the mechanism of favored occurrence of a reaction.

In complicated polyatomic molecules, many occupied and unoccupied levels exist other than HOMO and LUMO. The difference of energy level can not discriminate HOMO and LUMO only. The reason why the particular MO interaction has a decisive significance in many reactions will first be clear with the explanation presented above.

The particular role of HOMO and LUMO in chemical reactions is thus evident. But the level of each MO is influenced by many factors. Also an energetically highest occupied MO happens to be unable to overlap with LUMO of the other molecule e.g. on account of symmetry relation in order to cause a given reaction. In such cases, the reaction will generally be not smooth, but when another occupied MO satisfying the necessary condition lies in the vicinity of the HOMO level, it is possible that this MO will play the part of HOMO. The same can be said with respect to LUMO. In these cases, the particular MO's can be *high-lying* occupied MO's and *low-lying* unoccupied MO's.

All the discussions made so far in this section are referred to the singele-center interaction in which only one bond is newly forming between two reactants A and B. The interaction of three species has been treated simply (Fujimoto *et al.*, 1973). In the reaction in which two bonds are simultaneously formed, the circumstances are more complicated. Here, it is recalled to mind that the double single-center reactions mentioned in Section 4B(1) should be distinguished from the 'concerted' two-center reactions, which is the object of immediate discussion. The electron distribution in the intermolecular region is almost dominated by the HOMO–LUMO interaction, as has been explained before. But it is evident that, if the HOMO–LUMO overlappings in the vicinity of these two centers of reaction are comparable to each other and have different signs, all the effects of HOMO–LUMO interaction contributing to the favorableness of the reaction coordinate hitherto discussed are cancelled, and other repulsive effects will be controlling. Therefore, the condition of actual occurrence of a concerted two- (or, more generally, many-) center reaction is that the overlap density has the same sign in the centers of bond formation. This conclusion agrees with the result already pointed out (Fukui, 1964).

The arguments so far made can essentially be applied also to the *intramolecular* reaction. In this case, the HOMO–LUMO criteria are employed with respect to the two parts of the molecule appropriately selected by the consideration of the essential relation to that reaction (Fukui, 1971a).

6. Theoretical Tracing of Reaction Paths

The locus of the intrinsic reaction coordinate is given by the differential equation, Equation (1), in which the denominators are replaced by Equation (5):

$$\cdots = \frac{M_\alpha \, dX_\alpha}{\frac{\partial V_{nn}}{\partial X_\alpha} + \int \varrho(1) \frac{\partial V_n(1)}{\partial X_\alpha} \, dv(1)} = \frac{M_\alpha \, dY_\alpha}{\frac{\partial V_{nn}}{\partial Y_\alpha} + \int \varrho(1) \frac{\partial V_n(1)}{\partial Y_\alpha} \, dv(1)} =$$

$$= \frac{M_\alpha \, dZ_\alpha}{\frac{\partial V_{nn}}{\partial Z_\alpha} + \int \varrho(1) \frac{\partial V_n(1)}{\partial Z_\alpha} \, dv(1)} = \cdots. \tag{35}$$

If the nuclear configuration of a point on the reaction coordinate is given, the direction of displacement of each nucleus in the usual three-dimensional space is obtained so as to determine the 'next' point. Repetition of such procedure results in a locus. The determination of positions of nuclei of this starting point is always possible by finding the geometry of minimum energy, by varying the geometry of the total system, with a given, definite distance between the centers of gravity of two reactant molecules. An appropriate electronic device might be able to visualize even a quasi continuous motion of nuclei along the intrinsic reaction coordinate. Such a theoretical tracing of reaction path may be called 'reaction ergodography', and realization of this would be of practical significance from the standpoint of instructive puposes. How to overcome the difficulties expected from the accumulation of error due to successive plots and the essential error accompanying the Hellmann–Feynman usage of nonaccurate wave functions will be the point.

7. Conclusion

Nonempiricism having been introduced recently into chemistry, which is originally an 'empirical' science where division was the means of control, seems to serve in enabling generalization and synthesis of chemistry and chemical concept. Now, for instance, tolerable methods of nonempirical calculation for chemically interacting systems are available to seek for a unique common criterion for judging the favorableness of chemical interactions, whether or not interior intramolecular reactions of various types, heterolytic or homolytic, singlecentered or multicentered, and also molecule or molecular complex formations, and irrespective of the kind of the molecules interacting, inorganic or organic, saturated or unsaturated, and radical or excited molecules. However, it must be premised that such a criterion should be employable for the 'chemical' design by chemists. In this sense, the concept of particular orbital interaction between molecular species is expected to be qualified as a candidate for such a criterion, since this is based on the presumably immutable chemical concept of 'molecule', and is probably suitable for the understanding of usual chemists who can hardly get rid of the usage of isolated molecule concept.

Acknowledgements

The author thanks Mr Shinichi Yamabe, Mr Shigeki Kato, Mr Tsutomu Minato, and Mrs Chizuko Tanaka for their cooperation.

References

Anh, N. T.: 1970, *Les Règles de Woodward-Hoffmann*, Edscience, Paris.
Baba, H., Suzuki, S., and Takemura, T.: 1969, *J. Chem. Phys.* **50**, 2078.
Bader, R. F. W.: 1962, *Can. J. Chem.* **40**, 1164.
Fujimoto, H. and Fukui, K.: 1972, *Adv. Quant. Chem.* **6**, 177.
Fujimoto, H., Yamabe, S., and Fukui, K.: 1971a, *Tetrahedron Letters*, pp. 439, 443.
Fujimoto, H., Yamabe, S., and Fukui, K.: 1971b, *Bull. Chem. Soc. Jap.* **44**, 2936.
Fujimoto, H., Yamabe, S., and Fukui, K.: 1972a, *Bull. Chem. Soc. Jap.* **45**, 1566.
Fujimoto, H., Yamabe, S., and Fukui, K.: 1972b, *Bull. Chem. Soc. Jap.* **45**, 2424.
Fujimoto, H., Katata, M., Yamabe, S., and Fukui, K.: 1972c, *Bull. Chem. Soc. Jap.* **45**, 1320.
Fujimoto, H., Kato, S., Yamabe, S., and Fukui, K.: 1973, *Bull. Chem. Soc. Jap.* **46**, 1071.
Fujimoto, H., Kato, S., Yamabe, S., and Fukui, K.: 1974, *J. Chem. Phys.* (in press); compare also the references cited therein.
Fujimoto, H., Yamabe, S., Minato, T., and Fukui, K.: 1972d, *J. Am. Chem. Soc.* **94**, 9205.
Fukui, K.: 1964, in *Molecular Orbitals in Chemistry, Physics, and Biology* (ed. by P.-O. Löwdin and B. Pullman), Academic Press, New York, p. 513.
Fukui, K.: 1965a, in Modern Quantum Chemistry, Istanbul Lectures (ed. by O. Sinanoğlu), Academic Press, New York, Part I, p. 49.
Fukui, K.: 1965b, *Tetrahedron Letters*, p. 2009.
Fukui, K.: 1970a, in *Sigma Molecular Orbital Theory* (ed. by O. Sinanoğlu and K. B. Wiberg), Yale University Press, New Haven, Conn., U.S.A., p. 121.
Fukui, K.: 1970b, *Fortschr. Chem. Forsch.* **15**, 1.
Fukui, K.: 1970c, *J. Phys. Chem.* **74**, 4161.
Fukui, K.: 1971a, *Accounts Chem. Res.* **4**, 57.
Fukui, K.: 1971b, in *XXIIIrd Int. Congr. Pure Appl. Chem. (IUPAC)*, Vol. 1, Butterworths, London, p. 65.
Fukui, K. and Fujimoto, H.: 1968, *Bull. Chem. Soc. Japan* **41**, 1989; and compare the references cited therein.
Fukui, K. and Fujimoto, H.: 1969a, in *Mechanisms of Molecular Migrations* (ed. by B. S. Thyagarajan), Vol. 2, Wiley-Interscience, New York, p. 117.
Fukui, K. and Fujimoto, H.: 1969b, *Bull. Chem. Soc. Jap.* **42**, 3399.
Fukui, K., Yonezawa, T., and Shingu, H.: 1952, *J. Chem. Phys.* **20**, 722.
Fukui, K., Yonezawa, T., Nagata, C., and Shingu, H.: 1954a, *J. Chem. Phys.* **22**, 1433.
Fukui, K., Yonezawa, T., and Nagata, C.: 1954b, *Bull. Chem. Soc. Jap.* **27**, 423.

Fukui, K., Fujimoto, H., and Yamabe, S.: 1972, *J. Phys. Chem.* **76**, 232.

Hoffmann, R.: 1968, *J. Am. Chem. Soc.* **90**, 1475.

Hoffmann, R., Gleiter, R., and Mallory, F. B.: 1970, *J. Am. Chem. Soc.* **92**, 1460.

Hoyland, J. R.: 1971, *Theor. Chim. Acta* **22**, 229.

Kato, S., Fumimoto, H., Yamabe, S., and Fukui, K.: 1974, *J. Am. Chem. Soc.*, in press.

Klopman, G.: 1968, *J. Am. Chem. Soc.* **90**, 223.

Klopman, G. and Hudson, R. F.: 1967, *Theor. Chim. Acta* **8**, 165.

Salem, L.: 1968, *J. Am. Chem. Soc.* **90**, 543, 553.

Salem, L.: 1969a, *Chem. Phys. Letters* **3**, 99.

Salem, L.: 1969b, *Chem. Brit.* **5**, 449.

Woodward, R. B. and Hoffmann, R.: 1965, *J. Am. Chem. Soc.* **87**, 395, 2511.

Woodward, R. B. and Hoffmann, R.: 1969a, *Angew. Chem.* **81**, 797.

Woodward, R. B. and Hoffmann, R.: 1969b, *The Conservation of Orbital Symmetry*, Academic Press, New York.

Yamabe, S., Minato, T., Fujimoto, H., and Fukui, K.: 1973a, *Theor. Chim. Acta*, in press.

Yamabe, S., Kato, S., Fujimoto, H., and Fukui, K.: 1973b, *Bull. Chem. Soc. Jap.* **46**, 3619.

PART IV

FORMATION AND EVOLUTION OF MOLECULAR EXCITED STATES

Chairman: Robert Parr

The Johns Hopkins University, Baltimore, U.S.A.

A plenary session.

PREPARATION AND DECAY OF EXCITED MOLECULAR STATES

JOSHUA JORTNER and SHAUL MUKAMEL

Department of Chemistry, Tel-Aviv University, Tel-Aviv, Israel

Abstract. This paper reviews the current state of the art of theoretical understanding of the radiationless decay channels in electronically excited states of polyatomic molecules.

1. Introductory Comments

The theoretical chemist dines and wines well at the theoretical physicist's table. The borderline between theoretical physics and quantum chemistry is fuzzy and not well defined. The theoretical techniques employed are identical, only the goals and the nature of the specific questions often differ. The theory of relaxation phenomena in excited electronic states of polyatomic molecules, which is the subject matter of the present paper, draws heavily on work performed in the fields of radiation theory, collision theory, nuclear reactions and even elementary particle physics. This is not surprising as the decaying electronically excited states of a molecular system are amenable to theoretical descriptions in terms of compound states or resonances. Thus, there is a set of general features common to a wide class of physical systems. The decay characteristics of metastable states in atomic, molecular, solid state, nuclear and elementary particle physics should be described in terms of a unified theoretical picture.

Let us now attempt to specify and classify the field of non-radiative transitions. From the experimentalist's point of view, radiationless processes involve any 'transition' between the 'states' (i.e. electronic, vibrational, rotational) of a system (i.e. an atom, a molecule or a solid) which do not involve absorption or emission of radiation. These processes encompass a wide class of phenomena, which can be classified in the following manner:

A. Atoms: (A1) Atomic autoionization [1]
B. Molecules: (B1) Molecular autoionization [2]
                   (B2) Predissociation [2]
                   (B3) Thermally induced predissociation
                   (B4) Electronic relaxation between different states of a large molecule. Internal conversion and intersystem crossing [3–15]
                   (B5) Vibrational relaxation [16]
                   (B6) Photochemical rearrangements [17–19]
C. Solids: (C1) Thermal ionization of impurities [20]
                   (C2) Thermal electron capture [20]
                   (C3) Electronic relaxation in impurity states
                   (C4) Electronic energy transfer [21–23]
                   (C5) Autoionization of metastable excitons [24]
                   (C6) Electronic relaxation of exciton states [24]

R. Daudel and B. Pullman (eds.), The World of Quantum Chemistry, 145–209. All Rights Reserved

D. Solutions: (D1) Thermal electron transfer [25]
                (D2) Electronic energy transfer [21]
                (D3) Dynamics of electron localization [26].

From the theoretician's point of view, this broad definition is fraught with difficulties, involving some hidden assumptions and pitfalls. First, and most important,
the concept of a 'state of the system' has to be specified. Obviously, if we specify the
Hamiltonian of the system in terms of the molecular Hamiltonian, all time dependent
transitions between the stationary states of this molecular Hamiltonian are radiative
in nature [27]. Thus, in order to exhibit nonradiative evolution, the system has to be
'prepared' by some experiment, via optical, electron impact, thermal or collisional
excitation in a nonstationary state of the system's Hamiltonian. The resulting
'metastable state' will subsequently exhibit time evolution, where some of its decay
channels may be nonradiative in nature. Second, the radiative decay channels (which
involve a change in the occupation of the photon field) and the nonradiative decay
channels (which conserve the number of photons) in an excited state cannot be separated. A large bulk of physical information now available [28–33] originates from
optical studies of optical line shapes, radiative decay characteristics of electronically
excited states and quantum yield measurements. A complete theoretical understanding
of radiationless processes in excited electronic states of molecules should emerge from
the proper description of their radiative decay.

The relevant relaxation processes in excited molecular states can be classified as follows:
 (a) Radiative decay.
 (b) Direct decomposition, i.e. photodissociation and photoionization.
 (c) Indirect decomposition, i.e. predissociation and autoionization.
 (d) Nonradiative electronic relaxation in excited states of large molecules.
 (e) Vibrational relaxation.
 (f) Unimolecular photochemical rearrangement reactions.

These processes, listed above, provide the main decay channels which can be
encountered in excited electronic states. Processes (a)–(c) which obviously occur in an
'isolated', collision-free, molecule are not expected to be modified by an external
medium. Process (d) occurs in an isolated large molecule which corresponds to the
'statistical limit', while it may be induced by medium perturbations in a small molecule. Process (e) exclusively originates from medium perturbations. Processes of type
(f) are very complex and may involve a combination of processes (b)–(d).

We would like to discuss some of the results of recent work [34–68] regarding the
fate of electronically excited molecular states, in an attempt to provide a unified
picture for the interplay between the various basic decay channels which involve
radiative decay, direct and indirect decomposition, electronic and vibrational relaxation.

2. Experimental Observables

From the experimentalist's point of view, the following spectroscopic information is of fundamental importance for the elucidation of the decay characteristics of excited molecular states:

(1) Decay characteristics of electronically excited state. The most detailed information originates from the time and energy resolved pattern of excited electronic states. In the simplest common case, the decay pattern is exponential and the excited state is characterized by a single lifetime. More complex decay patterns which involve a superposition of exponentials were also recorded. Finally, the decay may (in principle) exhibit an oscillatory behavior, which originates from interference between closely spaced discrete levels. This phenomenon of quantum beats which is well known in level crossing atomic spectroscopy, was not yet conclusively established in large molecules.

(2) Cross sections for photon scattering from molecules. These involve both elastic photon scattering to the ground electronic-vibrational state and resonance Raman scattering to the ground electronic configuration. We shall refer to these processes as 'resonance fluorescence'.

(3) Optical absorption line shapes.

(4) Cross sections for photodissociation for molecules undergoing direct photodecomposition.

(5) Quantum yields for resonance fluorescence.

(6) Quantum yields for photodissociation.

These experimental observables fall into two different categories. In general, two classes of experiments which will be referred to as 'short excitation' and 'long excitation' processes can be utilized to extract physical information concerning the decay of electronically excited states of large molecules. When the temporal duration of the exciting photon field is short relative to the reciprocal width of the molecular resonance, it is feasible to separate the excitation and the decay processes and to consider the decay pattern of the metastable state. This experimental approach involves a 'short excitation' process. The study of the decay pattern of an 'initially' excited state corresponds to such a 'short excitation' experiment. On the other hand, when the exciting photon field is characterized by a high energy resolution, being switched on for long periods (relative to the decay time) the excitation and the decay processes cannot be separated and one has to consider resonance scattering from large molecules within the framework of a single quantum-mechanical process. Such 'long excitation' experiments involve the determination of optical line shapes, cross sections for resonance fluorescence, for intramolecular electronic relaxation and for photodissociation. Emission quantum yields can be obtained both from 'short time' excitation experiments (by the integration of the decay curve) or from 'long time' experiments (which yield the energy dependence of the quantum yield). In the special case of an isolated resonance, the information regarding the resonance width, its decay time and the corresponding energy independent quantum yield obtained from 'long time' excitation experiments should be identical to that obtained from a 'short time' excitation. However, for more complex physical situations one cannot get away by considering just the resonance width. When interference effects between resonances are exhibited, the decay curve in the 'short excitation' experiment is nonexponential, the quantum yield is energy-dependent, and the information obtained from 'short time' and 'long time' experiments is complementary but not identical.

In the discussion of 'short time' and 'long time' experiments, we have focused attention on the nature of the optical excitation process. One can subsequently consider two extreme types of photon detection. Broad band detection which admits all emitted photons and narrow band detection which spans a narrow energy region. It is important to note that the detection process is independent of the excitation mode. Both detection methods are useful for specific purposes.

3. Models for Relaxation of Electronically Excited States

During the past few years several theoretical models for the decay characteristics of excited states were developed and solved at various levels of sophistication. These models provide a schematic description of the energy levels of a zero-order Hamiltonian which should also include the radiation field, while the residual interactions couple the zero-order states. Clearly, the choice of the basis set is merely a matter of convenience and does not affect any observable quantities. Let us consider the conventional dissection of the molecular Hamiltonian, H, for a molecular system including the radiation field,

$$H = H_M + H_{rad} + H_{int},$$
$$H_M = H_{MO} + H_V,$$

$$(3.1)$$

where the molecular Hamiltonian, H_M, consists of H_{MO}, the zero-order molecular Hamiltonian and H_V which corresponds to the nonadiabatic intramolecular interaction which involves the interstate coupling via nuclear kinetic energy or spin orbit coupling. There has been lately a lively controversy regarding the nature of these interactions [52–53, 69–77]. We shall avoid a detailed discussion of this problem and just point out, at the risk of triviality, that any untruncated complete molecular zero order basis set is adequate for the specification of H_M. It was recently emphasized [52–53, 74, 77] that the Born–Oppenheimer (BO) basis set is superior to the crude adiabatic basis set as it minimizes off-resonance coupling terms between different electronic configurations. Thus the utilization of the BO basis set (i.e. $H_{MO} = H_{BO}$) and the identification of H_V with the breakdown of the BO approximation provides the ideological basis for the basic model systems for electronic relaxation which usually involve a two-electronic level system for the excited state. To complete the definitions in Equation (3.1), H_{rad} is the Hamiltonian for the free electromagnetic field, while H_{int} is the radiation-matter interaction term.

The electronically excited eigenstates of H_{MO} will be labelled as follows: (1) Discrete levels $|s\rangle$, $|r\rangle$, etc. which correspond to low lying vibronic components of excited electronic configurations. (2) A manifold of levels $\{|l\rangle\}$ corresponding to a lower electronic configuration and which are quasidegenerate to the $|s\rangle$ (and/or to the $|r\rangle$) level. The levels of types (1) and (2) are sufficient for the description of electronic relaxation. For direct and indirect photodissociation we have to consider in addition: (3) an intramolecular dissociative continuum $\{|l\rangle\}$. The electronic ground state of the system will be labelled by the vibronic components $|gv\rangle$ where $v=0$ refers to the vibrationless level while $v\neq0$ represents excited vibronic components. Note that for the low lying ground state $|gv\rangle$ can be considered as eigenfunctions of H_0 as well as of H_0+H_V, as off-resonance nonadiabatic corrections for these states are negligible. The eigenfunctions of H_{rad} will be given by the zero photon state $|vac\rangle$ and by one-photon states $|ke\rangle$ where k and e are the wave vector and the polarization vector of a photon, respectively.

A possible choice of the zero-order Hamiltonian, being often used is

$$H_0 = H_{MO} + H_{rad} \equiv H - V,$$
$$V = H_V + H_{int}. \tag{3.2}$$

The eigenstates of H_0 consists of zero-photon states $|s, vac\rangle$, $|r, vac\rangle$, $\{|l, vac\rangle\}$ etc., and of one-photon states $|gv, ke\rangle$.

It is important to emphasize that the separation of the Hamiltonian as expressed by Equation (3.1) is by no means unique and this can be accomplished in a variety of ways. Another useful approach is to adopt the molecular eigenstates basis $|n\rangle$, which diagonalizes the total electronic Hamiltonian H_M [7, 34] whereupon

$$H_0 = H_M + H_{rad},$$
$$V = H - H_0 = H_{int}. \tag{3.2a}$$

The eigenstates of H_0 consist now of the one-photon states $|gv, ke\rangle$ and the zero-photon states $|n, vac\rangle$. The hierarchy of basis sets useful for the description of decay channels in excited molecular states is summarized in Table I.

It should be noted that we have neglected the contribution of multiphoton states of the radiation field such as $|gv, ke, k'e'\rangle$, however, in general (at least for conventional excitation sources), the contribution of such states is negligibly small. We shall now proceed to portray the energy level diagrams for the excited states of the relevant physical systems.

In figure 1a we present a highly idealized level scheme which provides a universal model for radiationless transitions in a large 'isolates' molecule [7, 34, 78]. A zero-order vibronic level $|s\rangle$ of a higher electronic state, which carries all the oscillator strength from the ground state, is quasidegenerate with an intramolecular manifold $\{|l\rangle\}$ of bound levels which correspond to a lower electronic state. The $\{|l\rangle\}$ manifold is devoid of oscillator strength. In the case of a large molecule when the energy gap between the electronic origins of $|s\rangle$ and the $\{|l\rangle\}$ states is reasonably large (~1 eV) we have large densities of $\{|l\rangle\}$ states which are quasidegenerate with $|s\rangle$. The $|s\rangle$ state plays a special central role as it is optically accessible from the ground state. This situation is analogous to a 'doorway state' in nuclear scattering where a single excitation can be reached via the incident channel [81–82]. This physical system will not exhibit a truly intramolecular nonradiative relaxation but rather a Poincaré cycle [34]. For the limiting case of a large density of states, the recurrence time, t_r, for

TABLE I

Hierarchy of basis sets for description of electronic relaxation processes

Basis set	Major properties	Applicability
$\|\chi\rangle$	Diagonalizes H (with zero- and one-photon states)	Proofs of general theorems for the properties of the decay amplitudes
$\|j, \text{vac}\rangle$	a. Radiative decay provides the only dissipative channel b. Defined in $\hat{P}$ subspace c. Diagonalizes H_{eff} d. Nonorthogonal e. Specifies independently decaying levels	Time evolution of discrete electronically excited states
$\|J, \text{vac}\rangle$	a. System characterized by two parallel decay channels, radiative and nonradiative b. (e) as for $\|j, \text{vac}\rangle$	Parallel radiative and nonradiative decay of excited states
$\|n, \text{vac}\rangle$	Diagonalizes H_M	Radiative decay of small molecules and intermediate-type states of large molecules
Born–Oppenheimer basis	a. Diagonalizes $H_{MO} = H_{BO}$ b. Off-resonance interactions with higher excited states are negligible	Description of the statistical limit

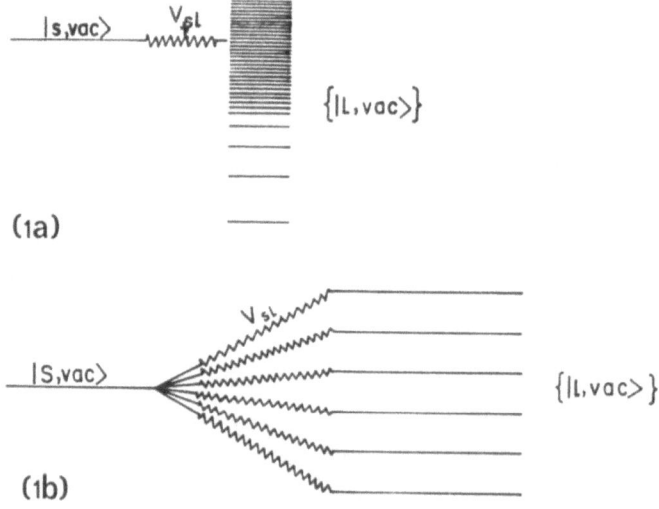

Figs. 1a–b.   Useful molecular decay models. — (a) Interstate coupling and nonradiative decay in an isolated large (statistical) molecule. (b) Interstate coupling in a small molecule.

the simple Bixon–Jortner model [34] (see Section 13) is $t_r = \hbar \varrho_l$ where ϱ_l corresponds to the density of states in the $\{|l\rangle\}$ manifold. This insures irreversible decay characterized by the decay time [34, 83–84] $\tau_{nr} = \hbar (2\pi |V_{sl}|^2 \varrho_l)^{-1}$ on the time scale, t, of interest, i.e. $\tau_{nr} \lesssim t \ll t_r$. Subsequent consecutive damping processes (which were disregarded in this simple scheme) of the $\{|l\rangle\}$ manifold, such as infrared radiation [34], or photon emission in the case of internal conversion [41, 79], will deplete these levels (see Section 13) insuring the occurrence of irreversible intramolecular radiationless processes in an isolated molecule. This situation is commonly referred to as the statistical limit [35]. If the large molecule is subjected to perturbations by an external medium [4, 64–65] the decay lifetime τ_{nr} is not affected in the statistical limit but the occupied $\{|l\rangle\}$ levels will be again depleted via vibrational relaxation insuring the irreversibility of the electronic relaxation process.

The basic energy levels scheme presented in Figure 1a is by no means restricted to a large molecule. The flexibility of molecular systems allows us to change the density of the $\{|l\rangle\}$ background states at will, by considering different molecules characterized by different numbers of degrees of freedom and by varying the electronic energy gap. A level scheme appropriate for a triatomic 'small' molecule [80] is presented in Figure 1b. In this context one has to be careful to distinguish between the implications of interstate coupling and intramolecular relaxation [34–35]. Intramolecular interstate nonadiabatic coupling is exhibited both in 'small' and in 'statistical' molecules. However, when the level density of the background states is low, no intramolecular relaxation is encountered in the small molecule. In Figure 1b we present the physical situation appropriate for level scrambling in a small molecule, where again a single $|s\rangle$ state corresponds to a doorway state.

The simple schemes (1a) and (1b) are grossly oversimplified, as the effect of the radiation field was not yet considered. In Figure 1c we present the appropriate level scheme (for the eigenstates of $H_0 = H_{BO} + H_{rad}$) corresponding to a large molecule. Now the doorway state $|s\rangle$ is simultaneously coupled to the radiation field $\{|g, \mathbf{ke}\rangle\}$ and to the intramolecular continuum $\{|l\rangle\}$. The analogous situation for the case of a small molecule is portrayed in Figure 1d. In this case two alternative descriptions which rest on different choices of the zero-order Hamiltonian are illuminating. We may proceed as before choosing $H_0 = H_{BO} + H_{rad}$ whereupon the BO doorway state $|s, \text{vac}\rangle$ is coupled to a sparse manifold $|l, \text{vac}\rangle$ and to the radiation field. Alternatively [40–41], one may look for the molecular eigenstates $|n\rangle$ which diagonalize the electronic Hamiltonian H_M. The zero-order states of $H_0 = H_M + H_{rad}$ correspond to $|n, \text{vac}\rangle$ which are coupled to the one-photon states. If the spacing between the molecular eigenstates considerably exceeds their radiative widths, each of these $|n\rangle$ levels will decay (and will be excited) independently. We thus encounter a situation occurring in atomic physics where a manifold of well separated levels (corresponding to the molecular eigenstates) is coupled to the radiation field. It should be noted, however, that accidental degeneracy between a pair of molecular eigenstates may result in interference effects which will exhibit quantum beats [38] in the radiative decay.

As it is common in physical problems, situations intermediate between the small

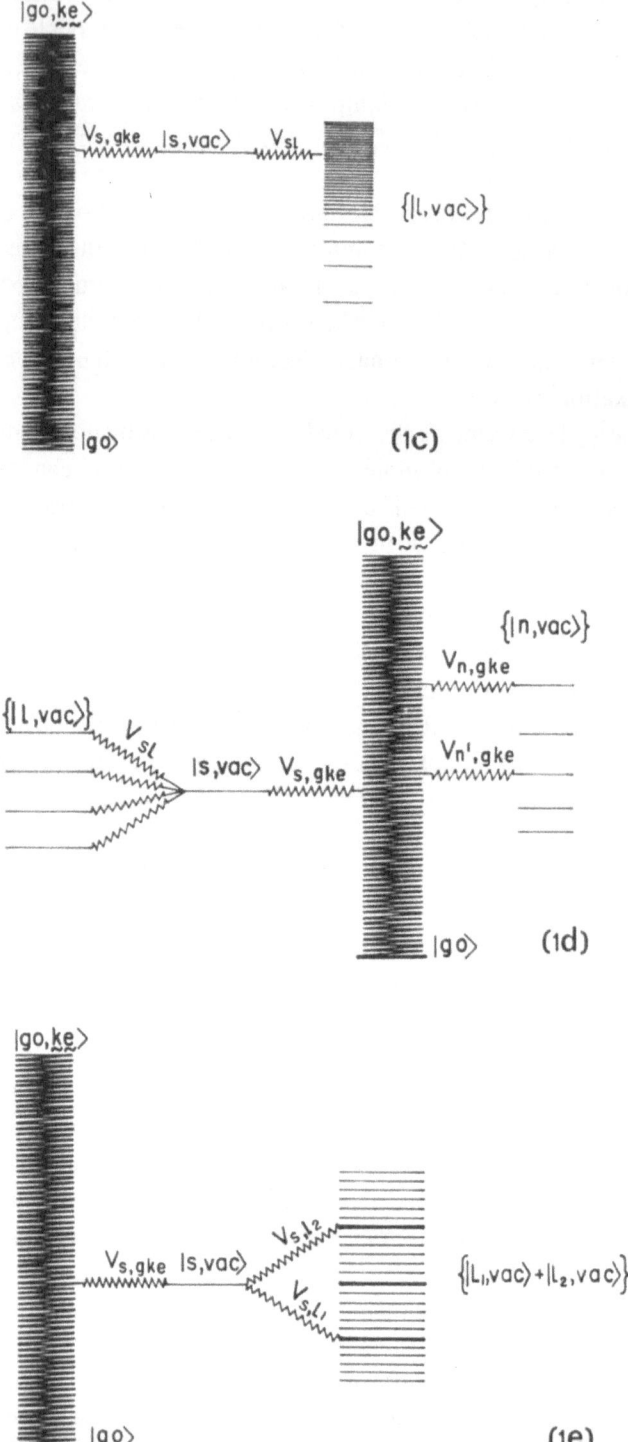

Figs. 1c–e.   (c) Radiative and interstate coupling in a statistical large molecule.
(d) Radiative coupling in a small molecule. (e) Intermediate level structure in a large molecule.

molecule and the statistical limit are encountered in real life. When the electronic energy gap between the electronic origins of two excited states of a large molecule is relatively small (2000–3000 cm^{-1}) the density of background $\{|l\rangle\}$ levels corresponding to the lower electronic configuration is not too high. This physical situation is referred to as the intermediate case [9–10, 47]. Now in such a case the details of the coupling strengths for various levels in the $\{|l\rangle\}$ manifold become crucial. Obviously, not all these levels are effectively coupled to the doorway state, and the coupling strength $(H_V)_{sl}$ varies substantially from one state to another. In the statistical limit the density of these effectively coupled accepting levels is still very high and the variation of their coupling strength is immaterial. In the intermediate case (Figure 1e) we expect simultaneous coupling of the doorway state $|s\rangle$ with a sparse manifold which will result in the decay characteristics of a small molecule, exhibited by an excited state of a large molecule which corresponds to the intermediate case.

Up to this point we have been concerned with simple coupling schemes where the intramolecular $\{|l\rangle\}$ manifold is not coupled to any additional decay channels. We have now to consider sequential decay processes where the doorway state (coupled to the radiative continuum) is also coupled to the (sparse or dense) $\{|l\rangle\}$ manifold, which in turn is coupled to a final dissipative continuum. We shall consider first the physical situation where each of the intermediate $\{|l\rangle\}$ levels is coupled to a different final continuum, and will thus exhibit non-interfering sequential decay. Relevant physical processes in this category are: (a) Sequential decay of the $\{|l\rangle\}$ quasicontinuum in an isolated statistical molecule due to infrared emission [34, 85]. Thus each of the $|l, \text{vac}\rangle$ levels is coupled to a separate radiative continuum $|l', \mathbf{k}_{ir}\mathbf{e}\rangle$ (see Figure 1f), where $|\mathbf{k}_{ir}\mathbf{e}\rangle$ correspond to an infrared photon. (b) Internal conversion in large molecules [51]. In the case of internal conversion from a highly excited singlet state the $\{|l, \text{vac}\rangle\}$ levels are electronically excited singlets, which are in turn radiatively coupled to highly excited (nontotally symmetric) vibrational levels $|g_\omega, \mathbf{k}'\mathbf{e}'\rangle$ of the ground electronic state (Figure 1g). (c) Vibrational relaxation of the $\{|l\rangle\}$ manifold of a statistical molecule embedded in a medium [64] (Figure 1h). In this case each $|l, \text{vac}\rangle$ level is separately coupled to a $|l'_{\{\omega_p\}}, \text{vac}\rangle$ continuum containing a collection of medium phonon modes, characterized by the frequencies $\{\omega_p\}$. (d) Sequential electronic-vibrational relaxation of a small molecule in a dense medium [65] (Figure 1i). As we have already pointed out, an isolated small molecule does not exhibit intramolecular electronic relaxation. However, when such a molecule is embedded in a dense medium each individual level in the sparse manifold $|l', \text{vac}\rangle$ can subsequently decay via medium-induced vibrational relaxation to a lower level $|l_{\{\omega_p\}}, \text{vac}\rangle$ thus providing a pathway for electronic-vibrational radiationless process. This process can be envisioned in terms of vibrational relaxation of the molecular eigenstates, which are well separated relative to their total (i.e. radiative and vibrational) relaxation widths. (e) Sequential decay via a single level [50]. This is a model system where strong coupling is exhibited between the doorway state $|s, \text{vac}\rangle$ and one of the $|l, \text{vac}\rangle$ levels (Figure 1j). The special $|l\rangle$ level is subsequently coupled to an internal continuum due to vibrational relaxation. Such a physical situation is encountered when an excited state of a

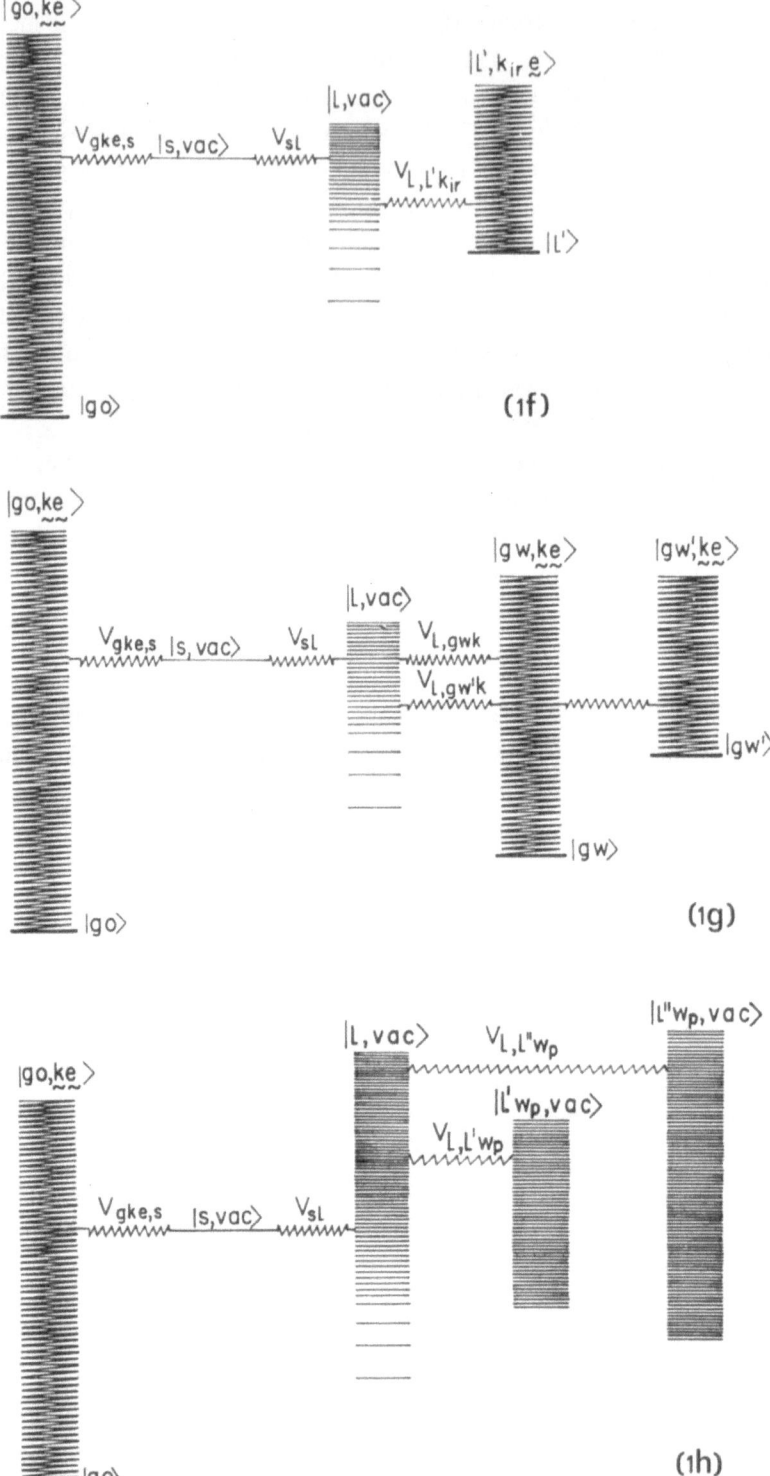

Figs. 1f–h.   (f) Sequential decay of intramolecular continuum by IR emission. (g) Internal conversion. (h) Vibrational relaxation in the intramolecular manifold due to medium perturbation.

large molecule, which corresponds to the intermediate case previously described in Figure 1e is subjected to medium induced collisional perturbation [50]. In this case of nearly degenerate two discrete levels interference effects in the decay process will be exhibited.

Finally we have to consider sequential decay processes involving interference where all the intermediate states $\{|l\rangle\}$ are coupled to the same continuum. The simplest example (Figure 1k) involves the coupling between a radiative (g, ke) and the non-radiative dissociative continuum $\{|l\rangle\}$ which provides the simplest case of direct photodissociation.

This state of affairs corresponds to 'elastic' photon scattering from and into a single radiative continuum. We can easily extend this picture taking into account (Figure 1l) the role of other radiative channels $|gv, ke\rangle$ corresponding to the vibrationally excited ground state levels. This system will now exhibit both elastic (Rayleigh-

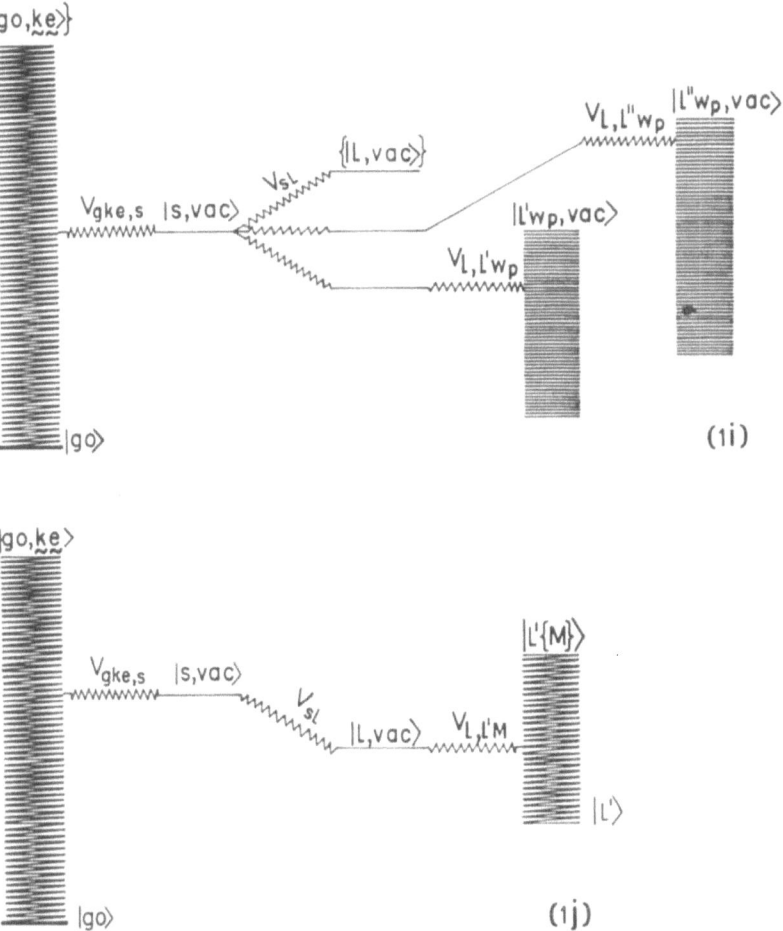

Figs. 1i–j.   (i) Electronic-vibrational relaxation of a small molecule in a medium. (j) Sequential and parallel decay of two discrete levels.

type) and Raman scattering [86]. Another interesting situation involves indirect photodissociation where a discrete excited state is involved being in turn coupled to a dissociative continuum [68]. This situation prevails for molecular predissociation. When the dissociative continuum is devoid of oscillator strength (Figure 1m with $V_{go,\text{ke};\, l_c\,\text{vac}} = 0;$) we have a situation which bears close analogy to radiationless transitions in the statistical limit (Figure 1c). An interesting state of affairs is encountered when the dissociative continuum is radiatively coupled (Figure 1m) to $|go, \text{ke}\rangle$ where interference effects will be exhibited. Finally, we consider direct and indirect photofragmentations of large polyatomic molecules. Here one has to consider (Fig-

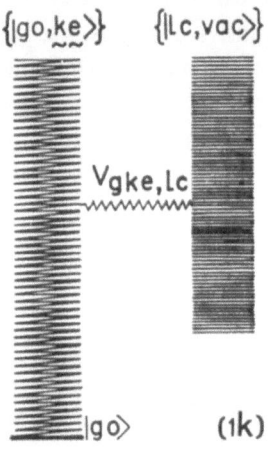

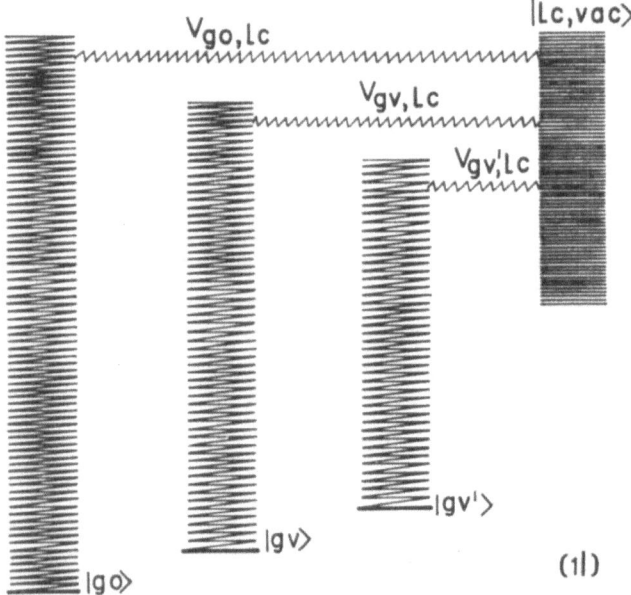

Figs. 1k–l.   (k) Direct photodissociation. (l) Coupling of dissociative continuum with a manifold of radiative continua.

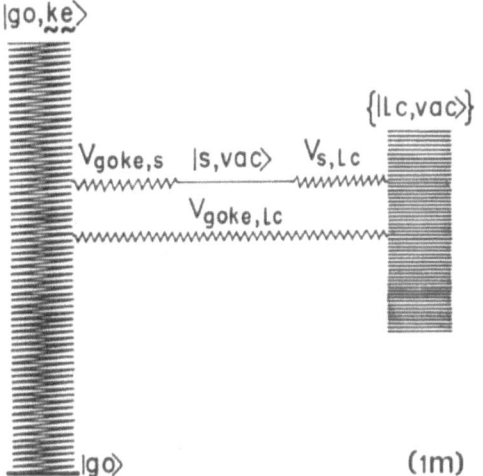

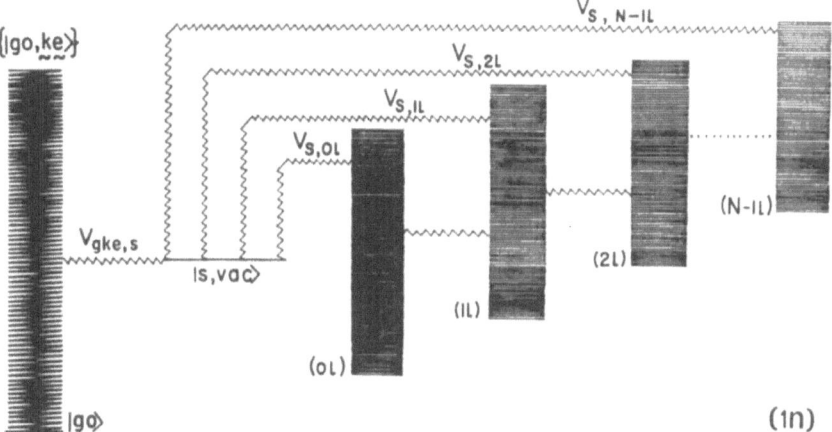

Figs. 1m–n. (m) Molecular predissociation. (n) Indirect molecular photofragmentation.

ure 1n) a set of dissociative continua each corresponding to a given vibrational state $|v_f\rangle$ of the fragments within the framework of the (reasonable) harmonic approximation coupling occurs only between adjacent continua.

These energy level schemes presented above provide the starting point for the theoretical study of the evolution of excited molecular states. The nature of the resulting physical information depends, of course, on the details of the experiment. We shall now proceed to a study of the problems.

4. Time Evolution of Excited Electronic States

We shall now consider excitation and decay processes in a general system consisting of an isolated single molecule and the radiation field. We shall segregate the eigenstates of the system's zero-order Hamiltonian into molecular excited states, $|m, \text{vac}\rangle$, i.e. $|s, \text{vac}\rangle$, $\{|l, \text{vac}\rangle\}$ etc., $\varepsilon\{|m, \text{vac}\rangle\}$ using the Born–Oppenheimer representation, or $|n, \text{vac}\rangle$ using the molecular eigenstates picture, and one-photon states $|g, \text{ke}\rangle$. In most general terms we can specify the state of the system $\Psi(0)$ at time $t = 0$ in terms of a superposition of these zero-order states

$$\Psi(0) = \sum_m A_m |m, \text{vac}\rangle + \sum_{\text{ke}} A_{\text{ke}} |g, \text{ke}\rangle. \tag{4.1}$$

The two sets of coefficients $\{A_m\}$ and $\{A_{\text{ke}}\}$ are referred to as the *preparation ampli-tudes* for the system, being determined by the special experimental conditions, which were not yet specified. Obviously, the state (4.1) is a nonstationary state of the system's Hamiltonian H (Equation (3.1)) and will exhibit time evolution. As we have chosen the Hamiltonian (3.1) in a time-independent representation, the evolution operator is simply

$$U(t, 0) = \exp(-iHt). \tag{4.2}$$

Thus the state of the system at time t is just

$$\Psi(t) = U(t, 0)\,\Psi(0) = \exp(-iHt)\,\Psi(0). \tag{4.3}$$

Expansion of $\Psi(t)$ in terms of the complete set of the eigenstates of H_0 (Equation (3.2)) results in

$$\Psi(t) = \sum_m |m\rangle\, d_m(t) + \sum_{\text{ke}} |g, \text{ke}\rangle\, d_{\text{ke}}(t), \tag{4.4}$$

where the general time-dependent amplitudes $\{d(t)\}$ of the relevant zero-order states are

$$d_m(t) = \langle m|\exp(-iHt)|\Psi(0)\rangle, \tag{4.5a}$$

$$d_{\text{ke}}(t) = \langle g, \text{ke}|\exp(-iHt)|\Psi(0)\rangle. \tag{4.5b}$$

The probability p_e of the system to be in any excited state at time t is given by

$$P_e(t) = \sum_m |\langle m, \text{vac} \,|\, \Psi(t)\rangle|^2 = \sum_m |d_m(t)|^2, \tag{4.6}$$

while the probability of the system to be found in any one-photon ground electronic state is

$$P_g(t) = \sum_{\text{ke}} |\langle g, \text{ke} \,|\, \Psi(t)\rangle|^2 = \sum_{\text{ke}} |d_{\text{ke}}(t)|^2. \tag{4.7}$$

The normalization condition for $\Psi(t)$ implies the conservation law

$$P_e(t) + P_g(t) = 1 \tag{4.8}$$

for all t. Finally, to bring this general treatment down to earth we have to relate these results to some experimental observables. The photon counting rate which monitors the number of photons emitted per unit time is just

$$\dot{P}_g(t) = dP_g/dt, \tag{4.9}$$

which by Equation (4.8) is given by

$$\dot{P}_g(t) = -\dot{P}_e(t).$$

(4.9a)

The total number of emitted photons at $t = \infty$ provides us with the quantum yield, Y,

$$Y = P_g(\infty) = 1 - P_e(\infty).$$

(4.10)

This general description of the time evolution of the system results in very cumbersome expressions for the time-dependent amplitudes $\{d(t)\}$ which are explicitly given in the form:

$$
\begin{aligned}
d_m(t) &= \sum_{m'} C_{mm'}(t) A_{m'} + \sum_{ke} C_{m, gk'e'}(t) A_{k'e'}, \\
d_{ke}(t) &= \sum_{m'} C_{gke, m'} A_{m'} + \sum_{k'e'} C_{gke, gk'e'}(t) A_{k'e'},
\end{aligned}
$$

(4.11)

where the time-dependent amplitudes $C_{\alpha\beta}(t)$ with α and $\beta \equiv |m, \text{vac}\rangle$ or $|g, \text{ke}\rangle$ are given by

$$C_{\alpha\beta}(t) = \langle \alpha | U(t, 0) | \beta \rangle \equiv \langle \alpha | \exp(-iHt) | \beta \rangle.$$

(4.12)

The (time-dependent) matrix elements $C_{\alpha\beta}(t)$ of the evolution operator between the zero-order states of H_0 are referred to as the *decay amplitudes* of the system. These include all the molecular information about the decay channels of the system. Thus the time evolution of the system, described in terms of the time-dependent amplitudes $\{d(t)\}$ (Equations (4.4) and (4.5)), can be completely specified by a superposition of products of the preparation amplitudes and of the decay amplitudes. The superposition (4.11) provides a mental, formal type separation of the initial conditions of the system (expressed in terms of the $\{A\}$ amplitudes) from the molecular radiative and nonradiative decay processes (expressed via the $\{C\}$-type amplitudes). Whether one can consider excitation followed by subsequent decay, or alternatively a single quantum photon scattering process depends on the specific experimental conditions. However, it should be realized that the formalism developed up to this point is applicable both for 'short excitation' and for 'long excitation' experiments. From the point of view of the formal theory we can now proceed to consider separately the decay amplitudes and the preparation amplitudes. The former are invariant with respect to the nature of the optical excitation, while the specification of the latter will determine the nature of the excitation process.

5. Formal Expressions for the Decay Amplitudes

When the evaluation of the decay amplitudes is involved it is more convenient and practical to express the matrix elements of the evolution operator in terms of the Fourier transform of the Green operator [87–90]

$$G(E) = (E - H + i\eta)^{-1},$$

(5.1)

where $\eta \to 0^+$. Thus all integrations over (5.1) will be performed over a contour which runs from $-\infty$ to ∞ just above the real E axis. Let us introduce at this point complete set of eigenfunctions $|\chi\rangle$ of the total Hamiltonian H. We realize that these can be determined in real life only for simple systems, but for the present general argument their detailed form is immaterial. Thus this basis set satisfies the Schrödinger equation

$$H|\chi\rangle = E_\chi|\chi\rangle \tag{5.2}$$

and the completeness condition

$$\sum_\chi |\chi\rangle\langle\chi| = 1. \tag{5.3}$$

We now write down immediately the time evolution operator (Equation (4.2)) and the Green operator (Equation (5.1)) in terms of their spectral representations

$$\exp(-iHt) = \sum_\chi |\chi\rangle \exp(-iE_\chi t)\langle\chi| \tag{5.4}$$

and

$$G(E) = \sum_\chi \frac{|\chi\rangle\langle\chi|}{E - E_\chi + i\eta}. \tag{5.5}$$

We note that expressions (5.4) and (5.5) are not valid for any arbitrary basis set, but just for the special basis set $|\chi\rangle$. Utilizing Equations (5.4) and (5.5) we can formally recast the time evolution operator in terms of the Fourier transform of the Green's function

$$\exp(-iHt) = \frac{1}{2\pi i} \int_{-\infty}^{\infty} \exp(-iEt)\, G(E)\, \mathrm{d}E, \tag{5.6}$$

where the conventional methods of residue integration have been utilized. Finally, utilizing the formal representation (5.6) the decay amplitudes (4.12) are explicitly expressed in the form

$$C_{\alpha\beta}(t) = \frac{1}{2\pi i} \int_{-\infty}^{\infty} G_{\alpha\beta}(E) \exp(-iEt)\, \mathrm{d}E \tag{5.7}$$

of

$$G_{\alpha\beta}(E) \equiv \langle\alpha|\, G(E)\, |\beta\rangle, \tag{5.7a}$$

i.e. the matrix elements of the Green's function between the zero-order states. Thus the evaluation of the decay amplitudes boils down to the evaluation of the matrix elements of the Green's function.

These general expressions for the decay amplitudes have many attractive features. From the point of view of general methodology, as we are dealing with a large number of levels, such general approach is most useful. From the point of view of mathematical convenience these expressions are quite easy to evaluate. In particular it is a simple matter to relate the matrix elements of $G(E) = (E - H + i\eta)^{-1}$ to those of the corresponding Green's function for the 'unperturbed' zero-order system $G^0(E) = (E - H_0 + i\eta)^{-1}$, where $H_0 = H - V$, via the Dyson operator identity [87]

$$G(E) = G^0(E) + G^0(E) VG(E) = G^0(E) + G(E) VG^0(E).$$

Some more formal and powerful techniques for the evaluation of the relevant matrix elements of the Green's function are available. We note that these matrix elements $G_{\alpha\beta}(E)$ are of three types: (1) $\langle g, \text{ke}| G(E) |m, \text{vac}\rangle$, (2) $\langle m, \text{vac}| G(E) \times \times |m', \text{vac}\rangle$, and (3) $\langle g, \text{ke}| G(E) |g\mathbf{k}'e'\rangle$. We shall now partition the Hilbert space as follows [88–90]:

$$\hat{P} = \sum_m |m, \text{vac}\rangle \langle m, \text{vac}|, \tag{5.8}$$

$$\hat{Q} = \sum_{\text{ke}} |g, \mathbf{ke}\rangle \langle g, \mathbf{ke}|, \tag{5.9}$$

where the subspace $\hat{P}$ contains the excited zero-photon levels while the subspace $\hat{Q}$ contains the one-photon zero-order states. Provided that we disregard the contribution of zero photon ground state and multiphoton excited states, which will yield only off-resonance terms, the completeness condition requires that

$$\hat{P} + \hat{Q} = 1. \tag{5.10}$$

We immediately notice that the matrix elements of type (1) combine the $\hat{Q}$ subspace with the $\hat{P}$ subspace, while the matrix elements of type (2) combine the $\hat{P}$ subspace with itself.

Thus, in view of the orthogonality of the subspaces $\hat{P}$ and $\hat{Q}$, evaluation of matrix elements of type (1), (2) and (3) requires the operators $\hat{Q}G\hat{P}$ and $\hat{P}G\hat{P}$ and $\hat{Q}G\hat{Q}$, respectively. The explicit forms for these operators are [88–90]

$$\hat{Q}G(E)\hat{P} = (E - \hat{Q}H_0\hat{Q} + i\eta)^{-1} \hat{Q}R(E)\hat{P}(E - H_0 - \hat{P}R(E)\hat{P})^{-1}, \tag{5.11}$$

$$\hat{P}G(E)\hat{P} = (E - \hat{P}H_0\hat{P} - \hat{P}R(E)\hat{P})^{-1}\hat{P}, \tag{5.12}$$

$$\hat{Q}G(E)\hat{Q} = \hat{Q}(E - \hat{Q}H\hat{Q} + i\eta)^{-1} + \\ + \hat{Q}(E - \hat{Q}H\hat{Q} + i\eta)^{-1} V\hat{P}G(E)\hat{P}V(E - \hat{Q}H\hat{Q} + i\eta)^{-1}\hat{Q}. \tag{5.13}$$

Being expressed in terms of the level shift operator

$$R(E) = V + V\hat{Q}(E - \hat{Q}H\hat{Q})^{-1}\hat{Q}V. \tag{5.14}$$

It is important to notice that the level shift operator consists of two contributions; a direct coupling, V, and a relaxation contribution.

These general formal expressions are of great value both for practical evaluation of the matrix elements and, more important, also for the choice of the most convenient basis set to describe the decay of the system. To conclude this exposition of the mathematical methods we would like to point out that the partitioning (5.8) and (5.9) is again not unique, as it is common for many intermediate steps in the formal theory. For some specific systems it will be convenient (see Section 10) to adopt an alternative partitioning procedure, retaining in subspace $\hat{P}$ just the discrete excited molecular states and throwing into subspace $\hat{Q}$ both the radiative continuum and the molecular (zero-photon) continua (or quasicontinua).

From the physical point of view the Green's function method is indeed very clear and transparent (at least after one has crossed a psychological barrier and became acquainted with these techniques). We have noted that the poles of the Green's function in the $|\chi\rangle$ representation (see Equation (5.5)) provide us with the energy spectrum of the system. It is important to note that in this case the poles $E = E_\chi$ are located on the real E axis, thus the system in an eigenstate of H is characterized by a real energy and does not exhibit a decay.

When we consider the matrix elements of $G(E)$ between the eigenstates of H_0 the poles of $G(E)$ will have imaginary parts of the form $E = E_e - \frac{1}{2}i\Gamma_e$. In general a large number of poles can be exhibited, whose imaginary components Γ_e provide the contributions to the decay rate of the nonstationary state.

6. Time Evolution Resulting from Wave-Packet Excitation

We have now to provide a physically realistic model for the excitation process. The most natural way to excite the system is to switch on a photon wave packet at the time $t = 0$, and then utilize the techniques of Section 3 to follow the time evolution of the system. It is important to note that by this general experiment we have not necessarily 'prepared' the system in an initial metastable decaying state, and that this formulation is general, and can be used both for 'short excitation' and for 'long excitation' experiments, as well as for intermediate excitation conditions.

Let us regress for a moment and consider some properties of a photon wave packet. The molecular system is now absent so that $H = H_{\text{rad}}$. A general representation of a photon wave packet is

$$\sum_{n_1, n_2 \ldots} a_{n_1 n_2 \ldots} \prod_i |n_i, \mathbf{k}_i \mathbf{e}_i\rangle, \tag{6.1}$$

where n_i is the population number of the photon state $|\mathbf{k}_i \mathbf{e}_i\rangle$ which is characterized by the energy $\varepsilon_i = k_i$ (where in this section we shall use the sloppy units $\hbar = c = 1$).

For moderately weak fields $n_i = 0$ or 1 for all i, under these conditions we may consider an initial state of the field $\Psi_p(t = 0)$ consisting of a wave packet of one-photon states [90]

$$\Psi_p(0) = \sum_{\mathbf{k}} a_{\mathbf{k}} |\mathbf{k}\mathbf{e}\rangle, \tag{6.2}$$

where $a_{\mathbf{k}}$ is the initial amplitude of the state $|\mathbf{k}\mathbf{e}\rangle$, while the summation $\sum_k$ represents integration over photon energies, over spatial directions and summation over all polarization directions. In what follows we shall use fixed propagation and polarization directions and sum only over the photon energies. The time evolution of the wave packet (in the absence of the molecular system) is

$$\Psi_p(t) = \exp(-iHt)|\Psi(0)\rangle = \sum_k a_k \exp(-ikt)|k e\rangle. \qquad (6.3)$$

It will be useful to define at this point the Fourier transform, $\varphi(t)$ of the wave-packet amplitudes

$$\varphi(t) = \sum_k a_k \exp(-ikt), \qquad (6.4)$$

which from (6.3) is just

$$\varphi(t) = \sum_{k'} \langle k'e \mid \Psi_p(t)\rangle. \qquad (6.5)$$

We shall refer to $\varphi(t)$ as the time-dependent field amplitude. The time evolution of the wave packet is expressed by the function

$$F(t) = |\langle \Psi_p(0) \mid \Psi_p(t)\rangle|^2 = \sum_k |a_k|^2 \exp(-ikt), \qquad (6.6)$$

which just corresponds to the Fourier transform of $|a_k|^2$. The power spectrum $|a_k|^2$ of the photon wave packet is the (inverse) Fourier transform of $F(t)$

$$|a_k|^2 = \frac{1}{2\pi} \int\limits_0^\infty dt \, \exp(ikt) \, F(t). \qquad (6.7)$$

We shall now present a specific example of a photon wave packet which will be subsequently utilized in the study of molecular excitation processes. Consider a wave packet whose amplitudes are given in terms of a coherent Lorentzian distribution [90]

$$a_k = \frac{A_N}{k - \bar{k} + \frac{1}{2}i\gamma_p}, \qquad (6.8)$$

where $\bar{k}$ is the center of the distribution, γ_p its width and A_N is a normalization factor. It must be stressed that this choice (6.8) of the energy spread of the wave packet is by no means unique, and will be used for the sake of mathematical convenience. Other shapes (i.e. Gaussian) can be used leading to similar results. The field amplitude for the Lorentzian wave packet is (see Equation (6.4))

$$\varphi(t) = \begin{cases} 2\pi i A_N \exp(-i\bar{k}t - \frac{1}{2}\gamma_p t) & t > 0 \\ 0 & t < 0 \end{cases} \qquad (6.9)$$

while the time evolution of the wave packet (expressed in terms of the Fourier transform of the power spectrum, Equation (6.6)) exhibits a simple exponential decay

$$F(t) = \frac{2\pi |A_N|^2}{\gamma_p} \exp[-\gamma_p |t|]. \qquad (6.10)$$

It will be useful at this stage to consider two limiting extreme situations for (6.8).

(a) An ideal 'long time' excitation experiment is characterized by a narrow wave packet

$$a_{\mathbf{k}} = \delta(\mathbf{k} - \bar{\mathbf{k}}) \tag{6.11}$$

and consequently

$$|\varphi(t)| = 1, \tag{6.12}$$

In this case the photon wave packet is well defined in energy. Note that this definition is general and does not depend on the specific form (6.8).

(b) When 'short time' excitation conditions are considered we require that $\varphi(t) \propto$ $\propto \delta(t)$ and consequently $a_{\mathbf{k}} = \mathrm{const}$. This limit may be obtained by choosing $A_N = \frac{1}{2}i\gamma_p$ in (6.8) and taking large γ_p values, i.e. $\gamma_p \to \infty$. Then for the relevant values of $\mathbf{k}$ $|\mathbf{k} - \bar{\mathbf{k}}| \ll \gamma_p$ and thus $a_{\mathbf{k}} \to 1$.

We note in passing that for different purposes we have to choose different normalization conditions for the initial photon wave packet. If we want to obtain the long time case (a) as a limiting form of a Lorentzian wave-packet excitation we should normalize the excitation amplitudes

$$\sum_{\mathbf{k}} a_{\mathbf{k}} = 1 \tag{6.13}$$

and then set $\gamma_p \to 0$. For case (b) $\varphi(t)$ should be normalized to a constant, (see Section 8)

$$\int_0^\infty \varphi(\tau)\, \mathrm{d}\tau = \mathrm{const}. \tag{6.14}$$

In order to follow the general time evolution of the system it is often convenient to normalize as the power spectrum, i.e.

$$\sum_{\mathbf{k}} |a_{\mathbf{k}}|^2 = 1. \tag{6.15}$$

This latter normalization is adopted later on in this chapter.

We now return to the physical situation of interest and insert back the molecule in the system. At the time $t = 0$ the photon wave packet is introduced so that the initial state of the system is

$$\Psi(0) = \sum_{\mathbf{k}} a_{\mathbf{k}} |g, \mathbf{ke}\rangle. \tag{6.16}$$

This representation of the initial conditions is much more simple and physically transparent than the general expression (4.1) where we have now set for the preparation amplitudes $A_{\mathbf{ke}} = a_{\mathbf{k}}$ and $A_m = 0$. The time evolution of the system is now obtained from Equation (4.3) in the explicit form

$$\Psi(t) = \sum_{k}\sum_{m} |m, \mathrm{vac}\rangle\, C_{m,\,g\mathbf{ke}}(t)\, a_k + \sum_{k}\sum_{k'} |g, \mathbf{k'e'}\rangle\, C_{g\mathbf{k'e'},\,g\mathbf{ke}}(t)\, a_k. \tag{6.16a}$$

We can now immediately project out from (6.16a) either the vacuum states $|m, \text{vac}\rangle$ or the one-photon state $|g, \mathbf{k'e'}\rangle$. Thus the probability (4.6) for finding the system in any excited zero-photon state at time t is

$$P_e(t) = \sum_m |\sum_\mathbf{k} C_{m, g\mathbf{ke}}(t) a_\mathbf{k}|^2 . \qquad (6.17)$$

While the probability for finding the system at a time t in any one-photon ground electronic state is (Equation (4.7)):

$$P_g(t) = \sum_{\mathbf{k'}} |\sum_\mathbf{k} C_{g\mathbf{k'e'}, g\mathbf{ke}}(t) a_\mathbf{k}|^2 , \qquad (6.18)$$

which obviously satisfies the conservation law $P_e(t) + P_g(t) = 1$ for all t. One can ask and answer at this stage some other more detailed and specific questions, such as what is the probability of the population of a subset of the $|m, \text{vac}\rangle$ levels, which correspond to some specific zero-order molecular states, or what is the population of a subset of the one-photon states (i.e. characterized by certain energies, or certain spatial or polarization directions). The general theoretical scheme presented above is able to answer all such questions, however, for the sake of presentation of the general arguments such extensions are not necessary. Let us reflect at this stage what are the 'hidden approximations' in our treatment. First, we have considered only a single electronic ground state. Second, a wave packet consisting only of one-photon states was considered to interact with the system. Third, we are averaging over photons spatial and polarization directions. These approximations can be indeed easily relaxed which will just result in more complex expressions which will not affect the general argument.

Thus Equations (6.17) and (6.18) provide us with all the pertinent general information regarding the experimentally observable time evolution of the system. The probability and rate of photon emission is expressed in terms of products of the preparation amplitudes, now given in terms of the wave packet amplitudes and the decay amplitudes which we had considered in detail in Section 5. We are now in a position to provide exact criteria for the applicability of the concept of 'decay of an initially excited state' and its range of validity, and also to specify the general conditions for 'short excitation' and 'long excitation' experiments.

7. Time Dependence of the Population of the Excited States

We shall now proceed to derive explicit theoretical expressions for the probability function $P_e(t)$. For this purpose we have to utilize the formalism outlined in Section 5 for the evaluation of the decay amplitudes of the system. We have chosen to calculate the function $P_e(t)$ (Equation (4.6)) rather than $P_g(t)$ (Equation (4.7)) although the latter is really related to experimental observables. The reason for that is simply mathematical convenience, as the decay amplitudes (or rather the corresponding matrix elements of the Green's function) of the form $C_{m, g\mathbf{ke}}(t)$ are somewhat easier to evaluate than the $C_{g\mathbf{k'e'}, g\mathbf{ke}}(t)$ amplitudes. In any case, once we have evaluated $P_e(t)$ we have $P_g(t)$ from the basic conservation relation (4.8). We shall now consider two relevant physical situations. First, the simplest case where only a single excited level carries oscillator strength from the ground state. This is a common state of affairs for many interesting molecular systems which were discussed in Section 2. Second, we shall focus attention on the general case when an arbitrary number of discrete $|m, \text{vac}\rangle$ levels are radiatively connected to the $|g, \text{ke}\rangle$ radiative continuum.

When a single molecular level, say $|s, \text{vac}\rangle$, acts as a doorway state we can set for the radiative coupling matrix elements

$$\langle g, \text{ke}| V |m, \text{vac}\rangle = (H_{\text{int}})_{\text{gke}, s}\, \delta_{s, m} \equiv V_{\text{gke}, s}\, \delta_{s, m}, \tag{7.1}$$

To evaluate the matrix elements of the Green's function which determine the decay amplitudes $C_{s, \text{gke}}$ and $C_{m, \text{gke}}\,(m \neq s)$ in Equations (6.17), (6.18), we make use of the Dyson Equation (5.7) and get

$$G_{s, \text{gke}}(E) = G_{ss}(E)\, V_{s, \text{gke}} \frac{1}{E - E_g - k + i\eta}, \tag{7.2}$$

$$G_{m, \text{gke}}(E) = G_{ms}(E)\, V_{s, \text{gke}} \frac{1}{E - E_g - k + i\eta}, \tag{7.3}$$

where E_g is the energy of the electronic ground state $|g\rangle$, which can be taken as $E_g = 0$. The form for the matrix elements (7.2) and (7.3) is very convenient, as we shall be able now to relate the probability function $P_e(t)$ to the decay amplitudes $C_{ss}(t)$ and $C_{ms}(t)$ combining only the excited states. These decay amplitudes now take the form

$$C_{m, \text{gke}}(t) = \frac{1}{2\pi i} \int_{-\infty}^{\infty} dE \exp(-iEt)\, G_{ms}(E^+) \frac{1}{E - k + i\eta}\, V_{s, \text{gke}}, \tag{7.4}$$

for all $|m\rangle$ including $|s\rangle$. Invoking the customary assumption that the radiative coupling matrix elements (7.1) exhibit a weak energy dependence we have

$$C_{m, \text{gke}}(t) = \frac{1}{2\pi i}\, V_{s, \text{gke}} \exp(-ikt) \int dE\, G_{ms}(E) \frac{\exp[-i(E-k)t]}{E - k + i\eta}. \tag{7.4a}$$

Now making use of the trivial identity

$$\frac{\exp[-i(E-k)t]}{E-k} = (-i) \int_0^t \exp[-i(E-k)\tau]\, d\tau + \frac{1}{E-k}, \tag{7.4b}$$

and utilizing again the basic definition (7.4) we get

$$2\pi i\, C_{m, \text{gke}}(t) = -iV_{s, \text{gke}} \exp(-ikt) \times$$

$$\times \int_{-\infty}^{\infty} dE \int_0^t d\tau \exp[-i(E-k)\tau]\, G_{ms}(E) + C_{m, \text{gke}}(0).$$

As by definition $C_{m,gke}(0) = 0$ we get the final result for the decay amplitude

$$C_{m,gke}(t) = -iV_{s,gke} \exp(-ikt) \int_0^t d\tau \exp(ik\tau) C_{ms}(\tau). \qquad (7.5)$$

Now utilizing Equation (6.17) for $P_e(t)$ and the definition (6.4) for the (time-dependent) field amplitudes we obtain

$$P_e(t) = |V_{s,gke}|^2 \sum_m \left| \int_0^t \varphi(t-\tau) C_{ms}(\tau) d\tau \right|^2. \qquad (7.6)$$

Equation (7.6) provides us with the final general result for the time dependence of the population of all the excited electronic states, for the simple case of a single doorway state. It is important to realize at this stage that the decay characteristics are determined by the convolution of the decay amplitudes $C_{ms}(t)$ (combining the doorway state with the other excited states) and the field amplitudes.

It is a simple matter to generalize this result for the case of an arbitrary number of optically active excited states whereupon $\langle m, \text{vac}| H_{\text{int}} |g, \text{ke}\rangle \equiv V_{m,gke} \neq 0$; for several (or even for all) $|m\rangle$ states. In order to calculate the relevant decay amplitudes we need the matrix elements $G_{m,gke}(E)$ which by the Dyson equation take the explicit form

$$G_{m,gke}(E) = \sum_{m'} \langle m, \text{vac}| G(E) |m', \text{vac}\rangle \cdot \langle m', \text{vac}| H_{\text{int}} |g, \text{ke}\rangle \frac{1}{E - k + i\eta}. \qquad (7.7)$$

We can now define a *generalized doorway states* $|N\rangle$ by the relation

$$|N, \text{vac}\rangle = \frac{1}{\gamma_N} \sum_m |m, \text{vac}\rangle \langle m, \text{vac}| H_{\text{int}} |g, \text{ke}\rangle, \qquad (7.8)$$

where

$$\gamma_N^2 = \sum_{m'} \langle g, \text{ke}| H_{\text{int}} |m', \text{vac}\rangle \langle m', \text{vac}| H_{\text{int}} |g, \text{ke}\rangle. \qquad (7.9)$$

The definition (7.8) implies a very simple physical interpretation. The generalized doorway state is just a superposition of the excited molecular states each weighted by its coupling strength with the electronic ground-state-radiative continuum $|g, \text{ke}\rangle$. In the special case of the single doorway state

$$|N, \text{vac}\rangle = |s, \text{vac}\rangle. \qquad (7.10)$$

The concept of the generalized doorway state was previously invoked by Nitzan and Jortner [60, 62] using first order perturbation theory. They wrote $|N\rangle = \sum_m |m\rangle \mu_{gm}$ where μ is the transition moment operator. Definition (7.8) is more general. Finally, it is worthwhile noting that the generalized doorway state (7.8) can be expressed in terms of the projection operators (5.8) and (5.9) whereupon $|N, \text{vac}\rangle = (1/\gamma_N) \hat{P} H_{\text{int}} \times$

$\times |g, \text{ke}\rangle$. As $\hat{Q}H_{\text{int}}\hat{Q}=0$, then in view of the completeness condition (5.10), one gets the compact form

$$|N, \text{vac}\rangle = \frac{1}{\gamma_N} H_{\text{int}} |g, \text{ke}\rangle, \tag{7.8a}$$

$$|\gamma_N|^2 = \langle g, \text{ke}| H_{\text{int}}^2 |g, \text{ke}\rangle. \tag{7.9a}$$

This definition is very useful being independent of the basis set used to specify the excited states in the $\hat{P}$ subspace.

Returning now to the evaluation of the decay amplitudes in the general case, Equation (7.7) for the matrix elements of the Green's function now takes the form

$$G_{m, g\text{ke}}(E) = \gamma_N G_{m, N}(E) \frac{1}{E - k + i\eta}. \tag{7.11}$$

It is apparent that Equation (7.11) provides us with a generalization of the simple relations (7.2) and (7.3) as in the case of a single doorway state Equation (7.10) applies.

Utilizing Equation (7.11) we can immediately obtain a general expression for the corresponding decay amplitudes following the same methods as applied for the derivation of Equations (7.5) from (7.4). Thus we get

$$C_{m, g\text{ke}}(t) = - i\gamma_N \exp(- ikt) \int_0^t d\tau \exp(ik\tau) C_{mN}(\tau), \tag{7.12}$$

while the population of the excited state is obtained from Equations (6.17), (6.4) and (7.12) in the form

$$P_e(t) = |\gamma_N|^2 \sum_m \left| \int_0^t d\tau \, \varphi(t - \tau) C_{mN}(\tau) \right|^2, \tag{7.13}$$

where for the sake of clarity we redefine the relevant decay amplitudes

$$C_{m, N}(t) = \langle m, \text{vac}| U(t, 0) |N, \text{vac}\rangle =$$

$$= \frac{1}{2\pi i} \int dE \exp(- iEt) \langle m, \text{vac}| G(E) |N, \text{vac}\rangle. \tag{7.14}$$

Equations (7.13) and (7.14) provide us with the desirable general result for the time evolution of an excited state consisting of an arbitrary (dense or sparse) level structure and where an arbitrary number of these levels carry oscillator strength from the ground state. This general form (7.13) implies that the time evolution $P_e(t)$ is determined by a sum of amplitudes squared, where each amplitude is determined by the convolution of the (time-dependent) field amplitude and the decay amplitude, which combines the generalized doorway state with the various (zero-order) molecular states. We are able now to discuss the general features of both 'long time' and 'short time' excitation experiments as limiting cases of Equation (7.13).

We shall start with the short excitations. This will be conducted in two stages. First, we shall provide a rigorous definition of an 'initially prepared' state of a physical system (see Section 8). Subsequently, by introducing the concept of 'independently decaying states' (Section 9) we shall be able to provide explicit expressions for the time evolution of a system consisting of any number of closely spaced levels at an arbitrary excitation time scale. This will be accomplished in Section 9.

8. An 'Initially Prepared' Doorway State

The understanding of the nature of the 'initially prepared' optically excited state is crucial for the understanding of any short excitation decay experiment when one wants to formulate the precise conditions for validity of the separation of the excitation and decay processes. Early treatments of this problem [34, 83–84] considered the radiative excitation process to lowest order and accounted for the nonradiative decay occurring during the excitation process. These formulations of the excitation and nonradiative decay process were provided by considering the time evolution of excited molecular eigenstates utilizing the time evolution of the density matrix. The original treatments [34, 83–84] were grossly oversimplified as they disregarded the radiative decay channel. Rhodes [91] has demonstrated how to handle the excitation and both radiative and nonradiative decay using the density matrix formalism and how to follow the time evolution after the termination of the pulse. He has shown [91] that for long excitation times the density matrix assumes a partially diagonal form, however [60], this does not affect the decay characteristics in the statistical limit. In this context Freed [79] has treated the excitation-decay process by starting from the system at $t = 0$ in the state $|g, \mathbf{ke}\rangle$ and subsequently terminating the photon field after an arbitrary time. $\Psi \, (t=0)$ in his formalism is precisely defined in energy and thus this approach is adequate to long time excitations rather than for short excitation experiments. An unsatisfactory feature common to all these treatments mentioned above involves the termination of the exciting pulse after an arbitrary time. A way out of this difficulty is using a delta function excitation in time and treating this preparation process to low order [9, 35–36]. This approach is basically valid for model systems, although it is esthetically unattractive. The present approach adopted herein provides a self-consistent general solution to the problem of 'preparation' of metastable decaying states.

It should be borne in mind that the definition of an 'initially excited' state is essentially a theoretical problem. However, the nature of the time resolved decay pattern experimentally observed in a short time experiment requires this definition. In order to provide a meaningful definition for this concept two basic conditions have to be satisfied: (a) A single state has to be defined which is radiatively coupled to and carrying all the oscillator strength from the ground state. (b) The duration of the exciting pulse is appreciably shorter than all the (radiative and nonradiative) 'decay times' of the excited states. This second condition will be considered in two stages. First we shall consider excitation by a 'white' pulse containing all frequencies. This leads to an unrealistic state of affairs, as obviously the system contains also other excited states, so that such a pulse will excite all states including those which are of no interest to the experimentalist. We shall thus have to provide a more precise definition of the decay times for a complicated molecular system.

We consider first the idealized system which involves a small number (say, two) of excited electronic configurations and these constitute the entire energy spectrum of excited states. Obviously this system can involve an arbitrary number of vibronic levels corresponding to these two electronic configurations. This idealized system is now excited by a light pulse which contains all frequencies, thus the Lorentzian pulse (6.8) takes the limiting form $a_k = $ const. for all k while the field amplitude is given by $\varphi(t) = \phi \cdot \delta(t)$ where ϕ is a normalization factor to be determined later. This delta excitation function obviously satisfies condition (b). Thus the time evolution of a general excited state (Equation (7.13)) takes now the simple form

$$P_e(t) = |\gamma_N|^2 |\phi|^2 \sum_m |C_{mN}(t)|^2 , \qquad (8.1)$$

while for the special case of a single doorway state we have

$$P_e(t) = |\gamma_N|^2 |\phi|^2 \sum_m |C_{ms}(t)|^2 . \qquad (8.1a)$$

Equation (8.1) provides a proper specification of an 'initially prepared' state. From these results we conclude that

(a) Under the extreme conditions of broad band excitation, the time evolution of the excited states is equivalent to preparing the $\phi \gamma_N |N, \text{vac}\rangle$ state at $t = 0$ and this initial state exhibits time evolution. Choosing $\phi = 1/\gamma_N$ we obtain the normalized state $|N, \text{vac}\rangle$ at $t = 0$.

(b) Under these extreme excitation conditions the time evolution of the system is solely determined by the decay amplitudes, and provides only information concerning the molecular decay characteristics.

(c) In this case, the time-dependent amplitudes (4.5) are $d_m(t) = C_{m,N}(t)$ and $d_{ke}(t) = C_{gke,N}(t)$. Utilizing the basic definitions (4.6) and (4.7) and the conservation law (4.8), the population probability of the ground one photon states is now

$$P_g(t) = \sum_{ke} |C_{gke,N}(t)|^2 , \qquad (8.2)$$

or for the special case of a single doorway state

$$P_g(t) = \sum_{ke} |C_{gke,s}(t)|^2 . \qquad (8.2a)$$

(d) It will be useful to provide an alternative expression for the population of the ground state (Equation (8.2)), expressing it in terms of the diagonal decay amplitudes $C_{NN}(t)$ (or of $C_{ss}(t)$ for a single doorway state). We shall thus be able to relate the photon counting rate to the probability of the survival of the 'initially prepared' excited state at time t. Let us consider first the simple situation of the single doorway state.

In analogy with (7.5) we have

$$C_{gke,s}(t) = -iV_{gke,s}\exp(-ikt)\int_0^t d\tau \exp(ik\tau) C_{ss}(\tau).$$ (8.3)

Inserting this result into Equation (8.2a) results in

$$P_g(t) = \sum_{ke}|V_{gke,s}|^2 \int_0^t d\tau \int_0^t d\tau' \exp(ik(\tau-\tau')) C_{ss}(\tau) C_{ss}^*(\tau')).$$ (8.4)

The summation in Equation (7.4) implies after averaging on polarization derections

$$\sum_{ke} \rightarrow \int \varrho_r(E)\, dk,$$ (8.5)

where $\varrho_r(E)$ is the density of photon states at the energy $E=k$. We shall further define the radiative width of the $|s\rangle$ state by the common relation

$$\Gamma_s(E) = 2\pi\,|V_{gke,s}|^2\,\varrho_r(E).$$ (8.6)

Assuming that it is a slowly varying function of the energy around $E=E_s$ and we set $\Gamma_s(E) = \Gamma_s(E_s) = \Gamma_s$. Equation (8.4) takes the form

$$P_g(t) = \frac{\Gamma_s}{2\pi}\int_0^t d\tau \int_0^t d\tau' \int_0^\infty dk\, \exp(ik(\tau-\tau'))\, C_{ss}^*(\tau)\, C_{ss}(\tau') =$$

$$= \Gamma_s \int_0^t d\tau\, |C_{ss}(\tau)|^2,$$ (8.7)

and the photon counting rate for a single doorway state is just

$$\dot{P}_g(t) = \Gamma_s |C_{ss}(t)|^2.$$ (8.7a)

Equation (8.7a) was previously derived using the Wigner–Weisskopf method [39, 47].

Turning now to the case of the generalized doorway state we utilize the basic definitions (7.8) to write

$$\gamma_N C_{gke,N}(t) = \sum_m C_{gke,m}(t) V_{m,gke},$$ (8.8)

and from Equation (7.11) we have

$$iC_{gke,m}(t) = \gamma_N \exp(-ikt)\int_0^t d\tau \exp(ik\tau) C_{Nm}(\tau).$$ (8.9)

Thus

$$P_g(t) = \sum_{ke} \left| \sum_m \int_0^t d\tau \exp(ik\tau) C_{Nm}(\tau) V_{m,\,gke} \right|^2 =$$

$$= |\gamma_N|^2 \sum_{ke} \left| \int_0^t d\tau \exp(ik\tau) C_{NN}(\tau) \right|^2. \tag{8.10}$$

Making use of Equation (8.5) and performing the integration over k we get

$$P_g(t) = \Gamma_N^{(\text{rad})} \int_0^t |C_{NN}(\tau)|^2 \, d\tau \tag{8.11}$$

and

$$\dot{P}_g(t) = \Gamma_N^{(\text{rad})} |C_{NN}(t)|^2. \tag{8.11a}$$

In the derivation of (8.10), (8.11) we have neglected the weak $\mathbf{k}$ dependence of γ_N and we have defined

$$\Gamma_N^{(\text{rad})} = 2\pi |\gamma_N|^2 \, \varrho_r, \tag{8.11b}$$

and ϱ_r is the density of the radiation field states around the relevant energy. In analogy with (8.7) we see that $\Gamma_N^{(\text{rad})}$ is the radiative width of the $|N\rangle$ state. The simple compact form of Equation (8.11) should not mislead us, as this expression involves a large number of cross terms. These concise expressions ((8.7 and (8.11)) will be invoked again to relate the decay rate of the system (i.e. photon counting rate) to optical absorption line shape (see Section 17).

Up to this point we have been concerned with a single ground molecular state $|g\rangle$. Any real molecule is characterized by a vibrational manifold $\{|gv\rangle\}$ (where $v=0, 1,\ldots$ correspond to the collection of vibrational quantum numbers) in the ground state and for certain applications we have to consider radiative decay processes to different radiative continua $\{|gv, \mathbf{ke}\rangle\}$ $(v=0, 1, \ldots)$. For the case of a single molecular resonance in a large molecule (see Section 13) this extension is of little interest. However, in the case of internal conversion [47, 51] from a highly excited singlet state (see Figure 1g) the $|l, \text{vac}\rangle$ states are radiatively coupled to $|gv, \mathbf{ke}\rangle$ levels characterized by high v values, and such an extension of the theory is pertinent. The initially excited doorway state is now

$$|N_0, \text{vac}\rangle = \frac{1}{\gamma_{N_0}} H_{\text{int}} |go, \mathbf{ke}\rangle. \tag{8.12}$$

We can subsequently define a whole set of discrete states

$$|N_v, \text{vac}\rangle = \frac{1}{\gamma_{N_v}} H_{\text{int}} |gv, \mathbf{ke}\rangle, \tag{8.13}$$

where

$$|\gamma_{N_v}|^2 = \langle gv, \mathbf{ke} | H_{\text{int}} \cdot H_{\text{int}} | gv, \mathbf{ke} \rangle. \tag{8.14}$$

In the ideal short time experiment, where the $|N_0\rangle$ state is initially excited, the probability $P_g^v(t)$ for population of the $|gv\rangle$ ground state level at time t is given by

$$P_g^v(t) = \Gamma_{N_v}^{(\text{rad})} \int_0^t |C_{N_v, N_0}(\tau)|^2 \, d\tau, \tag{8.15}$$

where

$$\Gamma_{N_v}^{(\text{rad})} = 2\pi |\gamma_{N_v}|^2 \, \varrho_r. \tag{8.16}$$

The different radiative decay channels to different v states (or groups of v states) can be separated experimentally by monitoring the energy resolved decay spectrum. We have to distinguish at this point between the doorway state $|N_0\rangle$ and the 'escaping' states $|N_v\rangle$ (all v). In the simplest case there is only a single escape state $|N_0\rangle$. In more complicated situations there is a whole manifold of scape states. Under the latter circumstances the decay probabilities $P_g^v(t)$ will be characterized by a different time dependence for different v values.

We were able to obtain a general formal picture for the decay of an initially excited state prepared by a delta function field excitation amplitudes. These general results are not entirely satisfactory because of two reasons. First, the expressions obtained for the general time evolution of the system and for the case of the decay of initially prepared state are formal. To account for any real life situation we have to provide explicit expressions for the decay amplitudes. Second, the description of the excitation process is suitable only for the model system consisting of two electronic configurations. When the experimentalist will hit a real molecule by an extremely broad pulse including all frequencies, he will excite a multitude of electronic states, and the experimentally monitored photon counting rate will not be very informative. One has to find some weaker conditions than the delta function excitation to specify the decay of the excited states. We have stated in condition (b) that the pulse duration should not exceed the relevant 'decay times' of the system. We have thus to provide a proper description of these characteristic decay times, or rather decay widths, of the general molecular system. For a bunch of closely spaced levels the resulting decay pattern is complex, exhibiting several decay times and/or oscillatory interference terms. Subsequently we shall be able to define a short time experiment by the condition that the width of the wave packet exceeds all these characteristic widths. We now proceed to provide the necessary parameters which determine the molecular decay amplitudes.

9. An Effective Hamiltonian for Independently Radiatively Decaying Levels

In order to apply the general theory outlined above we shall introduce and explore an effective Hamiltonian which specifies the time evolution of the excited molecular states in the presence of the radiation field. The use of such effective Hamiltonians is common in fields such as magnetic resonance, where in handling relaxation problems one considers the time evolution of a small part of the system. In our problem we shall consider the time evolution of the subpart $\{|m, \text{vac}\rangle\}$ consisting of all discrete zero photon electronically excited states.

Radiationless transitions in a hypothetical system, in the absence of radiative decay, can be adequately described in terms of the molecular eigenstates which diagonalize the electronic Hamiltonian H_M. However, for a real physical system of closely spaced levels, the molecular eigenstates lose their general physical utility. Adopting the generalized Wigner–Weisskopf approximation, Bixon et al. [38] have demonstrated that the time evolution of the excited molecular states can be described in terms of an effective Hamiltonian. The same argument was provided by Freed and Jortner [41] in terms of the Green's function formalism.

The definition of the effective Hamiltonian for the excited states rests on the following observations:

(1) The Hilbert space is partitioned into the subspaces $\hat{P}$ and $\hat{Q}$ (see Equations (5.8)–(5.10)).

(2) The general time evolution of the excited states Equation (7.13) or Equation (7.6), is determined by decay amplitudes combining levels in the $\hat{P}$ subspace.

(3) Thus, the evaluation of the relevant decay amplitudes requires the matrix elements of $\hat{P}G\hat{P}$ between excited states.

(4) The operator $\hat{P}G\hat{P}$ (Equation (5.12)) will be rewritten in the form

$$\hat{P}G(E)\hat{P} = \hat{P}(E - H_{\text{eff}} + i\eta)^{-1} \hat{P}, \tag{9.1}$$

where the effective Hamiltonian in the $\hat{P}$ subspace is

$$H_{\text{eff}} = \hat{P}(H_0 + R)\hat{P}. \tag{9.2}$$

(5) The evolution operator in the $\hat{P}$ subspace can be formally represented utilizing Equation (5.6)

$$\hat{P}U(t, 0)\hat{P} = \int_{-\infty}^{+\infty} dE \exp(-iEt)\,\hat{P}G(E)\hat{P}. \tag{9.3}$$

Now, making use of (9.2) we get

$$\hat{P}U(t, 0)\hat{P} = \int_{\infty}^{\infty} dE\,\hat{P}\,\frac{\exp(-iEt)}{E - H_{\text{eff}} + i\eta}\,\hat{P} = \exp[-i\hat{P}(H_0 + R)\hat{P}t]\,\hat{P} =$$

$$= \hat{P}\exp[-iH_{\text{eff}}t]\,\hat{P}. \tag{9.3a}$$

(6) A set of states $\{|j, \text{vac}\rangle\}$ defined in the $\hat{P}$ subspace which diagonalize the effective Hamiltonian can be then used for the spectral representation of $\hat{P}G\hat{P}$ and of the evolution operator (9.3a) in the $\hat{P}$ subspace.

Let us now proceed to explore the general form of the effective Hamiltonian (9.1).

Utilizing the definition of the level shift operator (Equation (5.13)) we have

$$H_{\text{eff}} = \hat{P} \left(H + V\hat{Q} \frac{1}{E - \hat{Q}H\hat{Q} + i\eta} \hat{Q}V \right) \hat{P}. \tag{9.4}$$

As $\hat{P} (H_{\text{rad}} + H_{\text{int}}) \hat{P} = 0$ and $\hat{P}V\hat{Q} = \hat{P}H_{\text{int}}\hat{Q}$ we obtain the formal result

$$H_{\text{eff}} = \hat{P} \left(H_M + H_{\text{int}}\hat{Q} \frac{1}{E - \hat{Q}H\hat{Q}} \hat{Q}H_{\text{int}} \right) \hat{P}. \tag{9.5}$$

The effective Hamiltonian can be thus recast in the matrix form

$$H_{\text{eff}} = H_M + \Delta - \tfrac{1}{2}i\, \Gamma, \tag{9.6}$$

where it is understood that H_{eff} combines only $|m, \text{vac}\rangle$ states in the $\hat{P}$ subspace. The explicit forms for the level shift matrix Δ and for the decay matrix Γ are obtained from the relaxation contribution of the level shift operator Equation (5.14) to Equation (9.5).

$$\Delta (E) - \frac{i}{2} \Gamma (E) = \sum_{\mathbf{ke}} \frac{H_{\text{int}} |g, \mathbf{ke}\rangle \langle g, \mathbf{ke}| H_{\text{int}}}{E - E_g - k + i\eta} =$$

$$= \sum_{\mathbf{e}} \int \frac{H_{\text{int}} |g\mathbf{ke}\rangle \langle g\mathbf{ke}| H_{\text{int}}}{E - k + i\eta} \varrho_r (k)\, d\mathbf{k}, \tag{9.7}$$

where we have applied Equation (8.5) and set $E_g = 0$. Utilizing the well-known relation

$$\int \frac{f(k)}{E - k + i\eta}\, dk = PP \int \frac{f(k)}{E - k} - i\pi f(E), \tag{9.8}$$

where PP represents the Cauchy principal part of the integral, we obtain

$$\Delta (E) = PP \sum_{\mathbf{e}} \int \frac{H_{\text{int}} |g\mathbf{ke}\rangle \langle g\mathbf{ke}| H_{\text{int}}}{E - k} \varrho (\mathbf{k})\, d\mathbf{k}, \tag{9.9}$$

$$\Gamma (E) = 2\pi \sum_{\mathbf{e}} \int d\Omega\, H_{\text{int}} |g\mathbf{ke}\rangle \varrho_r (E) \langle g\mathbf{ke}| H_{\text{int}}. \tag{9.10}$$

The matrix elements of the level shift matrix are

$$\Delta_{mm'} = PP \sum_{\mathbf{e}} \int \frac{\langle m, \text{vac}| H_{\text{int}} |g, \mathbf{ke}\rangle \langle g, \mathbf{ke}| H_{\text{int}} |m', \text{vac}\rangle \varrho_r (\mathbf{k})}{E - k}\, d\mathbf{k}. \tag{9.11}$$

This is a generalization of the concept of the ordinary level shift of a single resonance (i.e. single level interacting with a continuum). The elements of the (real) level shift matrix (9.9) diverge when the integration over k is performed to infinity, as $\varrho_r (k) \propto k^2$. This well-known difficulty [92] of quantum electrodynamics is resolved by a renormalization procedure as is done in the theory of the Lamb shift. The diagonal and off-diagonal matrix elements of Δ are expected to be of the order of [86] $\Delta \sim L^H/n$ where

$L^H \sim 10^{-2}$ cm^{-1} is the hydrogenic Lamb shift, and n is the number of effectively coupled levels. Then Δ varies from 10^{-2} cm^{-1} for a small number of coupled levels to 10^{-6} cm^{-1} for a dense distribution. These terms will result in shifts of the (real part) of H_{eff} and they are of minor practical interest.

Of crucial importance is the damping matrix (9.11) which is explicitly given in the form

$$\Gamma_{mm} = 2\pi \sum_e \int d\Omega \langle m, \text{vac}| H_{\text{int}} |g, \text{ke}\rangle \langle g, \text{ke}| H_{\text{int}} |m', \text{vac}\rangle \varrho_r. \quad (9.12)$$

The following features of the damping matrix should be noted:

(a) Γ provides a generalization of the Fermi golden rule. For the decay of a single resonance, the resonance width (or the reciprocal decay time) is given by Γ_{mm}.

(b) Γ is, in general, nondiagonal.

(c) Γ is Hermitian.

(d) Γ is a slowly varying function of the energy in the range of interest.

(e) The offdiagonal matrix elements of Γ represent coupling between the $|m, \text{vac}\rangle$ states via the one-photon states,

$$|m, \text{vac}\rangle \rightarrow |g, \text{ke}\rangle \rightarrow |m', \text{vac}\rangle.$$

These off-diagonal contributions will be important only in the case of near degeneracy when these terms are comparable to the energy spacing between the energy levels, i.e.

$$\Gamma_{mm'} \sim |E_m - E_{m'}|. \quad (9.13)$$

To be more specific let us now provide some simple examples for the effective Hamiltonian (9.6) in different representations, neglecting the (small) contribution of the level shift matrix. Consider first the Born–Oppenheimer basis $|s\rangle$ and $\{|l\rangle\} E|m\rangle$ where $|s\rangle$ is the single doorway state. Then the effective Hamiltonian is

$$H_{\text{eff}} = \begin{pmatrix} E_s - \frac{1}{2}i\Gamma_s & (H_v)_{sl} & (H_v)_{sl'} \cdots \\ (H_v)_{l_s} & E_l & 0 & \cdots \\ (H_v)_{l'_s} & 0 & E_{l'} & \cdots \end{pmatrix}, \quad (9.14)$$

so that the electronic Hamiltonian is off-diagonal, while the damping matrix is diagonal, with a single finite diagonal term given by (7.7). In the molecular eigenstates representation $\{|n\rangle\} E\{|m\rangle\}$ where the molecular eigenstates $\{|n\rangle\}$ are characterized by the energies $E_1, E_2, ..., E_n$ etc. The effective Hamiltonian assumes the form

$$H_{\text{eff}} = \begin{pmatrix} E_1 - \frac{1}{2}i\Gamma_{11} & -\frac{1}{2}i\Gamma_{12} & \cdots \\ -\frac{1}{2}i\Gamma_{21} & E_2 - \frac{1}{2}i\Gamma_{22} \cdots \\ \vdots & \vdots \end{pmatrix}. \quad (9.15)$$

Thus the electronic Hamiltonian is diagonal, but we pay the price by having an off-diagonal damping matrix.

We shall now proceed to explore the general properties of the effective Hamiltonian (9.6):

(1) The effective Hamiltonian is nonhermitian. This can be rationalized by noting that we consider only a subspace of the Hilbert space consisting of the (discrete) zero-photon manifold. In fact, from the basic definition (9.6) and from the hermitian property of Γ we conclude that H_{eff} is in general a sum of a Hermitian matrix H_M and an antihermitian matrix, $-\tfrac{1}{2}i\Gamma$.

(2) When the effective Hamiltonian is nondiagonal within a given basis of zero-photon states, these states cannot be considered to decay independently, in view of the appearance of off-diagonal terms in H_{eff}. Thus the evolution operator (9.3a) in the $\hat{P}$ subspace will contain such off-diagonal contributions.

(3) One can find, in principle, for the general case and in practice for simple model systems, a basis of zero-photon states $\{|\,j, \mathrm{vac}\rangle\}$ which diagonalize the effective Hamiltonian. This basis set is obtained by the transformation:

$$\begin{pmatrix} j_1, \mathrm{vac} \\ j_2, \mathrm{vac} \\ \vdots \end{pmatrix} = D \begin{pmatrix} m_1, \mathrm{vac} \\ m_2, \mathrm{vac} \\ \vdots \end{pmatrix}. \tag{9.16}$$

(4) The effective Hamiltonian is diagonalized by the transformation

$$D\ H_{\mathrm{eff}}\ D^{-1} = \Lambda, \tag{9.17}$$

where Λ is diagonal, $\Lambda_{ij} = \Lambda_{ii}\delta_{ij}$. As H_{eff} is a sum of Hermitian matrix H_M and an antihermitian matrix $-\tfrac{1}{2}\Gamma$, D is nonunitary matrix. When we use a basis set $|m\rangle$ of real functions, H_{eff} is a complex symmetric matrix and it can be always diagonalized using an orthogonal (nonunitary) transformation matrix D.

(5) The matrix elements of the diagonal matrix Λ (9.17) are in general complex

$$\Lambda_{jj'} = (E_j - \tfrac{1}{2}i\Gamma_j)\,\delta_{jj'}. \tag{9.18}$$

(6) The effective Hamiltonian matrix in the $|\,j, \mathrm{vac}\rangle$ representation is not given by the usual scheme suitable for orthogonal basis sets, i.e.

$$\Lambda_{jj'} \neq \langle j, \mathrm{vac}|\,H_{\mathrm{eff}}\,|j', \mathrm{vac}\rangle = (E_{j'} - \tfrac{1}{2}i\Gamma_{j'})\cdot\langle j, \mathrm{vac}|j', \mathrm{vac}\rangle. \tag{9.19}$$

(7) The new basis set $|\,j, \mathrm{vac}\rangle$ which diagonalizes H_{eff} is characterized by complex energies $E_j - \tfrac{1}{2}i\Gamma_j$. It is natural to assign the real part of (9.19) to the energies of these states, i.e. $\mathrm{Re}\,\Lambda_{jj} = E_j$ while the imaginary parts $\mathrm{Im}\,\Lambda_{jj} = \tfrac{1}{2}\Gamma_j$ corresponds to the characteristic widths of the system in the presence of the radiation field.

(8) The diagonal sum rule applies to the transformation (9.17). Thus

$$\mathrm{Re}\,[\mathrm{Tr}\,\Lambda] = \mathrm{Re}\,[\mathrm{Tr}\,H_{\mathrm{eff}}] = \mathrm{Tr}\,[H_M], \tag{9.20}$$

$$\mathrm{Im}\,[\mathrm{Tr}\,\Lambda] = \mathrm{Im}\,[\mathrm{Tr}\,H_{\mathrm{eff}}] = -\tfrac{1}{2}\,[\mathrm{Tr}\,\Gamma]. \tag{9.21}$$

Equation (9.20) implies that

$$\sum_j E_j = \sum_m E_m, \tag{9.22}$$

which is the conventional diagonal sum rule, whereupon the sum of the real energies of the $|j, \text{vac}\rangle$ states is equal to the sum of the energies of any other basis states $|m, \text{vac}\rangle$. From Equation (9.21) we have the more interesting result

$$\sum_j \Gamma_j = \sum_m \Gamma_{mm}. \tag{9.23}$$

Thus the sum of the widths of the $|j, \text{vac}\rangle$ states is equal to the sum of the diagonal elements of the (nondiagonal) Γ matrix in any $|m, \text{vac}\rangle$ representation.

(9) The $|j, \text{vac}\rangle$ basis set is *not* orthogonal. This is a consequence of the antihermitian property of $\frac{1}{2}i\Gamma$ which causes the non-unitarity of D.

(10) In order to expand $\hat{P}$ in terms of diagonalized projections we shall now define the complementary basis set $|\bar{j}, \text{vac}\rangle$ by the relation:

$$\begin{pmatrix} |\bar{j}_1, \text{vac}\rangle \\ |\bar{j}_2, \text{vac}\rangle \\ \vdots \end{pmatrix} = (D^{-1})^\dagger \begin{pmatrix} |m_1, \text{vac}\rangle \\ |m_2, \text{vac}\rangle \\ \vdots \end{pmatrix}. \tag{9.24a}$$

In the special case when the $|m, \text{vac}\rangle$ basis set has a real representation the transformation is orthogonal, i.e. $D^{-1} = \tilde{D}$, and we get

$$(D^{-1})^\dagger = D^*, \tag{9.24b}$$

so that in this special case $|\bar{j}, \text{vac}\rangle = |j^*, \text{vac}\rangle$.

We can write for the general case

$$\hat{P} = \sum_j |j, \text{vac}\rangle \langle \bar{j}, \text{vac}|, \tag{9.25}$$

where $|\bar{j}, \text{vac}\rangle$ are obtained from the transformation (9.24). This relation is the consequence of the orthogonality of $\{|\ j\rangle\}$ and $\{|\bar{j}\rangle\}$, i.e.

$$\langle \bar{j}, \text{vac} \mid j', \text{vac}\rangle = \delta_{jj'}.$$

Finally we can derive a form of $\hat{P}$ in terms of the $|j, \text{vac}\rangle$ basis. From Equations (9.24) and (9.25) we get

$$|\bar{j}, \text{vac}\rangle = \sum_{j'} [(D^{-1})^\dagger D^{-1}]_{jj'} |j', \text{vac}\rangle \tag{9.26}$$

and

$$\hat{P} = \sum_{jj'} |j', \text{vac}\rangle [(D^{-1})^\dagger D^{-1}]_{jj'}^* \langle j', \text{vac}|. \tag{9.27}$$

In the special case when we use a real basis set,

$$(D^{-1})^{\dagger*} = D, \\ (D^{-1})^* = D^\dagger \tag{9.28}$$

and we get

$$\hat{P} = \sum_{jj'} |j, \text{vac}\rangle (DD^\dagger)_{jj'} \langle j', \text{vac}|. \tag{9.29}$$

(11) We shall now explore the most important feature of the basis sets. The time

evolution operator (Equation (9.3a)) is

$$\hat{P}U(t,0)\,\hat{P} = \hat{P}\exp\left(-iH_{\text{eff}}t\right)\hat{P} = \sum_j |j,\text{vac}\rangle \exp\left[-i\varLambda_{jj}t\right]\langle j,\text{vac}| =$$

$$= \sum_j |j,\text{vac}\rangle \exp\left[-iE_jt - \tfrac{1}{2}\varGamma_jt\right]\langle j,\text{vac}|. \qquad (9.30)$$

This general result implies that the decay amplitudes combining any pair of $|m,\text{vac}\rangle$ states will be expressed as a superposition of terms of the form $\exp\left[-\tfrac{1}{2}\varGamma_jt - iE_jt\right]$, i.e. a sum of independently decaying exponentials. Thus the basis set $|j,\text{vac}\rangle$ can be considered as the set of *independently decaying levels* characterizing the molecular system.

(12) To conclude this formal discussion we shall recast the Green's function in the $\hat{P}$ subspace in the spectral representation of the independently decaying levels $|j,\text{vac}\rangle$. Making use of Equations (9.1), (9.19) and (9.25) we get

$$\hat{P}G(E)\,\hat{P} = \sum_j \frac{|j\rangle\langle j|}{E - E_j + \tfrac{1}{2}i\varGamma_j}, \qquad (9.31)$$

which is of course nothing but the inverse Fourier transform of (9.30). Equation (9.31) will be useful in the study of optical lineshapes of a system with a large number of closely spaced levels (see Section 17).

10. Theoretical Results for Time Evolution of Excited Molecular States

We are now able to provide explicit expressions for the general time evolution of an excited molecular state. The decay amplitude (7.14) which describes the time evolution of the excited state with the aid of (9.30) takes the form

$$C_{mN}(\tau) = \sum_j \langle m,\text{vac}\,|\,j,\text{vac}\rangle \exp\left[-iE_j\tau - \tfrac{1}{2}\varGamma_j\tau\right]\langle j,\text{vac}\,|\,N,\text{vac}\rangle. \qquad (10.1)$$

Thus Equation (7.13) is

$$P_e(t) = |\gamma_N|^2 \sum_m \left| \int_0^t d\tau\varphi(t-\tau)\sum_j \langle m,\text{vac}\,|\,j,\text{vac}\rangle\cdot\exp\left[-iE_j\tau - \tfrac{1}{2}\varGamma_j\tau\right]\times \right.$$

$$\left. \times \langle j,\text{vac}\,|\,N,\text{vac}\rangle \right|^2. \qquad (10.2)$$

We now note that $\hat{Q}|j,\text{vac}\rangle = 0$ for each $|j,\text{vac}\rangle$ whereupon Equation (10.2) can be further simplified and recast to include only matrix elements of the $|N,\text{vac}\rangle$ state and the $|j,\text{vac}\rangle$ basis

$$P_e(t) = \int_0^t d\tau \int_0^t d\tau'\,\varphi(t-r)\,\varphi^*(t-\tau')\cdot\sum_{jj'} A_{j'j}\exp\left[-iE_j\tau - \tfrac{1}{2}\varGamma_j\tau\right]\times$$

$$\times \exp\left[iE_{j'}\tau' - \tfrac{1}{2}\varGamma_{j'}\tau'\right], \qquad (10.3)$$

where

$$A_{j'j} = \langle N | j', \text{vac} \rangle \langle j', \text{vac} | j, \text{vac} \rangle \langle j, \text{vac} | N, \text{vac} \rangle \cdot | \gamma_N |^2. \qquad (10.4a)$$

Utilizing the general formal definition (7.8a) of the doorway state we can express the coefficients $A_{j'j}$, in (10.3) in terms of the radiative coupling matrix elements with the ground state

$$A_{j'j} = \langle g, \mathbf{ke} | H_{\text{int}} | j', \text{vac} \rangle \langle j', \text{vac} | j, \text{vac} \rangle \langle j, \text{vac} | H_{\text{int}} | g, \mathbf{ke} \rangle. \qquad (10.4b)$$

Equation (10.3) may be rewritten in the following manner:

$$P_e(t) = \sum_{jj'} A_{j'j} F_{j'}^{p*}(t) F_j^p(t) = \sum_j A_{jj} | F_j^p(t) |^2 + 2 \text{Re} \sum_{j' > j} A_{j'j} F_{j'}^{p*}(t) F_j^p(t), \qquad (10.5)$$

where

$$F_j^p(t) = \int_0^t d\tau\, \varphi(t - \tau) \exp(-iE_j\tau) \exp(-\tfrac{1}{2}\Gamma_j\tau). \qquad (10.6)$$

We note that $\varphi(\tau)=0$; and $C_{jj}(\tau)=0$ for $\tau<0$. Thus the integral (10.6) can be rewritten as $\int_{-\infty}^{\infty} d\tau \varphi(t-\tau) C_{jj}(\tau)$. Utilizing the convolution theorem for Fourier transforms we get

$$F_j^p(t) = -i \int_{-\infty}^{\infty} dE \exp(-iEt) G_{jj}(E) a_k(E). \qquad (10.7)$$

To bring Equation (10.3) into a more tractable form let us utilize the Lorentzian photon wave packet (6.8) $a_k = A_N/(k - \bar{k} + \tfrac{1}{2}i\gamma_p)$ for optical excitation. Thus we obtain

$$F_j^p(t) = 2\pi A_N \frac{\exp[-iE_j t] \exp[-\tfrac{1}{2}\Gamma_j t] - \exp[-i\bar{k}t] \exp[-\tfrac{1}{2}\gamma_p t]}{\bar{k} - E_j + \tfrac{1}{2}i(\Gamma_j - \gamma_p)}, \qquad (10.8)$$

and the photon counting rate is

$$\dot{P}_g(t) = -\frac{d}{dt} \left[\sum_j \sum_{j'} A_{j'j} F_{j'}^p(t)^* F_j(t) \right]. \qquad (10.9)$$

Equations (10.5), (10.8) together with the definition (10.4) provide us with the desired general results concerning the time evolution of the excited state. From these results we can immediately draw some general conclusions for the time evolution of a system of closely spaced levels:

(a) The time evolution of the excited states is expressed in terms of cross products of the functions $F_j^p(t)$. It is important to notice that the matrix $A_{j'j}$ (Equation (10.4)) is not diagonal in view of the nonorthogonality of the basis set $| j, \text{vac} \rangle$.

(b) Each of the functions $F_j^p(t)$ incorporates dual information. It contains the molecular energies E_j and widths Γ_j of the independently decaying levels, together with relevant energy parameters $\bar{k}$ and γ_p which characterize the energy maximum and the width of the exciting pulse.

(c) The time-independent denominators of F_j^p provide the attenuation factor for absorption of the pulse energy by the $|j, \text{vac}\rangle$ level.

(d) In the *mathematical* limit $t \to \infty$, $F_j^p(t) \to 0$ for all j irrespective of the relation between γ_p and $\{\Gamma_j\}$. This implies that $P_e(\infty) = 0$. Thus for a physical system characterized by a *discrete* spectrum of excited states the total photon emission yield at $t = \infty$ will be unity, i.e. $P_g(\infty) = 1$. It is important to stress at this point that the procedure which led to the definition of the $|j, \text{vac}\rangle$ basis and the derivation of Equation (10.4) considered a discrete molecular spectrum. When the spectrum of H_{BO} (when the Born–Oppenheimer molecular basis is employed) or of H_M (when the molecular eigenstates are used) contains continuum states we should not incorporate them in the $\hat{P}$ subspace. Under these more complicated circumstances we have to include the zero photon continuum molecular states in the $\hat{Q}$ subspace while the $\hat{P}$ subspace will contain only discrete levels. Under these conditions the probability of the system to be in the (extended) $\hat{Q}$ space at $t = \infty$ will be still unity, however, the photon emission yield at $t = \infty$ may be lower than unity due to the branching between the radiative channels and the nonradiative continuum channels. It should be finally pointed out that these results do by no means contradict the idea of electronic relaxation (internal conversion or intersystem crossing) in an isolated molecule, where the concept of the statistical limit rests on the notion of (a) practical irreversibility at a time scale short relative to the (exceedingly long) Poincaré recurrence time and (b) the occurrence of sequential decay processes in the dense intramolecular manifold.

(e) In the limit of high energy resolution of the exciting pulse, $\gamma_p \ll \Gamma_j$ and the contribution to the $F_j^p(t)$ functions originating from the molecular lifetimes Γ_j will be masked out by the long decay time of the pulse. Under these circumstances the time resolved photon counting rate will not result in any relevant information regarding the 'molecular' widths. This situation corresponds to the 'long time' excitation experiment.

(f) In the limit of a broad excitation pulse we encounter the 'short excitation' experiment and the time resolved decay pattern provides us with pertinent information regarding the molecular decay widths.

11. 'Initially Prepared' Decaying State, Revisited

As we have already pointed out in Section 8 the description of an 'initially prepared' decaying state excited by a delta function field amplitude (6.16) has to be modified. We can now provide a less stringent realistic condition for the 'preparation' process by requiring that the energetic spread γ_p of the photon wave packet considerably exceeds the characteristic widths Γ_j for all the independently decaying levels, i.e.

$$\gamma_p \gg \Gamma_j \tag{11.1}$$

for all j. Under these circumstances the functions (10.8) take the form

$$F_j^p(t) = -2\pi A_N \frac{\exp[-iE_j t] \exp[-\tfrac{1}{2}\Gamma_j t]}{E_j - \bar{k} + \tfrac{1}{2} i \gamma_p}. \tag{11.2}$$

Using Equations (10.4) and (11.2) the time evolution of the excited states resulting from the realistic 'short time' excitation experiment can be now written as follows

$$P_e(t) = \sum_j \frac{\bar{A}_{jj}}{(E_j - \bar{k})^2 + (\tfrac{1}{2}\gamma_p)^2} \exp(-\Gamma_j t) +$$

$$+ 2\text{Re} \sum_{j<j'} \sum \frac{\bar{A}_{j'j} \exp[i(E_{j'} - E_j)t - \tfrac{1}{2}(\Gamma_j + \Gamma_{j'})t]}{[E_j - \bar{k} + \tfrac{1}{2}i\gamma_p][E_{j'} - \bar{k} - \tfrac{1}{2}i\gamma_p]}, \qquad (11.3)$$

where we have defined

$$\bar{A}_{j'j} = 4\pi^2 |A_N|^2 \, A_{j'j} = 4\pi^2 |A_N|^2 \, \langle N, \text{vac}| \, j', \text{vac}\rangle \times$$
$$\times \langle j', \text{vac} \mid j, \text{vac}\rangle \langle j, \text{vac} \mid N, \text{vac}\rangle \, |\gamma_N|^2 . \qquad (11.3a)$$

Separating the mixed coefficients Equation (11.3) into their real and imaginary parts

$$\frac{\bar{A}_{j'j}}{[E_j - \bar{k} + \tfrac{1}{2}i\gamma_p][E_{j'} - \bar{k} - \tfrac{1}{2}i\gamma_p]} \equiv R_{j'j} \exp[i\varphi_{j'j}] \qquad (11.3b)$$

we get

$$P_e(t) = \sum_j \frac{\bar{A}_{jj} \exp[-\Gamma_j t]}{(E_j - \bar{k})^2 + \tfrac{1}{4}\gamma_p^2} +$$

$$+ 2 \sum_{j<j'} R_{j'j} \exp[-\tfrac{1}{2}(\Gamma_j + \Gamma_{j'})]t \cos[(E_{j'} - E_j)t + \varphi_{j'j}]. \qquad (11.3c)$$

This result provides us with the proper description of the time evolution of the discrete excited states resulting from a realistic short time excitation.

Thus the time evolution of the excited state is solely determined by the molecular parameters E_j and Γ_j. This result differs from that obtained in Section 8 for the extremely broad excitation condition (i.e. $\gamma_p \to \infty$) only by the introduction of the time-independent numerical factors $[(E_j - \bar{k}) + \tfrac{1}{2}i\gamma_p]^{-1}$ in the denominators of all the terms in Equation (11.3). These attenuation factors account for the absorption strength of the exciting pulse by the various independently decaying levels $|j, \text{vac}\rangle$.

We can now introduce a second condition for the pulse width

$$\gamma_p \gg |E_j - \bar{k}| \qquad (11.4)$$

for all E_j which implies that the pulse width exceeds the energy spread of $|j, \text{vac}\rangle$. When both conditions (11.1) and (11.4) are simultaneously satisfied we get

$$P_e(t) = \frac{4}{\gamma_p^2} \sum_{j, j'} \bar{A}_{j'j} \exp[i(E_{j'} - E_j)t - \tfrac{1}{2}(\Gamma_j + \Gamma_{j'})t], \qquad (11.5)$$

which corresponds to the extremely broad excitation condition. Thus Equation (11.1) provides the necessary condition for a realistic broad band excitation. This condition is useful for the study of a sparse distribution of strongly coupled levels as is the case

for interstate coupling in small molecules. The combination of conditions (11.1) and
(11.4) provides us with the circumstances equivalent to a delta function excitation in
time, which are useful for the study of systems of closely spaced levels, i.e. a dense
level structure in the excited states of large molecules.

The time evolution, Equation (11.3), consists of two contributions: (a) a sum of
decaying exponentials; (b) a sum of cross terms, which contain oscillatory contributions for the time evolution of the excited states. These oscillatory terms characterized
by the periods $\hbar(E_{j'} - E_j)^{-1}$ may lead to the observation of quantum beats in the
radiative decay.

When we consider the excitation of the system by a single pulse, disregarding more
sophisticated techniques such as double resonance methods, the experimental information regarding the decay features originates from the photon counting rate. When
only condition (11.1) is satisfied we get from Equation (11.3) for the realistic short
time excitation experiment:

$$
\dot{P}_g(t) = -\dot{P}_e(t) = \sum_j \frac{\bar{A}_{jj}\Gamma_j \exp(-\Gamma_j t)}{(E_j - \bar{k})^2 + (\tfrac{1}{2}\gamma_p)^2} +
$$

$$
+ \sum_{j \neq j'} \frac{\bar{A}_{j'j}\left[i(E_{j'} - E_j) - \dfrac{\Gamma_j + \Gamma_{j'}}{2}\right]}{[E_j - \bar{k} + \tfrac{1}{2}i\gamma_p][E_{j'} - \bar{k} - \tfrac{1}{2}i\gamma_p]} \times
$$

$$
\times \exp[i(E_{j'} - E_j)t]\exp[-\tfrac{1}{2}(\Gamma_j + \Gamma_{j'})t]. \tag{11.6}
$$

Now, for a system of densely spaced excited levels, we can invoke the additional condition (11.4) whereupon Equation (11.6) is simplified to read

$$
\dot{P}_g(t) = \frac{4}{\gamma_p^2} \sum_{jj'} \bar{A}_{j'j}[\tfrac{1}{2}(\Gamma_j + \Gamma_{j'}) + i(E_{j'} - E_j)] \times
$$

$$
\times \exp[i(E_{j'} - E_j)t]\exp[-\tfrac{1}{2}(\Gamma_j + \Gamma_{j'})t], \tag{11.7}
$$

where the coefficients $\bar{A}_{j'j}$ are given by (10.4) and (11.3a). Equation (11.7) could have
been alternatively derived from the expression (8.11a) utilizing the form of the evolu-
tion operator (9.30) in the $|j\rangle$ representation

$$
\dot{P}_g(t) = 2\pi |\gamma_N|^2 \varrho_r |C_{NN}(t)|^2 =
$$

$$
= \sum_j \sum_{j'} B_{j'j} \exp[-i(E_j - E_{j'})t - \tfrac{1}{2}(\Gamma_j + \Gamma_{j'})t], \tag{11.8}
$$

where

$$
B_{j'j} = 2\pi |\gamma_N|^2 \varrho_r \langle N| j\rangle \langle j | N\rangle \langle N | j'\rangle \langle j' | N\rangle. \tag{11.9}
$$

In the short excitation limit we choose $A_N = i\gamma_p/4\pi\gamma_N$ (in order to obtain $\varphi(t) =$
$= (1/\gamma_N)\,\delta(t)$, see Section 8). Hence, the equivalence of Equations (11.7) and (11.8)
implies that

$$
\hat{A}_{j'j} = \hat{B}_{j'j}, \tag{11.10}
$$

where

$$\hat{A}_{j'j} = \langle j' \mid j \rangle \left[i \left(E_j - E_{j'} \right) + \tfrac{1}{2} \left(\Gamma_j + \Gamma_{j'} \right) \right] \qquad (11.10a)$$

and

$$\hat{B}_{j'j} = \langle N \mid j \rangle \langle j' \mid N \rangle \, 2\pi \left| \gamma_N \right|^2 \varrho_r . \qquad (11.10b)$$

This equality (11.10) can be easily proved by utilizing the general properties of H_{eff}. We have

$$\hat{A}_{j'j} = i \left[\langle j' \mid H_{\text{eff}} \mid j \rangle - \langle j^* \mid H_{\text{eff}}^* \mid j'^* \rangle \right] . \qquad (11.11)$$

Separating H_{eff} into Hermitian and an antihermitian parts i.e. $H_{\text{eff}} = H_M - (i/2)\,\Gamma$, where H_M and Γ are obviously Hermitian, we get

$$\hat{A}_{j'j} = i \left[\langle j' \mid H_M - i\Gamma \mid j \rangle - \langle j^* \mid H_M + i\Gamma \mid j'^* \rangle = 2 \langle j' \mid \Gamma \mid j \rangle . \qquad (11.11a)$$

Using the definition of Γ (Equation (9.10)), we get from (11.11a)

$$\hat{A}_{j'j} = 2\pi \left| \gamma_N \right|^2 \varrho_N \langle j' \mid N \rangle \langle N \mid j \rangle = \hat{B}_{j'j} . \qquad (11.12)$$

We note in passing that the equivalence of Equations (11.7) and (11.8) implies that for the extreme case of short time excitation, the generalized doorway state obeys the relation

$$\dot{P}_g(t) = \frac{d}{dt} \left[\langle N(t) \mid N(t) \rangle \right] = 2\pi \left| \gamma_N \right|^2 \varrho_r \left| \langle N(0) \mid N(t) \rangle \right|^2 . \qquad (11.13)$$

From these results we conclude that the radiative decay rate of a system of discrete excited levels exhibits the following features:

(a) The photon counting rate can be in general recast in terms of linear superposition of a sum of direct exponentials and of a sum of oscillatory terms.

(b) The feasibility of the observation of the oscillatory pattern of the decay is crucially determined by the nature of the physical system.

(c) When the spacings between the $\mid j \rangle$ levels considerably exceed their radiative widths, i.e. $\Gamma_{j'}, \Gamma_j \ll \mid E_j - E_{j'} \mid$ for all $\mid j \rangle$ and $\mid j' \rangle$ the oscillatory term will exhibit extremely fast oscillations on the time scale Γ_j^{-1} or $\Gamma_{j'}^{-1}$ which will average out to zero. Thus for a system of coarsely spaced $\mid j \rangle$ levels no oscillatory contributions to the decay in (11.6) or (11.7) will be exhibited and the radiative decay rate will be determined by the first sum in (11.6), i.e. a linear superposition of decaying exponentials. This situation prevails for strong coupling between a sparse distribution of levels in a small molecule.

(d) Consider now the opposite extreme case of a dense level distribution with a single (zero-order) $\mid s \rangle$ level acting as a doorway state, as is the situation in a large isolated statistical molecule. In this case there is a large number of cross terms in Equation (11.6) or rather in (11.7). These oscillatory terms will lead to a destructive interference effect resulting in shortening of the radiative decay time on the experimentally relevant time scale (see Section 13).

(e) Interference effects, i.e. quantum beats in the radiative decay of an isolated molecule can be experimentally observed only for a system characterized by a small number of closely spaced $|j, \text{vac}\rangle$ levels, where $\Gamma_{j'}$, $\Gamma_j \sim |E_{j'} - E_j|$. This situation requires effective coupling between a small number of zero order molecular levels corresponding to two electronic configurations. In real life it may be possible [47] to observe quantum beats in the decay of an excited state of a large molecule which corresponds to the intermediate level structure (see Figure 1e and discussion in Section 15).

(f) From the point of view of general methodology it is important to notice that the oscillatory terms which may result in observable quantum beats are exhibited both in the probability for population of the excited state, $P_e(t)$, and of the ground state, $P_g(t)$. Thus the phenomenon of quantum beats in the radiative decay rate originates from the oscillations of the system between its electronically excited zero-photon levels. An attempt was made [93] to consider 'recurrence oscillations' in the excited state as distinguished from quantum beats in the radiative decay. The general treatment presented herein demonstrates that such a distinction is not acceptable.

12. Parallel Decay of Metastable States

We have demonstrated in Section 10 that for a discrete spectrum of zero-order excited states (which result in a discrete manifold $|j, \text{vac}\rangle$), $P_e(t) \rightarrow 0$ and $P_g(t) \rightarrow 1$, in the mathematical limit $t \rightarrow \infty$, whereupon the quantum yield for emission at $t = \infty$ is unity and no discrete excited levels are populated at the distant future. The situation is drastically different when the excited molecular states contain a continuum $|lc, \text{vac}\rangle$ characterized by the density of states ϱ_c, for example, a dissociative intramolecular continuum in the case of predissociation. We shall consider now the simplest physical common situation where the radiative continuum $|g, \text{ke}\rangle$ and the molecular zero photon continuum $|lc, \text{vac}\rangle$ are not directly coupled, i.e.

$$\langle g, \text{ke}| \ V \ |l_c, \text{vac}\rangle = 0. \tag{12.1}$$

We shall partition the Hilbert space as follows: the $\hat{P}$ subspace will contain the discrete zero photon excited states, as before

$$\hat{P} = \sum_m |m, \text{vac}\rangle \langle m, \text{vac}|, \tag{12.2}$$

while the $\hat{Q}$ subspace will contain the two continua, that is

$$\begin{aligned}
\hat{Q} &= \hat{Q}_r + \hat{Q}_l, \\
\hat{Q}_r &= \sum_{\text{ke}} |g, \text{ke}\rangle \langle g, \text{ke}|, \\
\hat{Q}_l &= \sum_{lc} |lc, \text{vac}\rangle \langle lc, \text{vac}|.
\end{aligned} \tag{12.3}$$

Under the conditions of short-time excitation satisfying conditions (11.1) and (11.4) together with (12.1) the initially excited state is still $|N, \text{vac}\rangle$ (Equation (7.8)) the time evolution of the discrete levels is geven by (8.1) while the rate of photon counting is (8.11a). For most practical purposes we can consider the decay amplitudes which combine only the (discrete) states in the $\hat{P}$ subspace. Now the relevant coupling matrix elements combining the $\hat{P}$ and $\hat{Q}$ subspaces are $\langle g, \text{ke}| H_{\text{int}}|m, \text{vac}\rangle$ and $\langle l_c, \text{vac}| H_v|m, \text{vac}\rangle$. We can again define a generalized effective Hamiltonian for the $\hat{P}$ subspace which will incorporate the effects of both radiative decay into $|g, \text{ke}\rangle$ and nonradiative decay into the dissipative continuum $|l_c, \text{vac}\rangle$. Equation (9.4) is still applicable however now the dissipative part of the level shift operator includes two contributions for parallel decay:

$$H_{\text{eff}} = \hat{P}\left(H_M + H_{\text{int}}\hat{Q}_r \frac{1}{E - \hat{Q}_r H \hat{Q}_r} \hat{Q}_r H_{\text{int}} + H_v \hat{Q}_l \frac{1}{E - \hat{Q}_l H \hat{Q}_l} \hat{Q}_l H_v \right)\hat{P}.$$
(12.4)

Thus the effective Hamiltonian takes the explicit form, which is a generalization of (9.6)

$$H_{\text{eff}} = \hat{P}\left(H_M + \delta - \tfrac{1}{2}i\gamma \right)\hat{P},$$
(12.5)

where δ and γ are the generalized level shift and decay matrices, now given in the explicit form

$$\delta_{mm'} = PP \sum_e \int dk \frac{\langle m, \text{vac}| H_{\text{int}}|g, \text{ke}\rangle \langle g, \text{ke}| H_{\text{int}}|m', \text{vac}\rangle}{E - k} \varrho_r(\mathbf{k}) +$$
$$+ PP \int dE_l \frac{\langle m, \text{vac}| H_v|l_c, \text{vac}\rangle \langle l_c, \text{vac}| H_v|m', \text{vac}\rangle \varrho_c(E_l)}{E - E_{l_c}}$$
(12.6)

and

$$\gamma_{mm'} = 2\pi \sum_e \int d\Omega \langle m, \text{vac}| H_{\text{int}}|g, \text{ke}\rangle \langle g, \text{ke}| H_{\text{int}}|m, \text{vac}\rangle \varrho_r +$$
$$+ 2\pi \langle m, \text{vac}| H_v|l_c, \text{vac}\rangle \langle l_c, \text{vac}| H_v|m', \text{vac}\rangle \varrho_c.$$
(12.7)

We can proceed as in Section 9 to find the basis set $|J, \text{vac}\rangle$ which diagonalizes the effective Hamiltonian (12.5), i.e.

$$(H_{\text{eff}})_{JJ'} = (E_J - \tfrac{1}{2}i\gamma_J)\delta_{JJ'},$$
(12.8a)

while for the complementary basis set $|\bar{J}, \text{vac}\rangle$ we have

$$(H_{\text{eff}}^*)_{JJ'} = (E_J + \tfrac{1}{2}i\gamma_J)\delta_{JJ'}.$$
(12.8b)

Now the decay widths of the independently decaying levels $|J, \text{vac}\rangle$ contain both radiative and nonradiative contributions. Finally, the evolution operator in the $\hat{P}$ subspace is

$$\hat{P}U(t, 0)\hat{P} = \sum_J |J, \text{vac}\rangle \exp\left[- iE_Jt - \tfrac{1}{2}\gamma_Jt \right]\langle \bar{J}, \text{vac}|.$$
(12.9)

It is a simple matter to extend the formalism presented in Sections 10 and 11 to include the role of the additional decay channel. To obtain the time evolution of the discrete states one has just to replace E_j by E_J and the radiative widths Γ_j by the total widths γ_J in Equations (10.3)–(10.6) for the general excitation and in Equations (11.3)–(11.7) for the 'short time' excitation. It is important to realize that $P_e(t)$ represents the time-dependent population of the excited discrete states and not of all excited states, so that $\dot{P}_g(t) \neq -\dot{P}_e(t)$ but rather $\dot{P}_g(t) = -\dot{P}_e(t) - \dot{P}_c(t)$ where $P_c(t)$ is the occupation probability of the $\{|l_c\rangle\}$ continuum. To gain some insight into the nature of the modification introduced by the presence of additional intramolecular decay channels let us write the photon counting rate for the excitation which satisfies both conditions (11.1) and (11.4). From Equations (8.11) and (10.9) we have

$$\dot{P}_g(t) = \sum_J \sum_{J'} B_{J'J} \exp\left[i\left(E_{J'} - E_J\right)t\right] \exp\left[-\tfrac{1}{2}\left(\gamma_J + \gamma_{J'}\right)t\right], \tag{12.10}$$

where

$$B_{J'J} = \langle N \mid J\rangle \langle J \mid N\rangle \langle N \mid J'\rangle \langle J' \mid N\rangle \cdot 2\pi \, |\gamma_N|^2 \, \varrho_r. \tag{12.11}$$

From these results we conclude that for the simplest case of parallel radiative and nonradiative decay:

(a) The time-dependent decay pattern is determined by the total widths γ_J of the independently decaying states. When the effective Hamiltonian (12.4) is nondiagonal these total widths have to be obtained from the general procedure outlined herein.

(b) Interference effects in the time evolution and in the photon counting rate of a system consisting of a small number of discrete coupled zero order excited states undergoing parallel decay may be exhibited. Quantum beats will be observed provided that the spacings between the small number of $|J, \text{vac}\rangle$ levels are comparable to their total widths, i.e.

$$\gamma_J, \gamma_{J'} \sim |E_J - E_{J'}|$$

for all J and J'. Quantum beats will not be observed for (1) a dense manifold of a large number of levels; (2) for extremely broadened levels manifold where

$$\gamma_J, \gamma_{J'} \gg |E_J - E_{J'}|.$$

(c) For a system of a small number of levels undergoing parallel decay it may be possible to vary continuously the γ_J widths via external pertubations and consequently modify the decay into the nonradiative relaxation channel [50]. Then interference effects will be exhibited for a narrow range of γ_J values.

(d) The total emission yield at $t = \infty$ can be obtained by integrating Equation (12.10), which results in the occupation probability of the radiative continuum in the distant future.

$$P_g(\infty) = \sum_J \sum_{J'} \frac{B_{J'J}}{i\left(E_J - E_{J'}\right) + \tfrac{1}{2}\left(\gamma_J + \gamma_{J'}\right)}. \tag{12.12}$$

As for $t = 0$ the initially excited $|N, \text{vac}\rangle$ state is normalized to unity thus $P_g(\infty)$ represents the emission quantum yield. The yield for decaying into the nonradiative continuum is

$$P_c(\infty) = 1 - P_g(\infty). \tag{12.13}$$

To lay the foundation for the discussion of the statistical limit we shall consider the simplest situation where a single level $|s, \text{vac}\rangle$ exhibits parallel decay into a radiative and a nonradiative continua. Now $\hat{P} = |s, \text{vac}\rangle \langle s, \text{vac}|$ while $\hat{Q}$ is given by (12.3). There is a single state in the $|J, \text{vac}\rangle$ manifold, i.e. $|J, \text{vac}\rangle \equiv |s, \text{vac}\rangle$. The time evolution of the excited state under the general conditions of wave-packet excitation is obtained from Equation (7.13) in the form

$$P_e(t) = |\gamma_N|^2 \left| \int_0^t \varphi(t - \tau) C_{ss}(\tau) \right|^2. \tag{12.14}$$

$C_{ss}(\tau)$ is the Fourier transform of

$$G_{ss}(E) = \frac{1}{E - E_s + \frac{1}{2}i\gamma_s}, \tag{12.15}$$

where the total width of the $|s, \text{vac}\rangle$ state is the sum of the radiative and nonradiative widths

$$\gamma_s = \Gamma_s + \Gamma_s^c \tag{12.16a}$$

$$\Gamma_s^c = 2\pi |\langle s, \text{vac}| H_v |l_c, \text{vac}\rangle|^2 \varrho_c, \tag{12.16b}$$

so that

$$C_{ss}(\tau) = \exp\left[-iE_s\tau - \frac{1}{2}\gamma_s\tau\right]. \tag{12.17}$$

Thus we get

$$P_e(t) = 4\pi^2 |A_N|^2 |\gamma_N|^2 \times$$
$$\times \frac{\exp(-\gamma_s t) + \exp(-\gamma_p t) - 2\exp\left(-\frac{1}{2}(\gamma_s + \gamma_p)t\right)\cos(E_s - \bar{k})t}{(E_s - \bar{k})^2 + \frac{1}{4}(\gamma_p - \gamma_s)^2}. \tag{12.18}$$

From this result we conclude that (1) the only molecular information originating from the time evolution of this system is the resonance width γ_s. Excitation characterized by different wave-packet widths (i.e. different excitation times) will not result in new information. (2) The trigonometric factor $\cos(E_s - \bar{k})t$ in (12.18) represents a 'ringing effect' between the field and the molecular system. (3) When $\gamma_p \gg \gamma_s$ we encounter the 'short excitation' condition,

$$P_e(t) \propto \exp(-\gamma_s t).$$

When both conditions (11.1) and (11.4) are obeyed the photon counting rate contains a single exponential decay

$$\dot{P}_g(t) = \Gamma_s \exp(-\gamma_s t). \tag{12.19}$$

Finally, the emission quantum yield is just the branching ratio between the radiative and the total width,

$$Y = \Gamma_s/\gamma_s = \Gamma_s/(\Gamma_s + \Gamma_s^c). \tag{12.20}$$

13. The Statistical Limit

When the background density of the vibronic levels in a large molecule is exceedingly high (see Figures 1a, c) one should enquire under what circumstances this intramolecular quasicontinuum can act as a practical decay channel. This question is central for understanding of electronic relaxation (i.e. internal conversion and intersystem crossing) in an isolated large molecule. This interesting problem imposes some conceptual difficulties. We have demonstrated that for a general discrete spectrum of the excited states the probability of the system to be in a discrete excited state is zero at $t = \infty$. Only when the system contains a real continuum the emission quantum yield is smaller than unity at $t = \infty$, i.e. the intramolecular continuum acts as a legitimate dissipative channel. However, we should note that the distinction between a 'real' (dissociative or ionization) continuum and an intramolecular dense quasicontinuum is not physical, as one can convert any 'real' continuum into a quasicontinuum by enclosing the system in a box. The experimental observables are not affected by the mathematical boundary conditions imposed on the system. We should now enquire what conditions should a quasicontinuum satisfy, to act, for all practical purposes, as a dissipative continuum [34, 78, 84].

Bixon and Jortner [34] have introduced the notion of practical irreversibility for the simple model of Figure 1a. In the absence of radiative decay the molecular eigenstates are adequate for the spectral representation of the Green's function,

$$G^+(E) = \sum_n \frac{|n\rangle\langle n|}{E^+ + E_n},\tag{13.1}$$

so that the time evolution of the doorway state $|s\rangle$ is

$$\langle s| U(t, 0)|s\rangle = \int_{-\infty}^{+\infty} \exp(-iEt)\, G_{ss}(E)\, dE =$$

$$= \sum_n \int dE \exp(-iEt) \frac{\langle s\mid n\rangle\langle n\mid s\rangle}{E - E_n + i\eta} =$$

$$= \sum_n |\langle s\mid n\rangle|^2 \exp(-iE_n t).\tag{13.2}$$

Thus the time evolution is described in terms of a Fourier sum which exhibits an oscillatory behavior. It was demonstrated that for the simple model system characterized by equal $\{|l\rangle\}$ level spacing $(1/\varrho_l)$ and constant V_{sl} coupling, the Fourier sum (13.2) exhibits an exponential decay on a time scale, t, which satisfies the condition

$$t \ll t_r = \hbar\varrho_l.\tag{13.3}$$

Thus Equation (13.3) establishes the time scale for the occurrence of an effective relaxation into a quasicontinuum. t_R corresponds to the recurrence time for the intramolecular nonradiative decay. This definition introduces the notion of a Poincaré recurrence cycle for the decay process. For excited states of large molecules characterized by a large electronic energy gap, $\hbar \varrho_l$ is exceedingly long compared to all relevant decay times.

In real life an 'isolated' large molecule cannot wait long enough to pass a Poincaré cycle. Under any realistic experimental conditions in the laboratory the population of the $\{|l\rangle\}$ manifold will be relaxed due to 'trivial' quenching processes such as wall collisions or kinetic collisions. Finally, it is important to realize that even an isolated large molecule in the outer space, in the absence of 'trivial' quenching mechanisms, will not exhibit a Poincaré cycle. We have focused attention just on the simple level scheme 1a or 1c. In a real molecule the $\{|l\rangle\}$ manifold will exhibit subsequent decay mechanisms such as: (a) Infrared emission to lower vibrational levels [34] as was indeed already observed by Drent and Kommandeur [85]. (b) Radiative decay to the highly vibrationally excited ground state levels in the case of internal conversion [41, 79] between high excited states. Thus, strictly speaking, all electronic relaxation processes in a large molecule involve noninterfering sequential decay (see qualitative discussion in Section 3). It is a simple matter to provide the time evolution for the physical systems portrayed in Figures 1f, g. The theory of noninterfering sequential decay for such level schemes was provided by Freed, Nitzan and Jortner [41, 64]. For a doorway state $|s, vac\rangle$ coupled in parallel to the radiative continuum $|g, \mathbf{ke}\rangle$ and to a quasicontinuum $|l, vac\rangle$, which in turn is coupled to a continuum $\{|l_c, vac\rangle\}$ one gets [64]

$$G_{ss}(E) = [E - E_s - \varDelta_s - \varDelta_s^{nr} + \tfrac{1}{2}i(\varGamma_s + \varGamma_s^{nr}(E))]^{-1}, \tag{13.4}$$

where $\varDelta_s$ is the radiative level shift, $\varDelta_s^{nr}$ is a nonradiative level shift function

$$\varDelta_s^{nr} = \sum_l \frac{2(E - \tilde{E}_l)|V_{sl}|^2}{(E - \tilde{E}_l)^2 + \tfrac{1}{4}\varGamma_l^2} \tag{13.5a}$$

and

$$\tilde{E}_l = E_l + pp \sum_{lc} \frac{|V_{l,\,lc}|^2}{E - E_{lc}}. \tag{13.5b}$$

$\varGamma_s$ is the radiative width of the doorway state while finally, and most important

$$\varGamma_s^{nr}(E) = \sum_l \frac{\varGamma_l |V_{ls}|^2}{(E - \tilde{E}_l)^2 + \tfrac{1}{4}\varGamma_l^2}, \tag{13.6}$$

where the width $\varGamma_l$ of each $\{|l\rangle\}$ level due to its coupling with the $\{|l_c\rangle\}$ continuum is

$$\varGamma_l = 2\pi |V_{l,\,lc}|^2 \varrho_{lc}(E_l). \tag{13.7}$$

The time evolution of a general system specified in terms of (13.4) may be very complex, as the Green's function may be characterized by a large number of poles. When the widths of the $\{|l\rangle\}$ levels considerably exceed their spacing, i.e.

$$\varGamma_l \gg |E_l - E_{l\pm1}| \simeq \varrho_l^{-1}. \tag{13.8}$$

The function $\Gamma_s^{nr}(E)$ (Equation (13.6)) is weakly varying with energy and can be considered to be constant. Only under these circumstances G_{ss} has a single pole at

$$E = E_s + \varDelta_s + \varDelta_s^{nr} - \tfrac{1}{2}i\left(\Gamma_s + \Gamma_s^{nr}\right), \tag{13.9}$$

and the time evolution of the excited states is characterized by the total width

$$\gamma_s = \Gamma_s + \Gamma_s^{nr}. \tag{13.10}$$

Conditions (13.3) and (13.8) provide us with the physical basis for the definition of the statistical limit in a large molecule. Each of these relations yields an independent necessary and sufficient condition for treating the intramolecular quasicontinuum as a legitimate dissipative continuum.

When condition (13.3) is satisfied and the widths Γ_l are very small (originating from infrared decay, as will be the case for intersystem crossing) we can set $\Gamma_l \to 0$ (in (13.6)) whereupon

$$\Gamma_s^{nr} = 2\pi \sum_l |V_{sl}|^2 \,\delta\left(E_s - \tilde{E}_l\right) \cong 2\pi \left\langle |V_{sl}|^2 \,\varrho_l \right\rangle, \tag{13.11}$$

which is the conventional expression for the nonradiative decay probability into a continuum [34] (note that the manifold of the delta functions, which enters as a bookkeeping device, has to be extremely dense). When only condition (13.8) is satisfied, as may be the case for internal conversion (or for electronic-vibrational relaxation of a small molecule in a medium) Equation (13.6) has to be used.

We have established the physical criteria for treating intramolecular quasicontinuum as a dissipative intramolecular channel. In the statistical limit we can factor the Hilbert space as follows

$$\hat{P} = |s\rangle \langle s| \tag{13.12}$$

for the discrete subspace and

$$\hat{Q} = \sum_{ke} |g, ke\rangle \langle g, ke| + \sum_l |l, vac\rangle \langle l, vac| \tag{13.13}$$

for the continuous part. The physical situation is that of parallel decay of a single discrete level into two noninteracting channels. Using the results of Section 11 we notice that in this case it is easy to satisfy the condition (11.1). When also condition (11.4) is obeyed we have for the photon counting rate

$$\dot{P}_g(t) = \Gamma_s \exp\left(-\gamma_s t\right), \tag{13.14}$$

where γ_s is given by (12.16) and for the quantum yield

$$Y = \frac{\Gamma_s}{\Gamma_s + \Gamma_s^{nr}}. \tag{13.15}$$

The major experimental characteristics of the statistical limit can be summarized as follows:

(1) Shortening of the radiative decay time. As $\gamma_s > \Gamma_s$ the experimental radiative decay time is shorter than expected on the basis of the integrated oscillator strength, which yields Γ_s.

(2) The decay resulting from short time excitation is a pure exponential.

(3) Reduction of the emission quantum yield, i.e. $Y < 1$. That implies that for any practical purpose the intramolecular quasicontinuum acts as a continuum.

(4) An inert medium will not in general modify the decay characteristics of a statistical molecule. Medium-induced vibrational relaxation will introduce a new contribution to the widths Γ_l. When $\Gamma_s^{nr}(E)$ is already a slowly varying function of the energy in the isolated molecule, this additional sequential decay is of minor importance.

(5) As the physical situation in the statistical limit is equivalent to that of a single discrete level exhibiting parallel decay into two continua, the only pertinent information is the resonance width γ_s. We cannot give new information about the decay characteristics of the system by changing the energetic width γ_p (or the duration) of the exciting photon wave packet.

14 Interstate Coupling in Small Molecules

In the small molecule limit [40–41] the interstate coupling matrix elements V_{sl} between the Born–Oppenheimer states are large while the density of states in the background manifold is low (see Figures 1b, d). The $\{|l\rangle\}$ levels are coarsely spaced, relative to their radiative widths. The sparse $\{|l\rangle\}$ manifold cannot act as a dissipative channel (in the isolated molecule) and we are encountered with the problem of the radiative decay of a set of discrete coupled levels, i.e. $|s, \text{vac}\rangle$ and $\{|l, \text{vac}\rangle\}$. In this case the molecular eigenstates basis, $|n, \text{vac}\rangle$ which diagonalizes H_M, is of great utility. The level distribution of $|n, \text{vac}\rangle$ is sufficiently sparse so that in the absence of accidental degeneracies we expect the off-diagonal matrix elements of the radiative decay matrix to be negligible compared to the level spacings, i.e.

$$\Gamma_{nn'} \ll |E_n - E_{n'}|. \tag{14.1}$$

So that H_{eff} in the $|n, \text{vac}\rangle$ representation (Section 3) is diagonal. Under these circumstances the molecular eigenstates are expected to provide a good description of the independently decaying levels $|j, \text{vac}\rangle$ (see Section 9). The corresponding complex energies are

$$(H_{\text{eff}})_{nn'} = (E_n - \tfrac{1}{2}i\gamma_n)\,\delta_{nn'}, \tag{14.2}$$

where the radiative widths of the molecular eigenstates are

$$\Gamma_n = \Gamma_s |\langle s, \text{vac} \mid n, \text{vac}\rangle|^2, \tag{14.3}$$

and Γ_s is the radiative width of the 'doorway state' (Section 7). We note in passing that for accidental degeneracies we have to diagonalize H_{eff} for these states. In this interference effects may be exhibited.

In view of the diagonal sum rule (9.23)

$$\Gamma_s = \sum_n \Gamma_n, \tag{14.4}$$

thus $\Gamma_n \ll \Gamma_s$ for all n. The overlap factors $|\langle s, \text{vac} \mid n, \text{vac}\rangle|^{-2}$ are of the order of the number of effectively coupled levels in the $\{|l\rangle\}$ manifold. We have thus provided an explanation for the anomalously long radiative decay times (as compared to what is expected on the basis of the integrated oscillator strength) of small molecules, reported by Douglas [80] (see Table II). The occurrence of interstate coupling in small molecules which results from the distribution of the absorption intensity of the doorway state and the dilution of its decay time among the molecular eigenstates, each of which is active in absorption and in emission. We also note that in this case of a discrete molecular spectrum we expect that $Y = 1$.

TABLE II

Long radiative lifetimes of small molecules [a]

Molecule	Transition	τ(exp.) sec	τ(integrated f) sec
NO_2	$^1B_2-^1A_1$ 4300 Å	44×10^{-6}	0.3×10^{-6}
SO_2	$^1B_1-^1A_1$ 3000 Å	42×10^{-6}	0.2×10^{-6}
CS_2	$^1\Sigma-^1\Sigma$ $^1\Pi-^1\Pi$ 3200 Å	15×10^{-6}	3×10^{-6}

[a] Experimental results for τ(exp.) from Douglas [80].

The detailed decay mode is determined by the pulse characteristics. We can easily satisfy condition (11.1) but not condition (11.4) in view of the large energy spread of the $|n\rangle$ levels. The decay law will now be (see Equation (11.6))

$$\dot{P}_g(t) \propto \sum_n \frac{\Gamma_n}{(\bar{k} - E_n)^2 + (\frac{1}{2}\gamma_p)^2} \exp\left(-\Gamma_n t\right). \tag{14.5}$$

Thus the decay mode is in general a superposition of exponentials. The constant coefficients in (14.5) just express the absorption strength of the wavepacket by the individual molecular eigenstates.

To conclude this discussion we would like to emphasize that as the small molecule case corresponds essentially to excitation and decay from the molecular eigenstates it is meaningless to consider nonradiative relaxation from $|s\rangle$ to $\{|l\rangle\}$ in the isolated small molecule. Only when such a small molecule is embedded in a medium, electronic-medium induced vibrational relaxation may result in non-radiative relaxation of the doorway state [65] (see Figure 1i).

15. Intermediate Level Structure

The statistical and the small molecule limits represent well defined, observable physical cases. Another potentially interesting situation involves the intermediate case when a small electronic energy gap exists between two electronic states of a large molecule [47]. It should be noted that now it is unjustified to use 'coarse graining' procedures employed in the statistical limit, which disregards the details of the variation of the interstate coupling terms and the level distribution in the background $\{|l\rangle\}$ manifold. These features have to be considered in detail for the intermediate case. The physical situation is closely related to the problem of intermediate structure in nuclear reactions [82] where the density of nuclear excitations is low and fine structure is exhibited in the nuclear scattering process.

As in the statistical limit, we can consider a single doorway state $|s, \text{vac}\rangle$ (see Figure 1e). In view of simple symmetry arguments, not all the states in the $\{|l\rangle\}$ manifold are coupled to $|s\rangle$ with the same efficiency. When the total density of the former states is relatively low, for small electronic energy gaps, say 10^3–10^4 cm^{-1} [9], only few of these levels will be effectively coupled to $|s\rangle$. We shall partition the $\{|l\rangle\}$ manifold into a small subset $\{|l_a\rangle\}$ of effectively coupled levels and another subset $\{|l_b\rangle\}$ which contains the majority of the levels, which are weakly coupled to $|s\rangle$. The $\{|l_b\rangle\}$ manifold may be considered as a statistical dissipative channel which leads to irreversible intramolecular decay on the relevant time scale. We should also incorporate in principle, other intramolecular statistical decay channels which correspond to dense vibronic manifold of even lower electronic configurations $|c, \text{vac}\rangle$ and of the ground state. This is a simple extension which was previously considered.

The subset of discrete states in the Born–Oppenheimer representation corresponds to the projection operator

$$\hat{P} = |s, \text{vac}\rangle \langle s, \text{vac}| + \sum_{l_a} |l_a\rangle \langle l_a| \tag{15.1}$$

and the projection into the remainder of the Hilbert space is

$$\hat{Q} = \sum_{l_b} |l_b\rangle \langle l_b| + \sum_c |c, \text{vac}\rangle \langle c, \text{vac}| + \sum_{ke} |g, \text{ke}\rangle \langle g, \text{ke}| . \tag{15.2}$$

The states in $\hat{P}$ constitute a sparse manifold of discrete levels, which bears a close analogy to the small molecule case, apart from the possibility of accidental degeneracies. We can now write the effective Hamiltonian $H_{\text{eff}} = \hat{P} H_0 \hat{P} + \hat{P} R \hat{P}$ for (15.1). Subsequently, it will be convenient to find the molecular eigenstates which diagonalize $\hat{P} H_M \hat{P}$.

The effective Hamiltonian Equation (9.4) is then

$$(H_{\text{eff}})_{nn'} = E_n \delta_{nn'} - \tfrac{1}{2} i \gamma_{nn'}, \tag{15.4}$$

where

$$\gamma_{nn'} = \Gamma_{nn'} + \Gamma^{nr}_{nn'}, \tag{15.5}$$

$$\Gamma_{nn'} = 2\pi \sum_e \int d\Omega \, \langle n, \text{vac}| H_{\text{int}} |g, \text{ke}\rangle \langle g, \text{ke}| H_{\text{int}} |n', \text{vac}\rangle \varrho_r \tag{15.6}$$

$$\Gamma_{nn'}^{nr} = 2\pi \langle n, \text{vac}| H_v |l_b, \text{vac} \rangle \langle l_b, \text{vac}| H_v |n' \rangle \, \varrho_{l_b} +$$
$$+ 2\pi \langle n, \text{vac}| H_v |c, \text{vac} \rangle \langle c, \text{vac}| H_v |n', \text{vac} \rangle \, \varrho_c, \qquad (15.7)$$

where ϱ_{l_b} and ϱ_c correspond to the densities of states in the intramolecular $\{|l_b\rangle\}$ and $\{|c\rangle\}$ manifolds. Radiative and non-radiative level shifts were neglected in (15.4). The physical situation corresponds to a parallel decay of a discrete manifold into radiative and nonradiative continua.

Two cases of increasing complexity will be considered:

(1) The molecular eigenstates in $\hat{P}$ are well separated relative to their total widths, i.e.

$$\gamma_{nn'} \ll |E_n - E_{n'}| \qquad (15.8)$$

for all n and n'. The situation is equivalent to that encountered in the small molecule case. The effective Hamiltonian is diagonal in the $|n, \text{vac}\rangle$ representation and the characteristic decay widths of the independently decaying levels are

$$\gamma_{nn'} = |\langle s, \text{vac} \mid n, \text{vac} \rangle|^2 \, (\Gamma_s + \Gamma_s^{nr}), \qquad (15.9)$$

where the radiative width Γ_s and nonradiative widths Γ_s^{nr} of the doorway state are obtained from (15.6) and (15.7) by replacing both n and n' by s.

The photon counting rate resulting from an excitation by a Lorentzian pulse is given by

$$\dot{P}_g(t) \propto \sum_n \frac{\gamma_{nn}}{(E_n - \bar{k})^2 + (\tfrac{1}{2}\gamma_p)^2} \exp(-\gamma_{nn}t), \qquad (15.10)$$

which is analogous to Equation (11.6) except that the radiative widths Γ_n are replaced by the total widths γ_{nn}.

As $|\langle s, \text{vac} \mid n, \text{vac} \rangle| \ll 1$ for all n then provided that $\Gamma_s^{nr} \sim \Gamma_s$ we expect that $\gamma_{nn} < \gamma_s$. The experimental decay width of the excited states $|n, \text{vac}\rangle$ now accessible by optical excitation will be reduced relative to the radiative width of the zero-order state obtained from the integrated oscillator strength. We expect a lengthening of the radiative decay times of a large molecule which corresponds to the intermediate case [47].

(2) When some of the molecular eigenstates in $\hat{P}$ are closely spaced relative to their total widths interference effects will be exhibited in the radiative decay [47]. The effective Hamiltonian (15.4) has to be diagonalized resulting in the $|j, \text{vac}\rangle$ states. The rate of radiative decay will be given by Equation (11.6).

The following experimental and theoretical features of the intermediate case have to be considered:

(a) Lenthening of the radiative decay times relative to those estimated from the integrated oscillator strength. Thus a state of a large molecule which corresponds to the intermediate level structure will exhibit the decay characteristics of a small molecule. This theoretical prediction was experimentally confirmed by Pertzepis et al. [95–97] (see Table III).

TABLE III

Anomalously long radiative decay times of some excited electronic states
of large molecules [a]

Molecule	Transition	Energy gap cm^{-1}	τ(exp.) sec	τ(integrated f) sec
3, 4-Benzopyrene	$S_2 \to S_0$	3800 ($S_2 - S_1$)	7×10^{-8}	1×10^{-8}
Naphthalene	$S_2 \to S_0$	3500 ($S_2 - S_1$)	4×10^{-8}	1×10^{-8}
Benzophenone	$S_1 \to S_0$	3000 ($S_1 - T_1$)	1×10^{-5}	1×10^{-6}

[a] Data from [95–97].

(b) The time resolved decay mode in case (1) above may exhibit a superposition of exponential decays (see Equation (15.10)) and vary with the mean excitation energy (if the exciting pulse sufficiently broad, i.e. $\gamma_p \gg \gamma_{nn}$).

(c) The $\{|l_a\rangle\}$ manifold is nondissipative. The strong interstate coupling between $|s\rangle$ and $\{|l_a\rangle\}$ does not provide a pathway for electronic relaxation in the isolated molecule.

(d) When the molecule is perturbed by an external medium a new relaxation channel is added to the $|s\rangle$ state: $|s\rangle \to \{|l_a\rangle\} \to \{|l_m\rangle\}$. Consecutive relaxation will occur as collisions or phonon coupling provide a vibrational relaxation decay channel. To provide a verification of these conclusions we note that the $|s\rangle$ state of the benzophenone molecule which is separated by 2800 cm^{-1} from T_1 and which does not exhibit fluorescence in solution [98] fluorescences in the low pressure gas phase [97].

16. 'Long Time' Excitation Experimental Observables

Up to this point we have been concerned with the interesting physical information which can be extracted from 'short time' excitation experiments exploring the conditions for and the consequences of separation between the preparation and the decay processes. The time has come to consider the second extreme situation of a 'long time' excitation where one has to consider photon scattering from large molecules as a single quantum-mechanical process. The exciting photon field can be now characterized by high energy resolution and we shall proceed to study the relevant cross sections (see Section 2) resulting from scattering of photons having the energy $E = \hbar ck$. We shall first follow the work of Nitzan and Jortner [62] and focus attention on the physical information which can be extracted from such 'long excitation' processes in a large molecule where the only nonradiative decay channel involves an intramolecular statistical quasicontinuum.

Scattering theory provides a powerful tool for the understanding of the interaction of a molecular system with the radiation field which is responsible for the absorption line shape and photon scattering processes. Nitzan and Jortner [62] proposed that 'long excitation' experimental observables pertaining to electronic relaxation in large molecules can be handled by considering a 'collision process' between a monochromatic wave train and the 'isolated' molecule within the framework of the Lippman–Schwinger equation, expressed in terms of the T matrix formalism, as was previously done for atomic autoionization [99]. At the distant past, the molecule is in the continuum state $|a\rangle = |go, \mathbf{k}e\rangle$ characterized by the energy E_a. The final (continuum) states resulting from photon scattering will be denoted by $|b\rangle = |gv, \mathbf{k}_f\mathbf{e}_f\rangle$ characterized by the energy E_b. Let us define [92] the transition probability per unit time from a continuum state $|a\rangle$ to a continuum state $|b\rangle$ as the increase (per unit time) of the probability that a system initially in the state $|a\rangle$ is found at time t to be in the state $|b\rangle$, i.e.

$$W_{ba} = \lim_{t_0 \to -\infty} \frac{\mathrm{d}}{\mathrm{d}t} |\langle b| \, U(t_0, t) \, |a\rangle|^2. \tag{16.1}$$

The probability W_{ba} is independent of t. To prove that assertion we quote here a result of scattering theory [92]

$$W_{ba} = \frac{2}{\hbar} \delta(b - a) \operatorname{Im} T_{aa} + \frac{2\pi}{\hbar} \delta(E_b - E_a) |T_{ba}|^2, \tag{16.2}$$

where the T matrix (the reaction operator) is defined by

$$T = V + VG(E^+)V. \tag{16.3}$$

Here we use the notation

$$E^+_{\eta \to 0^+} = E + i\eta, \tag{16.3a}$$

with $V = H_V + H_{\mathrm{int}}$. Equations (16.1) and (16.2) are the generalization of Fermi's golden rule, where the delta function insures energy conservation. The physically meaningful concept involved in Equation (16.1) is a transition to a group of final states within the energy interval $\mathrm{d}E_b$, so that when this equation is integrated over the final states one gets the familiar density of states ϱ_b in the final expression. The cross section for the process $a \to b$, $\sigma(a \to b)$, is obtained by dividing the transition probability by the photon flux $F = c/Q$, where c is the velocity of light and Q represents the volume, and we use box normalization for the radiation field. Thus, the cross section is:

$$\sigma(a \to b) = \frac{2\pi Q}{hc} |T_{ba}|^2 \, \delta(E_b - E_a). \tag{16.4}$$

The second general result we require is the rate of disappearance, W_a, of the initial state $|a\rangle$, which is given by the optical theorem of scattering theory

$$W_a = \frac{\mathrm{d}}{\mathrm{d}t} |\langle a| \, U(t_0, t) \, |a\rangle|^2 = -\frac{2}{\hbar} \operatorname{Im}(T_{aa}), \tag{16.5}$$

while the absorption cross section σ_a is given again by dividing (16.5) by the flux

$$\sigma_a = -\frac{2Q}{\hbar c}\, \text{Im}\, T_{aa}. \tag{16.6}$$

We can immediately apply these results by setting for the initial energy $E_a = = E(|go, \text{ke}\rangle) = E_{go} + E$ where E_{go} is the energy of the ground state vibrationless level and $E = k\hbar c$ is the incident photon energy, whereupon the absorption cross section is obtained from (16.6) in the form

$$\sigma_a(E) = -\frac{2Q}{\hbar c}\, \text{Im}\, \langle go, \text{ke}|\, T\, |go, \text{ke}\rangle =$$

$$= -\frac{2Q}{\hbar c}\, \text{Im}\, \langle go, \text{ke}|\, VG(E^+)\, V\, |go, \text{ke}\rangle. \tag{16.6a}$$

Strictly speaking Equation (16.6) represents the absorption cross section at zero temperature. At finite temperatures a proper thermal average has to be performed.

Consider now the cross section for resonance fluorescence. We focus attention on the photon scattering process $|go, \text{ke}\rangle \to |gv, \text{k}_f\text{e}_f\rangle$, which takes place between the initial state $|go, \text{ke}\rangle$ characterized by the energy $E_{go} + E = E_{go} + k\hbar c$ and the final states $|gv, \text{k}_f\text{e}_f\rangle$ characterized by the energy $E_{gv} + E_f = E_{gv} + k_f\hbar c$ (where by E_{gv} we denote the energy of the state $|gv\rangle$ and where the emitted photon which is characterized by the polarization e_f and momentum k_f is scattered into the spherical angle $\Omega_{k_f} - (\Omega_{k_f} + d\Omega_{k_f})$. Equation (16.4) results in

$$\sigma(go, \text{ke} \to gv, \text{k}_f\text{e}_f) = \frac{2\pi Q}{\hbar c} \times$$

$$\times \delta(E_{gv} + E_f - E_{go} - E)\, |\langle gv, \text{k}_f\text{e}_f|\, T\, |go, \text{ke}\rangle|^2. \tag{16.7}$$

The density of final states in the radiation field is

$$\varrho_r(k_f) = \frac{4\pi k_f^2 Q}{(2\pi)^3}, \tag{16.8}$$

and one has to take $k_f c = E_{gv} - E_{go} - kc$ to insure energy conservation. The resonance scattering cross section $\sigma_r^v(E)$ into the final molecular state $|gv\rangle$ will be obtained by summing up Equation (16.7) over all final spatial directions and polarization directions. This scattering cross section depends on the energy E of the initial photon. We consider a sample of randomly oriented (noninteracting) molecules, and provided that we are not interested in polarization measurements, then averaging over the initial polarization directions e, results in

$$\sigma_r^v(E) = \left\langle \sum_{e_f} \int d\Omega_{k_f} \sigma(go, \text{ke} \to gv, \text{k}_f\text{e}_f) \right\rangle, \tag{16.9}$$

where $\langle \ \rangle$ denotes averaging over molecular orientations with respect to photon polarization.

The total cross section for resonance fluorescence is obtained by monitoring all the emitted photons resulting from scattering into all the final molecular states $|gv\rangle$

$$\sigma_r(E) = \sum_v \sigma_r^v(E). \tag{16.10}$$

In a similar manner we can define a cross section $\sigma_{nr}(E)$ for effective scattering into the quasicontinuum $\{|l, \text{vac}\rangle\}$, which we consider to be an operational continuum. This is given by

$$\sigma_{nr}(E) = \left(\frac{2\pi Q}{\hbar c}\right) |\langle go, \mathbf{ke}| T(E) |l, \text{vac}\rangle|^2 \, \varrho_l(E). \tag{16.11}$$

The unitarity relations for the scattering matrix result in the optical theorem [99]

$$-\frac{1}{\pi} \operatorname{Im} T_{aa} = \sum_{\text{all } b} |T_{ba}|^2 \, \delta(E_a - E_b), \tag{16.12}$$

which leads to the conservation law

$$\sigma_a(E) = \sum_v \sigma_r^v(E) + \sigma_{nr}(E). \tag{16.13}$$

The (energy-dependent) quantum yield resulting from absorption of a photon of energy E leading to the molecular state $|gv\rangle$ is given by the ratio of the resonance scattering cross section Equation (16.9) and the absorption cross section Equation (16.6)

$$Y^v(E) = \sigma_r^v(E)/\sigma_a(E). \tag{16.14}$$

If the ground state energy levels are well spaced the different channels can be resolved. Finally, the total quantum yield for emission is given by

$$Y(E) = \sum_v Y^v(E) = \sigma_r(E)/\sigma_a(E). \tag{16.15}$$

In a similar way the quantum yield for electronic relaxation in a statistical molecule is

$$Y_{nr}(E) = \sigma_{nr}(E)/\sigma_a(E), \tag{16.16}$$

and Equation (16.13) implies that

$$Y(E) + Y_{nr}(E) = 1. \tag{16.17}$$

The general expressions for the absorption cross sections, for the resonance fluorescence cross sections and for the emission quantum yields in the 'statistical' molecular case will involve as 'open channels' not only the radiation continuum but also the intramolecular quasicontinuum $\{|l\rangle\}$ which for all practical purposes can be considered as an 'open' decay channel. In this case the unitarity relations for the scattering matrix do not imply that $Y(E)$ is equal to unity as intramolecular decay channels have to be considered, as is evident from Equation (16.13).

17. Relation between 'Short Excitation' and 'Long Excitation' Experiments

At first sight it may appear to the uninitiated reader that there is no direct relation between the theoretical treatment based on the time evolution of the molecular system resulting from wave-packet excitation (Sections 6–7) and the study of photon scattering exposed in Section 16. We have now to establish the connection between observables obtained under 'short time' and 'long time' excitation conditions. This treatment will result in a general useful definition of the emission quantum yields and to a new insight into the physical interpretation of the absorption line-shape function.

The reaction matrix, T, containing all the relevant information regarding 'long time' experiments can be obtained from time-dependent scattering formalism by taking appropriate limits for the evolution operator from the distant past to the far future. Formally, we can define the scattering matrix S [92]:

$$S = \lim_{\substack{t' \to -\infty \\ t'' \to +\infty}} U^I(t'', t'), \tag{17.1}$$

where the evolution operator $U^I(t'', t')$ in the interaction representation is

$$U^I(t'', t') = \exp(iH_0 t'') \, U(t'', t') \exp(-iH_0 t'). \tag{17.2}$$

The S and T matrices are related by [99]

$$S = I - 2\pi i \delta (E_i - E_f) T, \tag{17.3}$$

when I is the unity matrix.

In the description of time evolution of excited states we have utilized the evolution operator $U(t, 0)$. However, as $U(t'', t') = U(t'' - t', 0)$, we can use the equivalent expression

$$U(t, 0) = U(\tfrac{1}{2}t, -\tfrac{1}{2}t). \tag{17.4}$$

Thus the scattering matrix (17.1) can be obtained by a single rather than a double limiting process [90]

$$S = \lim_{t \to \infty} U^I(\tfrac{1}{2}t, -\tfrac{1}{2}t). \tag{17.5}$$

Making use of (17.4) we have

$$S = \lim_{t \to \infty} U^I(t, 0) \tag{17.6}$$

or making use of the relation (5.6) we get a relation between the S matrix and the Green's function

$$S = \frac{1}{2\pi i} \lim_{t \to \infty} \exp(iH_0 t) \int_{-\infty}^{\infty} dE G(E^+) \exp(-iEt). \tag{17.7}$$

This relation enables us to consider the relation between the decay of a 'prepared' state at the finite time $t = 0$ and photon scattering, while the conventional formulation would have required a state of affairs where the excited states is 'prepared' at $t = -\infty$ whereas one could not utilize the Green's function formalism to describe its time evolution. Equations (17.7) and (17.3) establish the formal connection between 'long' and 'short' excitation experiments.

The quantum yield (Section 16) in a given exit channel is defined in a general way as the number of photons or molecules scattered into that channel divided by the number of absorbed photons. The system contains as effective exit channels, the radiative continua $|gv, \mathbf{k}_f\mathbf{e}_f\rangle$ and the intramolecular quasicontinuum.

We proceed to calculate the probability for finding the system in the exit channels at $t = \infty$. Taking the initial state (6.2) and utilizing the form (17.7) for the S matrix we have

$$\Psi(\infty) = S\Psi(0) = \sum_{\mathbf{k}_f\mathbf{e}_f}\sum_v\sum_k |gv, \mathbf{k}_f\mathbf{e}_f\rangle \langle gv, \mathbf{k}_f\mathbf{e}_f| S |go, \mathbf{ke}\rangle a_k +$$
$$+ \sum_l\sum_k |l, \text{vac}\rangle \langle l, \text{vac}| S |go, \mathbf{ke}\rangle a_k +$$
$$+ \sum_m\sum_k |m, \text{vac}\rangle \langle m, \text{vac}| S |go, \mathbf{ke}\rangle a_k . \tag{17.8}$$

The probability of the system decaying radiatively into the final states $\{|gv, \mathbf{k}_f\mathbf{e}_f\rangle\}$ resulting in the molecular ground state $|gv\rangle$ at $t = \infty$ is

$$P_g^v(\infty) = \sum_{\mathbf{k}_f\mathbf{e}_f}\left|\sum_k a_k \langle gv, \mathbf{k}_f\mathbf{e}_f| S |go, \mathbf{ke}\rangle\right|^2 . \tag{17.9}$$

Making use of Equation (17.3) (and performing spatial integration and summation over polarization directions as in Section 16) we get

$$P_g^v(\infty) = 4\pi^2 \sum_{\mathbf{k}_f\mathbf{e}_f}\sum_k |a_k|^2 |\langle gv, \mathbf{k}_f\mathbf{e}_f| T |go, \mathbf{ke}\rangle|^2 \delta(E - k) . \tag{17.10}$$

In an analogous manner we obtain the probability for the population of the quasi-continuum at $t = \infty$

$$P_l(\infty) = 4\pi^2 \sum_l\sum_k |a_k|^2 |\langle l, \text{vac}| T |go, \mathbf{ke}\rangle|^2 \delta(E - E_l) . \tag{17.11}$$

Thus the quantum yield for emission into $|gv\rangle$ is

$$Y^v = P_g^v(\infty)\bigg/\left(\sum_v P_g^v(\infty) + P_l(\infty)\right), \tag{17.12}$$

and the total emission quantum yield

$$Y = \sum_v Y^v . \tag{17.12a}$$

The denominator of Equation (17.12) represents just the total probability of populating all the decay channels, which is just the probability of photon absorption.

From Equations (17.11) and (17.10) we have

$$P_l(\infty) + \sum_v P_g^v(\infty) = 4\pi^2 \sum_k |a_k|^2 \left[\sum_{k_f e_f} |\langle gv, \mathbf{k}_f \mathbf{e}_f| T |go, \mathbf{k}\mathbf{e}\rangle|^2 \delta(E - k_f) + \right.$$
$$\left. + \sum_l |\langle l, \text{vac}| T |go, \mathbf{k}\mathbf{e}\rangle|^2 \delta(E - E_l) \right]. \qquad (17.13)$$

Making use of the optical theorem (Equation (16.12)) we obtain

$$P_l(\infty) + \sum_v P_g^v(\infty) = -\sum_k |a_k|^2 \, \text{Im} \, \langle go, \mathbf{k}\mathbf{e}| T |go, \mathbf{k}\mathbf{e}\rangle \cdot 4\pi. \qquad (17.13a)$$

The emission quantum yields (17.12a) and (17.12) can be recast with the help of Equations (16.6), (16.7), (16.9) and (17.12a) into the final form

$$Y^v = \frac{\sum_k |a_k|^2 \, \sigma_r^v(E)}{\sum_k |a_k|^2 \, \sigma_a(E)} = \frac{\int_0^\infty |a(E)|^2 \, \sigma_r^v(E) \, dE}{\int_0^\infty |a(E)|^2 \, \sigma_a(E) \, dE} \qquad (17.14)$$

and

$$Y = \frac{\int_0^\infty |a(E)|^2 \, \sigma_r(E) \, dE}{\int_0^\infty |a(E)|^2 \, \sigma_a(E) \, dE}. \qquad (17.15)$$

In a similar way the yield for nonradiative decay is

$$Y_{nr}(E) = \frac{\int |a(E)|^2 \, \sigma_{nr}(E) \, dE}{\int |a(E)|^2 \, \sigma_a(E) \, dE}, \qquad (17.16)$$

where we have set $a_k \equiv a(E)$.

Equations (17.14)–(17.16) constitute the general expressions for the quantum yields, which can be expressed as the ratio of integrals involving the products of the relevant cross sections and the power spectrum of the exciting light pulse. These results are valid for all excitation conditions. Although for short time experiments one can evaluate $P_g(\infty)$ directly, the present method is more general and useful. From these results we conclude that:

(a) The quantum yields are determined in general by both the characteristics of the molecular system, expressed in terms of the cross sections $\sigma_r^v(E)$, $\sigma_{nr}(E)$ and $\sigma_a(E)$ and by the features of the source.

(b) Regarding the source characteristics we notice that the quantum yields are determined by the power spectrum of the source, $|a(E)|^2$, the only relevant feature is the energetic spread. Unlike the time-resolved decay pattern which is determined by the excitation amplitudes (Equation (7.13)) and thus possibly by phases of the radiation field, the quantum yields are just determined by the energetic spread of the pulse.

(c) In the 'long excitation' limit $|a_k|^2$ is sharply peaked around $\bar{k}$ and $\sigma_r^v(E)$ and $\sigma_a(E)$ vary slowly in the range where the power spectrum is finite. Then the quantum yields are obtained in terms of Equations (16.14)–(16.15).

(d) For the 'short excitation' limit $|a_k|^2$ is a slowly varying function of the energy, thus the emission quantum yields become

$$Y^v = \int_0^\infty \sigma_r^v(E)\,dE \Big/ \int_0^\infty \sigma_a(E)\,dE. \tag{17.17}$$

(e) Only in the long excitation and short excitation limits the quantum yields are solely determined by the molecular parameters.

(f) Only when both cross sections $\sigma_r^v(E)$ and $\sigma_a(E)$ exhibit the same dependence on the energy, E, the quantum yields will be independent of the pulse width being identical for long and for short excitation conditions as well as for intermediate situations. This situation is encountered in the simple case of a single molecular resonance where both cross sections are characterized by a Lorentzian energy dependence.

(g) In general, when interference effects are exhibited, the quantum yield differs for different energy conditions. Nitzan and Jortner [62, 50] have demonstrated this effect for the case of two overlapping resonances, which was applied to provide the only available interpretation for the decay characteristics of the S_1 state of biacetyl.

To pursue further the relation between short excitation and long excitation experiments we shall establish the relation between the scattering cross sections $\sigma_r^v(E)$ and $\sigma_a(E)$ and the independently decaying discrete molecular states (see Sections 9–12). As we are interested in parallel coupling of the discrete levels $|m, \text{vac}\rangle$ to radiative and nonradiative channels we shall consider the effective Hamiltonian H_{eff} (Equation (12.4)) characterized by the eigenstates $|J, \text{vac}\rangle$. The cross sections are determined by matrix elements of T involving one-photon states, so that $V = H_{\text{int}}$, which combines only states within the $\hat{Q}$ and $\hat{P}$ subspaces (see Equation (12.4)). Furthermore, we have just to evaluate the matrix elements of $H_{\text{int}}\hat{P}G\hat{P}H_{\text{int}}$ between one-photon states. Thus we get, apart from irrelevant numerical factors,

$$\sigma_r^v(E) \propto \left| \sum_J \frac{\langle gv, \mathbf{k}_f \mathbf{e}_f| H_{\text{int}} |J, \text{vac}\rangle \langle J, \text{vac}| H_{\text{int}} |go, \mathbf{ke}\rangle}{E - E_J + \frac{1}{2}i\gamma_J} \right|^2 \times$$

$$\times \varrho_r(k_f)\,\delta(E_{go} - E_{gv} + \hbar kc - \hbar k_f c), \tag{17.18}$$

$$\sigma_a(E) \propto \sum_J \text{Im} \frac{\langle go, \mathbf{ke}| H_{\text{int}} |J, \text{vac}\rangle \langle J, \text{vac}| H_{\text{int}} |go, \mathbf{ke}\rangle}{E - E_J + \frac{1}{2}i\gamma_J}, \tag{17.19}$$

where γ_J are the total widths of the independently decaying states. These results exhibit some interesting features. The photon scattering cross section $\sigma_r^v(E)$ will definitely involve interference effects provided that $|E_J - E_{J'}| < \gamma_J, \gamma_{J'}$. This is the analogue for long-time experiments of the quantum beats expected under these circumstances for short-time excitation experiments. The expression for the absorption cross section reveals formally a superposition of Lorentzians. This feature should not mislead us, as one has to bear in mind that the states $|J, \text{vac}\rangle$ and their complementary states $|\bar{J}, \text{vac}\rangle$ are characterized by complex expansion coefficients of $|m, \text{vac}\rangle$, whereupon the single sum (17.19) will exhibit interference effects of the absorption line shape for closely spaced (relative to γ_J) levels.

Finally we shall consider some features of the optical absorption line shape. The basic form (16.6a) for $\sigma_a(E)$ and the definition of the generalized doorway states $|N, \text{vac}\rangle$ (Equation (7.8)) imply that

$$\sigma_a(E) = -\left(\frac{2Q}{\hbar c}\right) |\gamma_N|^2 \, \text{Im} \langle N, \text{vac}| \, G(E^+) \, |N, \text{vac}\rangle. \qquad (17.20)$$

The decay rate of the 'initially prepared' doorway state $|N\rangle$ in the short excitation limit (Equation (8.11a)), as monitored by the photon counting rate is

$$\dot{P}_g(t) = 2\pi |\gamma_N|^2 \, \varrho_r \left| \frac{1}{2\pi i} \int_{-\infty}^{+\infty} dE \exp(-iEt) \langle N, \text{vac}| \, G(E^+) \, |N, \text{vac}\rangle \right|^2. \qquad (17.21)$$

Making use of the dispersion relation [87]

$$\langle N, \text{vac}| \, G(E^+) \, |N, \text{vac}\rangle = -\frac{1}{\pi} \int_{-\infty}^{\infty} \frac{\text{Im} \langle N, \text{vac}| \, G(E') \, |N, \text{vac}\rangle \, dE'}{E' - E - i\eta} \qquad (17.22)$$

we get for $t > 0$

$$\frac{1}{2\pi i} \int_{-\infty}^{\infty} dE \langle N, \text{vac}| \, G(E^+) \, |N, \text{vac}\rangle \exp(-iEt) =$$

$$= \frac{1}{2\pi i} \frac{1}{\pi} \int_{-\infty}^{\infty} dE \int_{-\infty}^{\infty} dE' \, \frac{\text{Im} \langle N, \text{vac}| \, G(E') \, |N, \text{vac}\rangle}{E - E' + i\eta} \exp(-iEt) =$$

$$= \frac{1}{\pi} \int_{-\infty}^{\infty} dE' \, \text{Im} \langle N, \text{vac}| \, G(E') \, |N, \text{vac}\rangle \exp(-iE't). \qquad (17.23)$$

Equations (17.20)–(17.23) result in

$$\dot{P}_g(t) = \frac{2}{\pi} \left(\frac{\hbar c}{2Q} \right)^2 \frac{\varrho_r}{|\gamma_N|^2} \left| \int_{-\infty}^{\infty} dE \exp\left(-iEt\right) \sigma_a(E) \right|^2. \tag{17.24}$$

The same result can be obtained by taking the Fourier transform of $\sigma_a(E)$ expressed in the $|J, \text{vac}\rangle$ representation, Equation (17.19).

Equation (17.24) provides us with a general result relating the decay rate of any excited state to the Fourier transform of the absorption line-shape function. This result was often derived and utilized for the simple case of a single resonance, where the lineshape is Lorentzian and the decay rate being exponential. The present discussion provides a proof which is valid for any level structure in the excited state.

18. Concluding Remarks

In this paper we have presented a unified theoretical scheme for the description of the diverse decay channels of excited electronic states of polyatomic molecules. We have focused attention on the dissipative channels in small, intermediate type and large molecules, bearing in mind that the same techniques utilized for electronic relaxation are applicable for direct and indirect photodissociation. We have made a conscientious attempt to focus attention on general theoretical schemes, rather than on specific applications, as many of the latter were already published.

Many applications of the theory were aimed towards the understanding of the decay characteristics in the statistical limit. As pointed out by Lin and Bersohn [100] and subsequently by Englman, Freed and Jortner [43–44], the nonradiative widths (13.6) can be considered as multiphonon processes displaying the transition probability as a Fourier transform of a generating function. Explicit solutions for a harmonic molecule were provided, which are amenable to numerical calculations and to analytic approximations. Two limiting cases were considered. The strong electron–phonon coupling limit, which corresponds to the Teller model [101] for crossing of potential surfaces and which reduces at high temperatures to an activated rate equation. The weak electron–phonon coupling situation which corresponds to most cases of internal conversion and intersystem crossing exhibits the energy gap law [43, 102] and the well-known deuterium isotope effect [103]. These calculations pertain to a 'harmonic' model molecule, and obviously unharmonicity corrections are important from the quantitative point of view. Unfortunately, in spite of recent efforts by Fisher et al. [104] there is no satisfactory way for handling unharmonicity corrections. This can be handled by displaying the Fourier integral in terms of a convolution, factoring out a small number of strongly unharmonic modes. These studies rested on the evaluation of the thermally averaged transition probability. An interesting related problem involves the optical selection studies [48, 54] where the nonradiative decay of a single zero-order vibronic level with excess vibrational energy above the electronic origin is considered.

The elegant Feynman operator technique can be applied for this problem, although brute force methods [105] resulted in somewhat more detailed information. Unfortunately, for molecules larger than the benzene molecule, sequence congestion effects [58] are important, so that the isolated molecule preserves the memory of the ground state Boltzmann population in the excited states, and a single vibronic level cannot be excited. Finally, when a molecule is externally perturbed, coupling between electronic and vibrational relaxation may be exhibited. Such processes are of cardinal importance for the understanding of some experimental results of picosecond spectroscopy [98] for ultrafast electronic relaxation in solution.

The statistical limit is well understood, and many of the theoretical results concur with experimental observations. The situation is different for the theory of interference effects in the radiative decay of large molecules, where the experimental information is meager. Three classes of interference effects can be distinguished:

(1) Interference between a small number of closely spaced levels (see Sections [14–15]). In spite of extensive experimental effort no conclusive evidence was yet obtained for the occurrence of this effect in excited states of large molecules which correspond to the intermediate case [95–97]. It is possible that sectroscopic sequence congestion effects mask this interesting feature of the decay.

(2) Interference between a zero-order discrete state and a dissociative continuum both of which are coupled to the radiative field. This problem originally solved by Fano [1] for $\sigma_a(E)$, may be of considerable interest for some molecular predissociation processes. We have recently derived [86] a general solution for the absorption, photon scattering and dissociation cross sections for the Fano problem incorporating radiative interactions to infinite order and elucidating the nature of interference effects between resonance and potential scattering.

(3) Interference between continua. The problem of sequential decay with interference was recently handled utilizing simple model systems [67]. We have recently provided a complete general solution for the coupling between the radiative continuum and a dissociative continuum [86]. This problem is central for the theoretical study of molecular photodissociation processes.

From the foregoing discussion of interference effects it is evident that classes (2) and (3) are relevant for a profound theoretical understanding of a variety of processes such as the dynamics of direct photofragmentation and indirect predissociation in large molecules. The same theoretical techniques are also directly applicable to the study of inverse radiative processes such as radiative recombination and inverse predissociation.

We hope that the present unified approach to the decay of excited states will be of general applicability for the elucidation of a variety of interesting photophysical processes.

Acknowledgements

This review contains results of research conducted in collaboration with Dr M. Bixon, Dr J. Dothan, Dr B. Englman, Dr K. F. Freed and Dr A. Nitzan, who have all profoundly contributed to the understanding of this interesting field. We are in particular indebted to Dr A. Nitzan for endless enlightening discussions and to Dr P. M. Rentzepis for exposing us to the experimental real facts of life.

References

[1] Fano, U.: 1961, *Phys. Rev.* **124**, 1866.
[2] For a review see Herzberg, G.: 1954, *Spectra of Diatomic Molecules*, Vol. II, Van Nostrand, Princeton, N.J.
[3] a. Kasha, M.: 1950, *Disc. Faraday Soc.* **9**, 14.
    b. Kasha, M.: 1960, *Radiation Res. Suppl.* **2**, 243.
[4] Robinson, G. W. and Frosh, R. P.: 1962, *J. Chem. Phys.* **37**, 1962; 1963: **38**, 1187.
[5] Seybold, P. and Gouterman, M.: 1965, *Chem. Rev.* **65**, 413.
[6] Lower, S. K. and El-Sayed, M. A.: 1966, *Chem. Rev.* **66**, 199.
[7] Henry, B. R. and Kasha, M.: 1968, *Ann. Rev. Phys. Chem.* **19**, 161.
[8] Jortner, J., Rice, S. A., and Hochstrasser, R. W.: 1969, in *Adv. Photochemistry*, Vol. 7 (ed. by W. A. Noyes, J. N. Pitts, and G. Hammond), Wiley, New York.
[9] Jortner, J.: 1970, *Pure Appl. Chem.* **24**, 165.
[10] Jortner, J.: 1970, *J. Chim. Phys.*, special issue 'Transitions non-radiatives dans les molécules', p. 1.
[11] Bixon, M. and Jortner, J.: 1968, *Israel J. Chem.* **48**, 715.
[12] Jortner, J.: 1971, in *Organic Solid State Chemistry*, Vol. 2 (ed. by M. D. Cohen), Butterworths, London.
[13] Schlag, E. W., Schneider, S., and Fischer, S. F.: *Ann. Rev. Phys. Chem.* (in press).
[14] Freed, K. F.: 1972, *Topics Current Chem.* **31**, 105.
[15] Henry, B. R. and Siebrand, W.: to be published.
[16] Montroll, E. W. and Shuler, K. E.: 1957, *J. Chem. Phys.* **26**, 454.
[17] See, for example, (a) Turro, N.: 1965, *Molecular Photochemistry*, Benjamin, New York; (b) Calvert, J. and Pitts, J. N.: 1966, *Photochemistry*, Wiley, New York.
[18] Gelbart, W. M. and Rice, S. A.: 1969, *J. Chem. Phys.* **50**, 4775.
[19] Gelbart, W. M., Freed, K. F., and Rice, S. A.: 1970, *J. Chem. Phys.* **52**, 2460.
[20] For a review see (a) Markham, J. J.: 1959, *Rev. Mod. Phys.* **31**, 956; (b) Perlin, Yu. E.: 1964, *Sov. Phys. Usp.* **6**, 542; (c) Rebane, K. K.: 1970, *Impurity Spectra of Solids*, Plenum Press, New York.
[21] Forster, T.: 1959, *Disc. Faraday Soc.* **27**, 7.
[22] Dexter, D. L.: 1953, *J. Chem. Phys.* **21**, 836.
[23] Soules, T. F. and Duke, C. B.: 1971, *Phys. Rev.* **B3**, 262.
[24] Jortner, J.: 1968, *Phys. Rev. Letters* **20**, 244.
[25] Levich, V. G.: 1970, in *Physical Chemistry – An Advanced Treatise* (ed. by H. Eyring), Academic Press.
[26] Jortner, J.: 1971, Berichte Bundesgesellschaft Physikalische Chem. **75**, 697.
[27] Harris, R. A.: 1963, *J. Chem. Phys.* **39**, 978.
[28] Birks, J. B.: 1970, *Photophysics of Aromatic Molecules*, Wiley, New York.
[29] McGlynn, S. P., Azumi, T., and Kinoshite, M.: 1969, *Molecular Spectroscopy of the Triplet State*, Prentice Hall, Englewood Cliffs, N.J.
[30] Becker, R. S.: 1969, *Theory and Interpretation of Fluorescence and Phosphorescence*, Wiley, New York.
[31] Lim, E. D. (ed.): 1969, *Molecular Luminescence*, Benjamin, New York.
[32] 1970, *J. Chim. Phys.*, special issue 'Transitions non-radiatives dans les molécules'.
[33] Williams, F. (ed.): 1970, *Proc. Int. Conf. Luminescence*, North Holland Publ. Co., Amsterdam.
[34] Bixon, M. and Jortner, J.: 1968, *J. Chem. Phys.* **48**, 715.
[35] Berry, R. S. and Jortner J.: 1968, *J. Chem. Phys.* **48**, 2757.
[36] Chock, D., Rice, S. A., and Jortner, J.: 1968, *J. Chem. Phys.* **49**, 610.
[37] Bixon, M. and Jortner, J.: 1969, *Mol. Crystals* **213**, 237.
[38] Bixon, M., Dothan, Y., and Jortner, J.: 1969, *Mol. Phys.* **17**, 109.
[39] Bixon, M. and Jortner, J.: 1969, *J. Chem. Phys.* **50**, 4061.
[40] Bixon, M. and Jortner, J.: 1969, *J. Chem. Phys.* **50**, 3284.

[41] Freed, K. and Jortner, J.: 1969, *J. Chem. Phys.* **50**, 2916.
[42] Morris, G. C. and Jortner, J.: 1969, *J. Chem. Phys.* **51**, 3689.
[43] Englman, B. and Jortner, J.: 1970, *Mol. Phys.* **18**, 145.
[44] Freed, K. and Jortner, J.: 1970, *J. Chem. Phys.* **52**, 6272.
[45] Gelbart, W. M., Spears, K. G., Freed, K. F., Rice, S. A., and Jortner, J.: 1970, *Chem. Phys. Letters* **6**, 345.
[46] Gelbart, M. M. and Jortner, J.: 1971, *J. Chem. Phys.* **54**, 2070.
[47] Nitzan, A., Rentzepis, P. M., and Jortner, J.: 1972, *Proc. Roy. Soc. London* **A327**, 367.
[48] Jortner, J.: 1971, *J. Chem. Phys.* **55**, 1355.
[49] Nitzan, A., Rentzepis, P. M., and Jortner, J.: 1971, *Chem. Phys. Letters* **8**, 445.
[50] Nitzan, A., Kommandeur, J., Drent, E., and Jortner, J.: 1971, *Chem. Phys. Letters* **9**, 273.
[51] Nitzan, A., Rentzepis, P. M., and Jortner, J.: 1972, *Mol. Phys.* **22**, 585.
[52] Jortner, J.: 1971, *Chem. Phys. Letters* **11**, 458.
[53] Nitzan, A. and Jortner, J.: 1972, *J. Chem. Phys.* **56**, 3360.
[54] Nitzan, A. and Jortner, J.: 1972, *J. Chem. Phys.* **56**, 2079.
[55] Nitzan, A. and Jortner, J.: 1972, *J. Chem. Phys.* **56**, 5200.
[56] Jortner, J.: 1972, *J. Chem. Phys.* **56**, 5742.
[57] Gebelein, H. and Jortner, J.: 1972, *Theor. Chim. Acta* **25**, 143.
[58] Nitzan, A. and Jortner, J.: 1972, *Chem. Phys. Letters* **13**, 466.
[59] Nitzan, A. and Jortner, J.: 1972, *Mol. Phys.* **24**, 109.
[60] Nitzan, A. and Jortner, J.: 1972, *Chem. Phys. Letters* **14**, 177.
[61] a. Nitzan, A. and Jortner, J.: 1973, *J. Chem. Phys.* **58**, 2412.
     b. Nitzan, A. and Jortner, J., 1972, *Chem. Phys. Letters* **15**, 350.
[62] Nitzan, A. and Jortner, J.: 1972, *J. Chem. Phys.* **57**, 2870.
[63] Nitzan, A. and Jortner, J.: 1973, *Mol. Phys.* **25**, 713.
[64] Nitzan, A. and Jortner, J.: 1973, *Theor. Chim. Acta* **30**, 217.
[65] Nitzan, A. and Jortner, J.: 1973, *Theor. Chim. Acta* **29**, 97.
[66] Nitzan, A. and Jortner, J.: 1973, *J. Chem. Phys.* **58**, 2669.
[67] Nitzan, A., Berne, B. J., and Jortner, J.: 1973, *Mol. Phys.* **26**, 281.
[68] Mukamel, S. and Jortner, J.: *Mol. Phys.* (in press).
[69] Sharf, B. and Silbey, R.: 1969, *Chem. Phys. Letters* **4**, 423.
[70] Sharf, B. and Silbey, R.: 1970, *Chem. Phys. Letters* **4**, 561.
[71] Burland, D. M. and Robinson, G. W.: 1970, *Proc. Nat. Acad. Sci.* **66**, 257.
[72] Sharf, B.: 1971, *Chem. Phys. Letters* **6**, 364; 1971, **8**, 238; 1971, **8**, 391.
[73] Sharf, B. and Silbey, R.: 1971, *Chem. Phys. Letters* **9**, 125.
[74] Lefebvre, R.: 1971, *Chem. Phys. Letters* **8**, 306.
[75] Orlandi, G. and Siebrand, W.: 1971, *Chem. Phys. Letters* **8**, 473.
[76] Siebrand, W.: 1971, *Chem. Phys. Letters* **9**, 157.
[77] Freed, K. F. and Gelbart, W. M.: 1971, *Chem. Phys. Letters* **10**, 187.
[78] Robinson, G. W.: 1967, *J. Chem. Phys.* **47**, 1967.
[79] Freed, K. F.: 1970, *J. Chem. Phys.* **52**, 1345.
[80] Douglas, A. E.: 1966, *J. Chem. Phys.* **45**, 1007.
[81] Feshbach, H.: 1958, *Ann. Phys. N.Y.* **5**, 357; 1967, **43**, 410; Estrada, L. and Feshback, H.: 1963, *Ann. Phys. N.Y.* **23**, 123.
[82] Feshback, H., Kerman, A. K., and Lemmer, R. H.: 1967, *Ann. Phys. N.Y.* **41**, 230.
[83] Rhodes, W., Henry, B. R., and Kasha, M.: 1969, *Proc. Nat. Acad. Sci.* **63**, 31.
[84] Rhodes, W.: 1969, *J. Chem. Phys.* **50**, 2885.
[85] Drent, E. and Kommandeur, J.: 1971, *Chem. Phys. Letters* **8**, 303.
[86] Mukamel, S. and Jortner, J.: *J. Chem. Phys.* (in press).
[87] Goldberger, M. L. and Watson, R. M.: 1969, *Collision Theory*, Wiley, New York.
[88] Mower, L.: 1966, *Phys. Rev.* **142**, 799.
[89] Mower, L.: 1968, *Phys. Rev.* **165**, 145.
[90] Cohen Tannoudji, C.:1972, *Théorie des processus élémentaires d'interaction entre les atomes et le champ électromagnétique.*
[91] Rhodes, W.: 1971, *Chem. Phys. Letters* **11**, 179.
[92] Schweber, S. S.: 1961, *An Introduction to Relativistic Quantum Field Theory*, Row, Peterson and Co., New York.

[93] Siebrand, W.: 1972, *Chem. Phys. Letters* **14**, 23.

[94] Nitzan, A. and Jortner, J.: 1973, *Theor. Chim. Acta (Berlin)* **29**, 97.

[95] Wannier, P., Rentzepis, P. M., and Jortner, J.: 1971, *Chem. Phys. Letters* **10**, 102.

[96] Wannier, P., Rentzepis, P. M., and Jortner, J.: 1971, *Chem. Phys. Letters* **10**, 182.

[97] Busch, G. E., Renzepis, P. M., and Jortner, J.: 1971, *J. Chem. Phys.* **56**, 361.

[98] Rentzepis, P. M.: 1970, *Science* **169**, 239.

[99] Shore, B. W.: 1967, *Rev. Mod. Phys.* **39**, 439.

[100] Lin, S. H. and Bersohn, R.: 1968, *J. Chem. Phys.* **48**, 2732.

[101] Teller, E.: 1937, *J. Chem. Phys.* **41**, 109.

[102] Robinson, G. W. and Frosch, R. P.: 1962, *J. Chem. Phys.* **37**, 1962; 1963, **38**, 1187.

[103] a. Hutchison Jr., C. A. and Magnum, B. W.: 1960, *J. Chem. Phys.* **32**, 1261.

    b. Wright, M. R., Frosch, R. P., and Robinson, G. W.: 1960, *J. Chem. Phys.* **33**, 934.

    c. de Groot, M. S. and van der Waals, J. H.: 1961, *Mol. Phys.* **4**, 189.

    d. Len, E. C. and Laposa, J. D.: 1964, *J. Chem. Phys.* **41**, 3257.

    e. Kellog, R. E. and Wyeth, N. C.: 1966, *J. Chem. Phys.* **45**, 3156.

    f. Hirota, N. and Hutchison Jr., C. A.: 1967, *J. Chem. Phys.* **46**, 1561.

    g. Martin, T. E. and Kalantes, A. H.: 1968, *J. Chem. Phys.* **48**, 4996.

    h. Johnson, D., Logan, L. M., and Ross, I. G.: 1964, *J. Mol. Spectrosc.* **14**, 198.

[104] Fischer, S.: 1971, *Chem. Phys. Letters* **11**, 577.

[105] Heller, D. F., Freed, K. F., and Gelbart, W. M.: 1972, *J. Chem. Phys.* **56**, 2309.

SOME ASPECTS OF UPS

E. HEILBRONNER

Physikalisch-chemisches Institut, Universität Basel, Switzerland

The primary process in photoelectron (PE) spectroscopy [1–2] is

$$M\left(\Gamma\right) \overset{h\nu}{\to} M^{+}\left({}^{2}\Psi_{j}\right) + e^{-}\left(T_{j}\right),\tag{1}$$

where Γ is the ground-state configuration of the closed-shell molecule M and ${}^{2}\Psi_{j}$ a doublet state of the radical cation M^{+}. Disregarding changes in the vibrational and/or rotational energies of M and M^{+}, the kinetic energy T_{j} of the ejected electron e^{-} is given by

$$T_{j} = h\nu - I_{j}.\tag{2}$$

Depending on whether the photon energy $h\nu$ corresponds to the far UV region (~ 10 eV) or the X-ray region ($\sim 10^{3}$ eV) one differentiates between UPS (ultraviolet photoelectron spectroscopy) [1] or XPS (X-ray photo-electron spectroscopy). The latter technique is also known as ESCA (electron spectroscopy for chemical analysis) [2]. Typical photon sources used in the two regions are He I ($\lambda = 584$ Å, $h\nu = 21.2$ eV) or He II ($\lambda = 304$ Å, $h\nu = 40.8$ eV) in UPS, and Al K$_{\alpha}$ ($\lambda = 8.3$ Å, $h\nu = 1486$ eV) in XPS. In the following we shall only be concerned with He I PE spectroscopy.

If $M(\Gamma)$ and $M^{+}({}^{2}\Psi_{j})$ assume their respective equilibrium structure and if both are in their vibrational and rotational ground state, then I_{j} will be the adiabatic ionization potential $I_{a,j}$, i.e. the position of the first vibrational component of the corresponding band in the PE spectrum (barring hot bands). On the other hand if $M^{+}({}^{2}\Psi_{j})$ is assumed to have the same structure as $M(\Gamma)$ (i.e. same internal coordinates), I_{j} will be the vertical ionization potential $I_{v,j}$. Necessarily $I_{v,j} \geqslant I_{a,j}$. In practice $I_{v,j}$ is assumed to correspond to the position of the fine structure component of highest intensity (largest Franck–Condon factor), that is if the vibrational fine structure of the PE band can be resolved. However, for larger molecules this is usually not possible, and one is forced to assume that $I_{v,j} \approx I_{m,j}$, i.e. that $I_{v,j}$ corresponds to the position of the maximum of the unresolved Franck–Condon envelope.

The ground state (configuration) of M will be written as a single determinantal wave function

$$\Gamma = \|\psi_{1}\bar{\psi}_{1} \ldots \psi_{j}\bar{\psi}_{j} \ldots \psi_{N}\bar{\psi}_{N}\|,\tag{3}$$

in which the ψ_{i} are the canonical SCF molecular orbitals of M which minimize the energy $\mathscr{E}(\Gamma)$ of M. Electronically allowed ionization processes (1) lead to states ${}^{2}\Psi_{j}$ of M^{+} which can again be written as single determinantal wave functions

$$\begin{aligned}\Psi_{j}^{\alpha} &= \|\psi_{1}'\overline{\psi_{1}'} \ldots \psi_{j}' \ldots \psi_{N}'\overline{\psi_{N}'}\|,\\\Psi_{j}^{\beta} &= \|\psi_{1}'\overline{\psi_{1}'} \ldots \overline{\psi_{j}'} \ldots \psi_{N}'\overline{\psi_{N}'}\|.\end{aligned}\tag{4}$$

R. Daudel and B. Pullman (eds.), The World of Quantum Chemistry, 211–235. All Rights Reserved

If the assumption is made that $\psi_i' \equiv \psi_i$, then it can be shown [3] that

$$I_{v,j} = -\varepsilon(\psi_j),\qquad\qquad\qquad\qquad\qquad\qquad\qquad (5)$$

where $\varepsilon(\psi_j)$ is the orbital energy associated with the molecular orbital ψ_j. This is known as Koopmans' theorem.

There is hardly any paper on photoelectron spectroscopy, be it UPS or XPS, which does not quote Koopmans' theorem. A survey of the literature will reveal that the corresponding references fall into two classes: (a) A majority which favors *Physica* **1** (1933) 104, and (b) a minority who prefers *Physica* **1** (1934) 104. The correct reference is quite definitely the latter. The mix-up is presumably due to the fact that Koopmans' paper is contained in the first issue of Vol. 1 of *Physica*, which was published in December 1933. However, the bulk of Vol. 1 appeared in 1934 and the front page of this volume carries the year 1934. Also Vol. 2 is 1935, etc.

Very few theoreticians or spectroscopists seem to be aware of what became of Koopmans. Tjalling C. Koopmans was born on August 28, 1910 at 's-Graveland in the Netherlands. He studied at the University of Utrecht, received his M. A. degree in physics and mathematics in 1933 and wrote two papers on quantum mechanics under the supervision of Prof. H. A. Kramers. One of these papers contains the formula $\varepsilon_k = E'(k) - E''(k)$. He went to Leiden where he got his Ph.D. in mathematical statistics with an important thesis 'Linear Regression Analysis of Economic Time Series'. From 1936 to 1938 he was lecturer at the Netherlands School of Economics and he accepted a position as an economist with the League of Nations in Geneva. 1940 found him as a research associate at Princeton University. From 1942 to 1944 he was statistician for the Combined Shipping Adjustment Board, and later Professor of Economics at the University of Chicago and Research Director of the Cowles Commission for Research in Economics. He is at present Alfred Cowles Professor of Economics at Yale. A collection of his many significant contributions to quantitative economics, many highly mathematical, has been published in honour of his 60th birthday in 1970 [4].

Especially in connection with the increasing 'orbital awareness' of the practising chemist, which has been stimulated by the work of Woodward and Hoffmann [5], the interpretation of the PE spectra in terms of Koopmans' theorem has become increasingly popular and almost all of the work published to date has been discussed with reference to orbital diagrams and orbital energies of the parent molecule M. As a consequence, statements such as the following one made by Mulliken [6], i.e. "PE spectroscopy... has given a new reality to the idea of molecular orbitals, by determining quantitative values for their binding energies, and also by giving information about their bonding characteristics", have been grossly misunderstood by the 'chemist in the street', in whose mind the theoretical construct of a molecular orbital and its associated energy have by now gained the status of (almost) observables.

A typical example for the application of (4) is shown in Figure 1, where the observed vertical ionization potentials $I_{v,j}$ of allene (**1**) [7–8], butatriene (**2**) [9], vinylacetylene (**3**) [9] and divinylacetylene (**4**) [9] have been plotted vs. the previously calculated

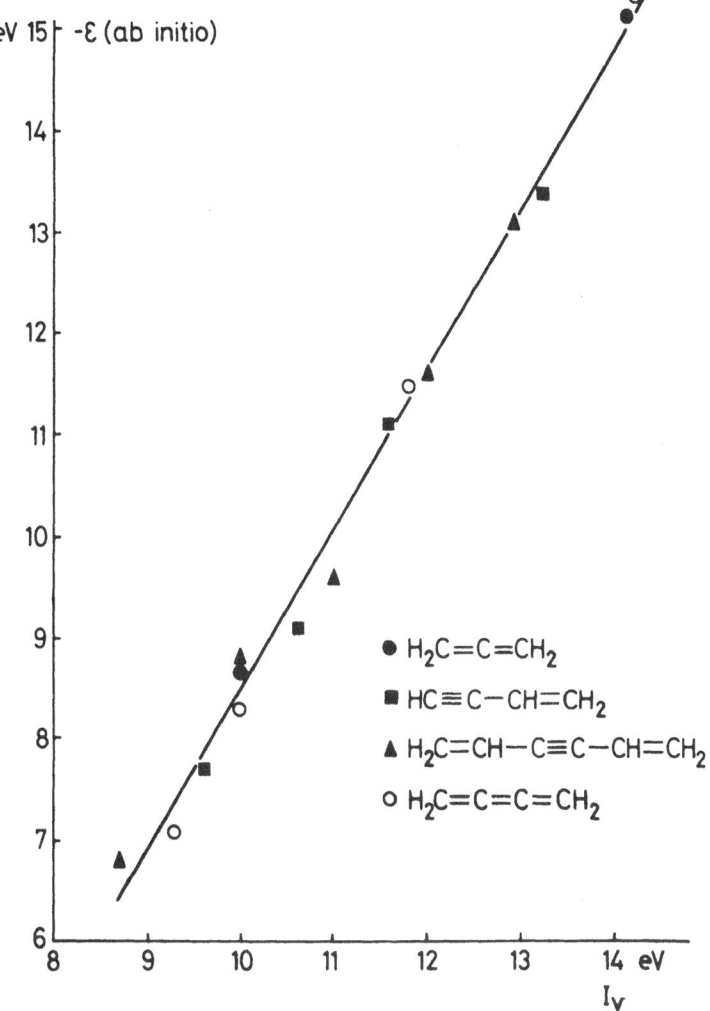

Fig. 1. Comparison of observed ionization potentials $I_{v,j}$ of **1**, **2**, **3** and **4** with computed *ab initio* orbital energies $\varepsilon(\Psi_j)$ [9].

ab initio [10] orbital energies $\varepsilon(\psi_j)$ obtained by Pople and Radom [11].

$$H_2C=C=CH_2 \qquad\qquad H_2C=C=C=CH_2$$
$$\textbf{1} \qquad\qquad\qquad\qquad \textbf{2}$$
$$H_2C=CH-C\equiv CH \qquad H_2C=CH-C\equiv C-CH=CH_2.$$
$$\textbf{3} \qquad\qquad\qquad\qquad \textbf{4}$$

The regression line

$$I_{v,j}(\text{pred.}) = [(4.643 \pm 0.157) - (0.632 \pm 0.015)\,\varepsilon(\psi_j)]\ \text{eV} \qquad (6)$$

is remarkably good, as evidenced by a correlation coefficient $r = 0.9963$ and a standard error $\text{SE}(I_{v,j}) = 0.154$ eV (degree of freedom $\phi = 14$). Note, however, that (6) does not go through the origin and that the slope differs from unity.

The question arises whether simpler models e.g. semiempirical SCF calculations [12] or Hückel-type procedures [13] yield similarly satisfactory results. This is indeed the case. As an example we examine briefly the application of the classical Hückel MO model to benzenoid hydrocarbons.

In Figure 2 are shown the PE spectra of the acenes [14–15] (a) benzene (**5**), naph-

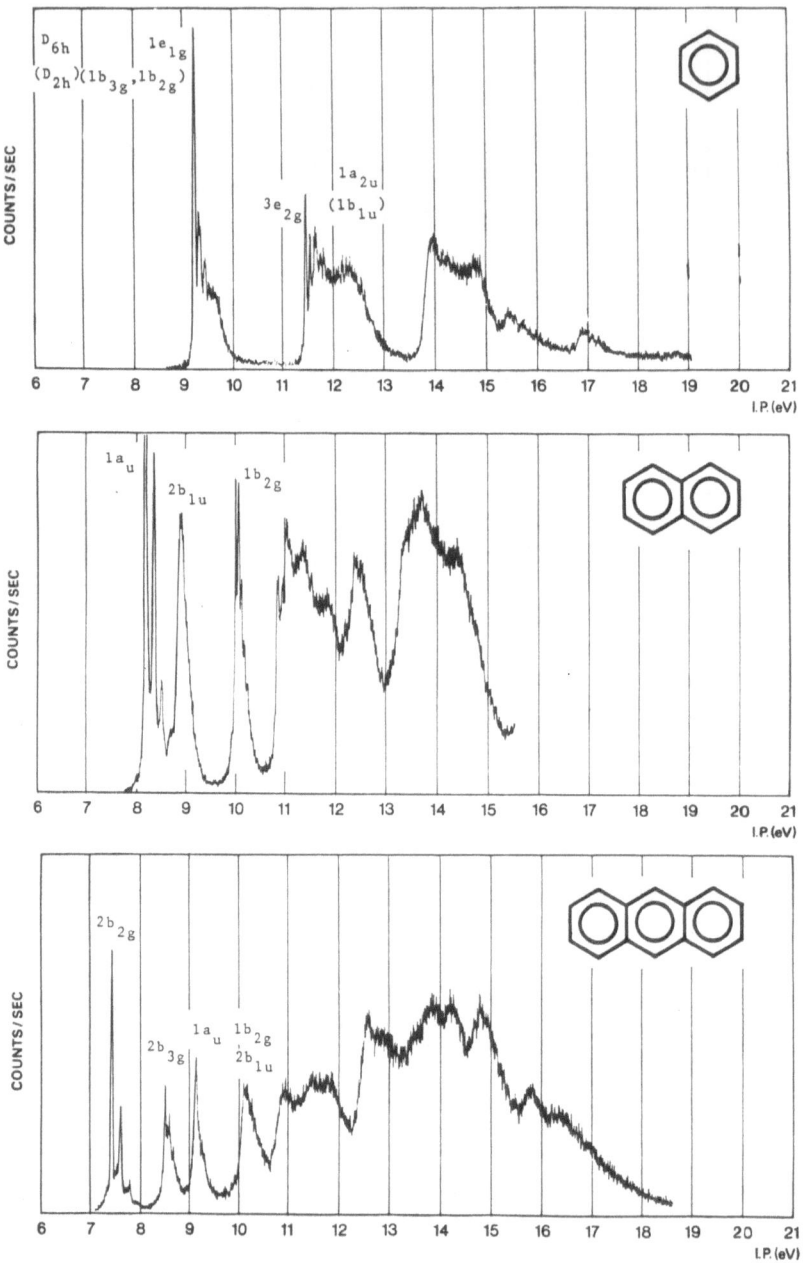

Fig. 2a.

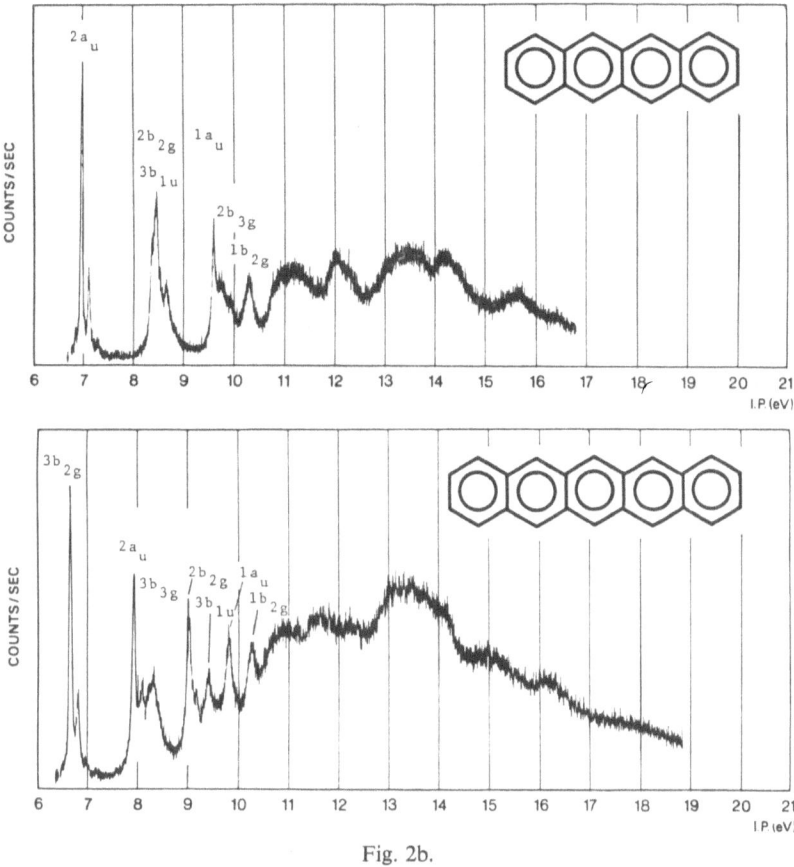

Fig. 2b.

Fig. 2.   PE spectra of *n*-acenes [14].

thalene (**6**), anthracene (**7**), (b) naphthacene (**8**) and pentacene (**9**). These spectra are

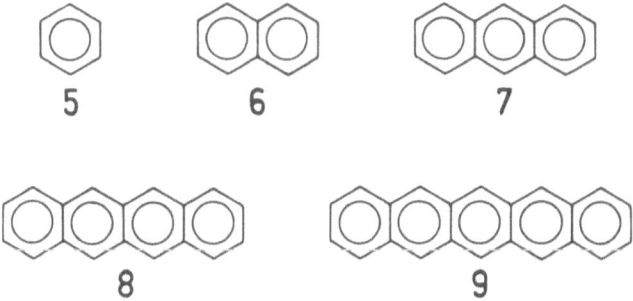

characterized by the fact that electron ejection (1) from a π-orbital gives rise to sharp bands with dominating $0 \leftarrow 0$ or $1 \leftarrow 0$ transition. According to Coulson [16] the HMO orbital energies ε_J^{HMO} of the bonding HMO's ψ_J^{HMO} of an *n*-acene (*n* is the number of annelated rings) are given by

$$\varepsilon_0^{HMO} = \alpha + \beta,$$
$$\varepsilon_J^{HMO} = \alpha + \beta (r_J + 1)/2, \qquad\qquad (7)$$

where $r_J = [9 + 8 \cos(\pi J/(n+1))]^{1/2}$ with $J = 1, 2, \ldots n$. Figure 3 shows a correlation diagram of these orbital energies in which those belonging to HMO's ψ_J^{HMO} of the same irreducible representations are joined by solid correlation lines. The n-independent orbital energy ε_0^{HMO} belongs alternatively to orbitals ψ_0^{HMO} of $B_{3g}(n=\text{odd})$ or $B_{1u}(n=\text{even})$ behaviour. The orbital energies (7) yield already a reasonably good representation of the observed data $I_{v,j}$, if (5) is assumed to be valid for the ε_J^{HMO}. However, the agreement can be much improved by including the corrections $\delta I_{v,J}$ which take care of the effect of first-order bond localization in the neutral n-acenes. This can be done by simple perturbation treatment [17], which yields

$$I'_{v,j} = I_{v,j} + \delta I_{v,j},$$
$$I'_{v,J} = -(\alpha + \beta x_J) + b \sum_{\mu, v} (p^+_{\mu v, J} - p_{\mu v}) (p_0 - p_{\mu v}), \qquad (8)$$

where (a) $\alpha + \beta x_J = \varepsilon_J^{HMO}$ is the orbital energy of the π-orbital ψ_J^{HMO} in the usual

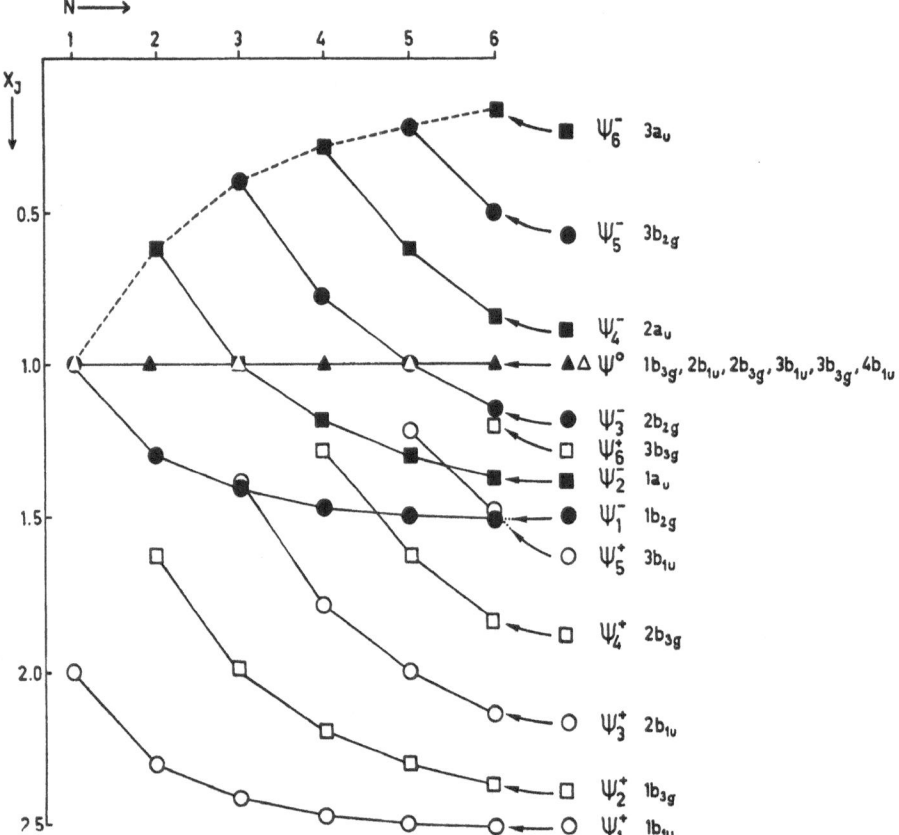

Fig. 3.　Orbital correlation diagram for n-acenes with $n = 1$ to 6. The HMO orbital energies $\varepsilon_J{}^{HMO}$ have been calculated from (6) [16].

Hückel approximation; (b) $p_{\mu v}$ is the bond order between the bonding centers μ, v of the neutral hydrocarbon M; (c) $p^+_{\mu v, J} = p_{\mu v} - c_{J\mu}c_{Jv}$ is the corresponding bond order in the radical cation $M^+ (^2\Psi_J)$, i.e. M^+ in the configuration in which ψ^{HMO}_J is only singly occupied; (d) $p_0 = 2/3$ is the standard bond order in benzene; (e) b is a factor dependent upon the force constants of the π- and σ-bonds and upon the derivative $d\beta/dR$ of the resonance integral β with respect to the length R of the π-bond. Summation is performed over all bonds. With $\alpha = (-5.864 \pm 0.110)$ eV, $\beta = (-3.196 \pm 0.107)$ eV and $b = (-7.859 \pm 1.475)$ eV the residual variance about the regression is $V(\varepsilon_J) = 0.0109$ eV^2, corresponding to a standard error SE$(\varepsilon_J) = 0.104$ eV.

The remarkable agreement between the observed $(I_{v,J})$ and calculated $(I'_J = -\varepsilon'_J)$ ionization potentials is evident from the correlation diagram of Figure 4. Apart from the general numerical precision, the most gratifying feature is that our perturbation model explains satifactorily the observed upwards trend of ε_0 as a function of N. Note that the observed accidental degeneracy of $3b_{1u}$ and $2b_{2g}$ for **8** is faithfully reproduced.

This result is not unexpected. Previous calculations based on formula (8) have shown that the correction δI_J leads to an equally convincing improvement of the

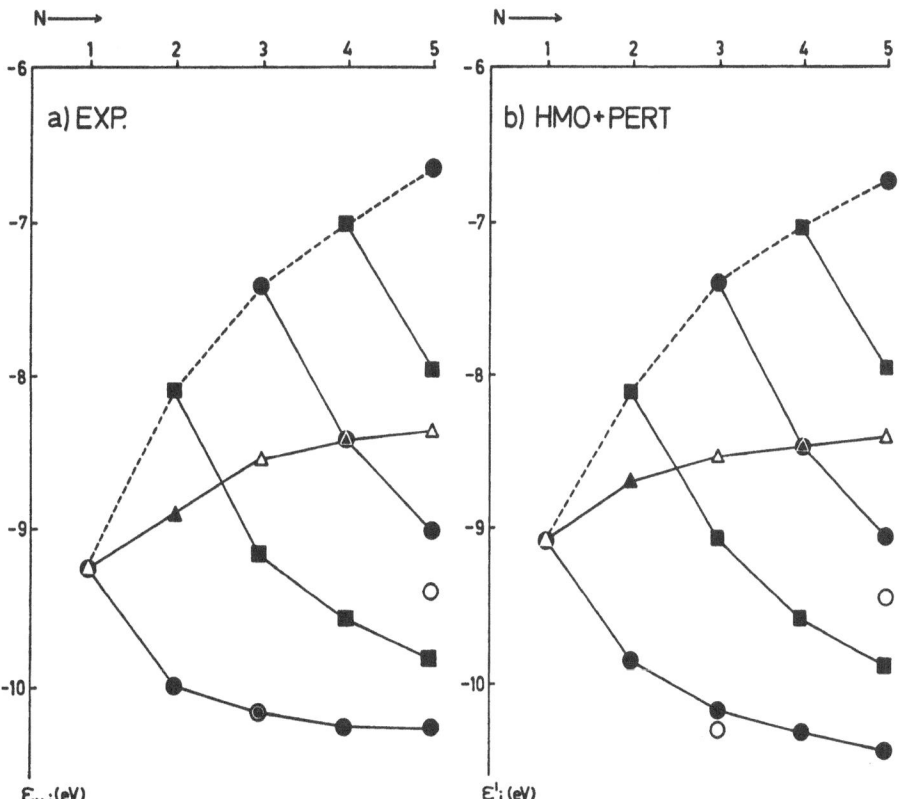

Fig. 4.   Comparison of 'observed' and computed π-orbital energies for the n-acenes. The 'observed' values have been obtained by applying (4) in reverse, the calculated ones (from (6)) have been corrected acconding to (7) for bond fixation in the neutral molecule M [17].

predicted π-orbital energies for linear unsaturated, non alternant and non benzenoid hydrocarbons [17] or for the five isomeric benzenoid molecules $C_{18}H_{12}$ [18]. In contrast, taking into account the uneven distribution of the excess positive charge in $M^+ (^2\Psi_J)$ (i.e. by a first-order perturbation treatment) does not yield a further significant improvement of the ε'_J values.

Obviously the same type of agreement can be obtained by applying any of the usual semiempirical [12] or *ab initio* methods (cf. Figure 1) to planar unsaturated hydrocarbons, if the correct structure parameters are used as input data.

We shall now discuss the limitation of Koopmans' theorem which is due to the neglect of the reorganization of the electrons in $M^+ (^2\Psi_j)$, a limitation which had already been recognized and clearly stated by Koopmans himself in his fundamental paper [3].

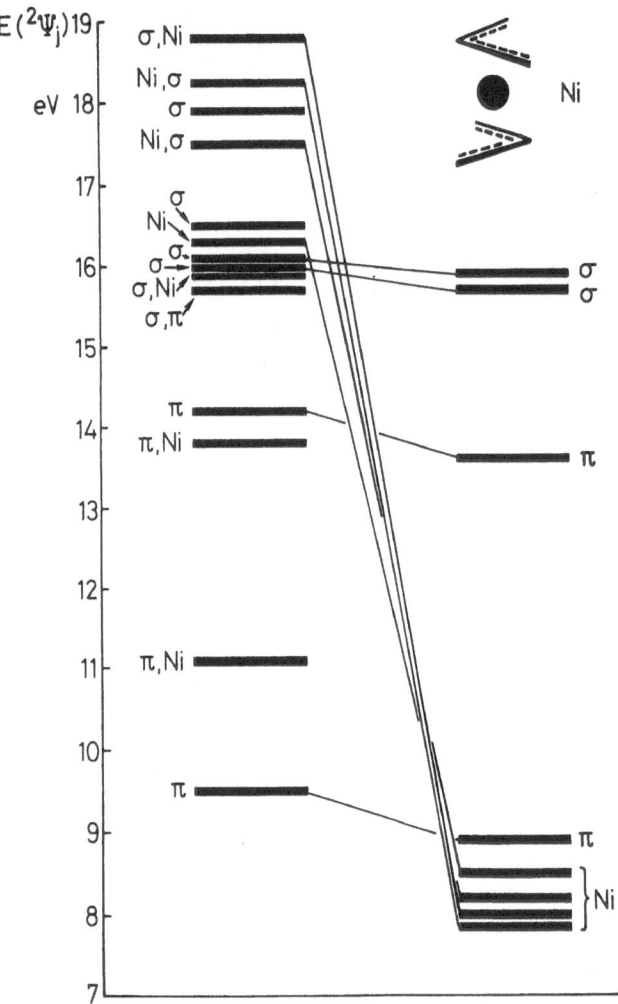

Fig. 5.   Stabilization of the electronic states of bis (π-allyl) nickel radical cation on electronic relaxation, relative to the state energies derived from Koopmans' theorem [21].

A dramatic and perhaps extreme example is shown in Figure 5. It concerns bis (π-allyl) nickel (**10**) (symmetry C_{2h}) for which an *ab initio* orbital calculation has been carried out by Veillard [19]. If (5) is applied to the set of orbital energies $\varepsilon(\psi_j)$, then the ground state $^2\Psi_0$ of the radical cation is predicted to correspond to the removal of an electron from the π-orbital $7a_u$ localized in the two allyl moieties of **10**, i.e.

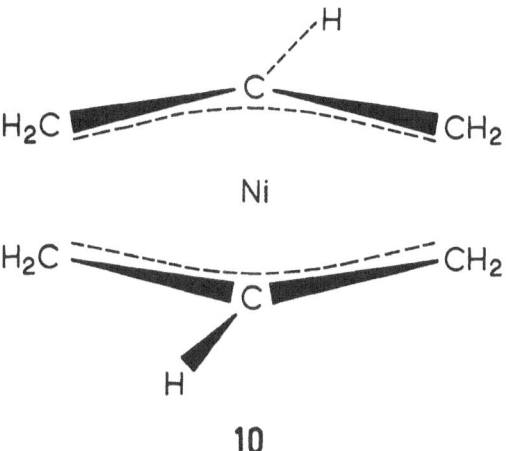

10

$^2\Psi_0 \equiv {}^2A_u$. In contrast, vacating one or the other of the orbitals $3b_g$, $9a_g$, $10a_g$ or $11a_g$ which are essentially $3d$-Ni in character, would lead to excited states of the radical cation of **10** which lie 7 to 9 eV above the presumed ground state ^2A_u. This result is contradicted by the experiment. Indeed Lloyd and Lynaugh have shown that the first band(s) in the PE spectrum of **10** corresponds – in terms of our model – to the ejection of an electron from a $3d$-Ni orbital [20]. This discrepancy is removed if the electrons of the radical cation of **10** are allowed to relax [21]. As shown in Figure 5 such a re-laxation process leads to a considerable stabilization of those 'Koopmans states' which correspond to the ejection of an electron from orbitals centred on the nickel atom (e.g. $3b_g$, $9a_g$, $10a_g$, $11a_g$) whereas only small changes in energy are predicted for the states obtained by removing an electron from σ- or π-orbitals located in the allyl subsystems [22]. Obviously, the size of the computed corrections is closely tied to the particular model that has been used and that other models would yield different orbital energies $\varepsilon(\psi_j)$ and thus different relaxation energies. However, the take-home-lesson would still be the same, i.e. that (6) has to be used with caution.

In the light of this and of similar results, expression (5) ought to be formally changed to

$$I_{v,j} = -\varepsilon(\psi_j) - R_j, \tag{9}$$

where R_j stands for the relaxation correction, i.e. the stabilization of $M^+\,(^2\Psi_j)$ which results when the 'frozen orbital' assumption underlying (5) is abandoned. It is general-ly believed that for hydrocarbon radical cations the corrections R_j is much smaller than in the above example **10** (i.e. only of the order of 1 to 2 eV) and that it will not depend strongly on j. In particular Koopmans' theorem is often used in the weaker perturbation form

$$\delta I_{v,j} = - \delta\varepsilon(\psi_j) \tag{10}$$

derived from (5), or from (6) under the assumption $\delta R_j=0$. This implies that Koopmans' theorem will at least allow an estimate of those changes $\delta I_{v,j}=I'_{v,j}-I_{v,j}$ in ionization potentials which are the consequence of a perturbation, e.g. alkyl substitution of benzene 5 to yield 11 or replacement of CH by N to yield pyridine 12. The orbital energy changes $\delta\varepsilon(\psi_j)$ that

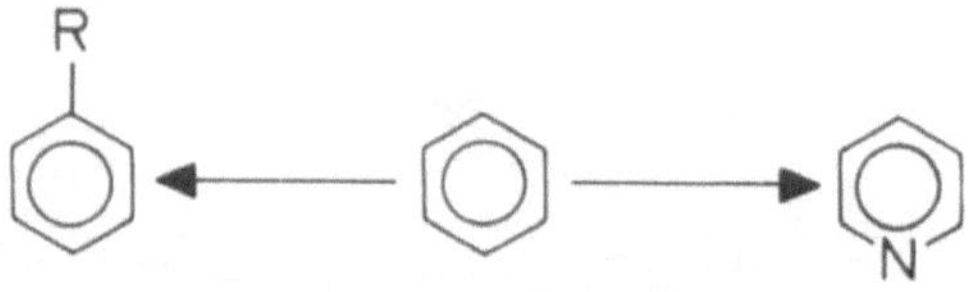

11   5   12

have to be inserted into formula (10) are calculated by way of the usual perturbation procedures. However, we shall now give an example which clearly indicates that even (10) has its severe limitations.

Figure 6 shows the PE spectrum of fulvene 13 (H, H) [23], the first two bands of

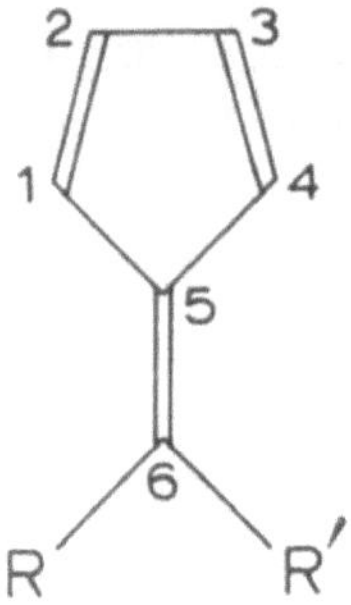

13 (R, R′)

which (① at 8.55 eV; ② at 9.54 eV) correspond to the ejection of an electron from the π-orbitals $1a_2(\pi)$ and $2b_1(\pi)$, respectively, whose nodal properties are given in (11):

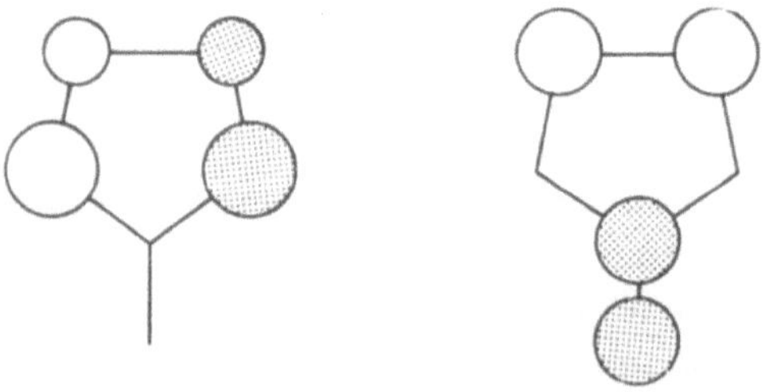

$1a_2(\pi)$ $2b_1(\pi)$

(11)

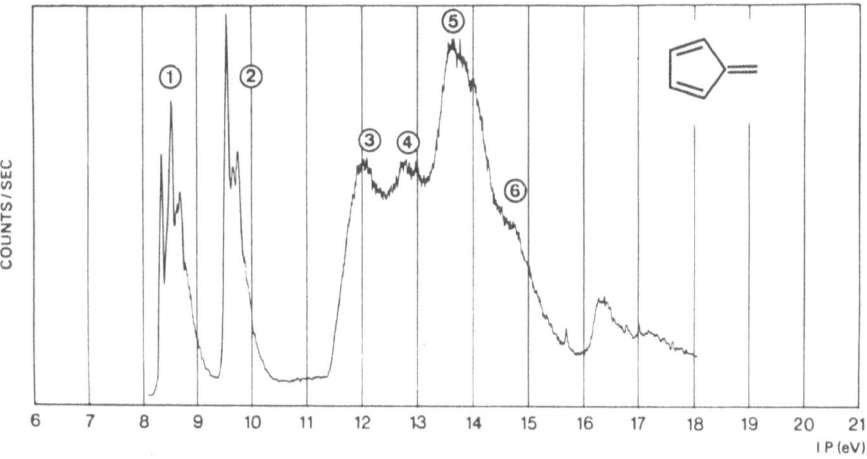

Fig. 6.   PE spectrum of fulvene **13** (H, H) [23].

Because of the nodal plane of $1a_2(\pi)$, which contains the centres 5 and 6 of **13** (H, H), we intuitively expect that $\delta I_{v,1}$ should be zero for all alkyl-substituted fulvenes **13** (R, R'), whereas $\delta I_{v,2}$ should be large and negative, a view strongly supported by the results of semiempirical SCF calculations (EHT, MINDO/2, SPINDO and INDO) [24]. However, this expectation is completely at variance with the experimental result summarized in Figure 7 and in Table I.

The linear regression

$$\delta I_{v,2} = (-0.069 + 0.030)\ eV + (1.503 + 0.071)\ \delta I_{v,1} \tag{12}$$

does not miss the origin significantly (95% security) so that we have for all derivatives **13** (R, R').

$$\delta I_{v,2}/\delta I_{v,1} = 1.5. \tag{13}$$

TABLE I

Shifts $\delta I_{v,1}$ and $\delta I_{v,2}$ of the bands ① and ② in the PE spectrum of fulvene and alkyl substitution

13(R, R')		$-\delta I_{v,1}$	$-\delta I_{v,2}$
R	R'	eV	eV
H	H	0.00	0.00
H	Me	0.24	0.45
H	Et	0.28	0.51
H	n-Pr	0.32	0.59
H	i-Pr	0.30	0.57
H	n-Bu	0.31	0.58
H	i-Bu	0.38	0.63
H	t-Bu	0.44	0.66
Me	Me	0.47	0.79
Et	Et	0.56	0.91
n-Pr	n-Pr	0.63	1.01
$-CH_2(CH_2)_2CH_2-$			

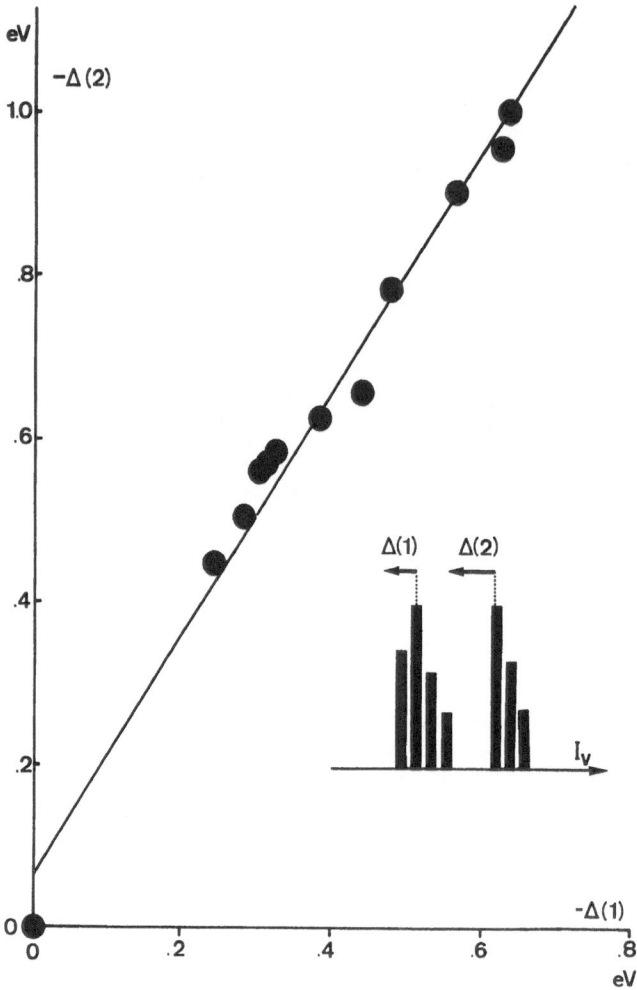

Fig. 7.   Regression of $\delta I_{v,2}$ on $\delta I_{v,1}$ for the shifts induced by alkyl substitution in the bands 1 and 2 of fulvene **13** (H, H). See Figure 5 and Table I.

The reason for the unexpected result $\delta I_{v,1} \approx \delta I_{v,2} \neq 0$ (rather than $\delta I_{v,1} = 0$) is immediately obvious if an SCF open-shell calculation is carried out for the radical cation of **13** (H, H) in the two states $^2\Psi_0 \equiv {}^2A_2$ and $^2\Psi_1 \equiv {}^2B_1$ [25]. In the diagrams of Figure 8 are given the electron drifts (in millielectrons) that have been calculated in the framework of an INDO model [26]. Ejection of an electron from $1a_2(\pi)$ or $2b_1(\pi)$ in a 'Koopmans process' with frozen orbitals ψ_j increases the atomic charges Q of neutral **13** (H, H) by the positive amounts indicated in the top row, i.e. amounts given by the squares of the orbitals $1a_2(\pi)$ and $2b_1(\pi)$. If the orbitals of the radical cation of **13** (H, H) are allowed to readjust i.e. $\psi_j \rightarrow \psi'_j$, the electron drifts indicated in Figure 8 are calculated in the framework of the model chosen. As a result, the atomic charges Q^+ given in the bottom row are obtained for the radical cation of **13** (H, H). The most noteworthy feature is, that ejection of an electron from $1a_2(\pi)$ places 0.23 of a positive charge in position 6 (as compared to 0.03 in neutral **13** (H, H)). It is gratifying, albeit somewhat accidental, that the ratio of the charges in position 6 which are calculated for an electron vacating $2b_1(\pi)$ or $1a_2(\pi)$ is 1.6, as compared to the shift ratio 1.5 given in (13).

From the foregoing it is obvious that any attempt to reparametrize semiempirical SCF models in the hope to obtain, according to (5), a perfect fit with the PE data of a large variety, even of simple hydrocarbons will necessarily be doomed to failure. Typically, none of the available semiempirical models predicts correctly the observed

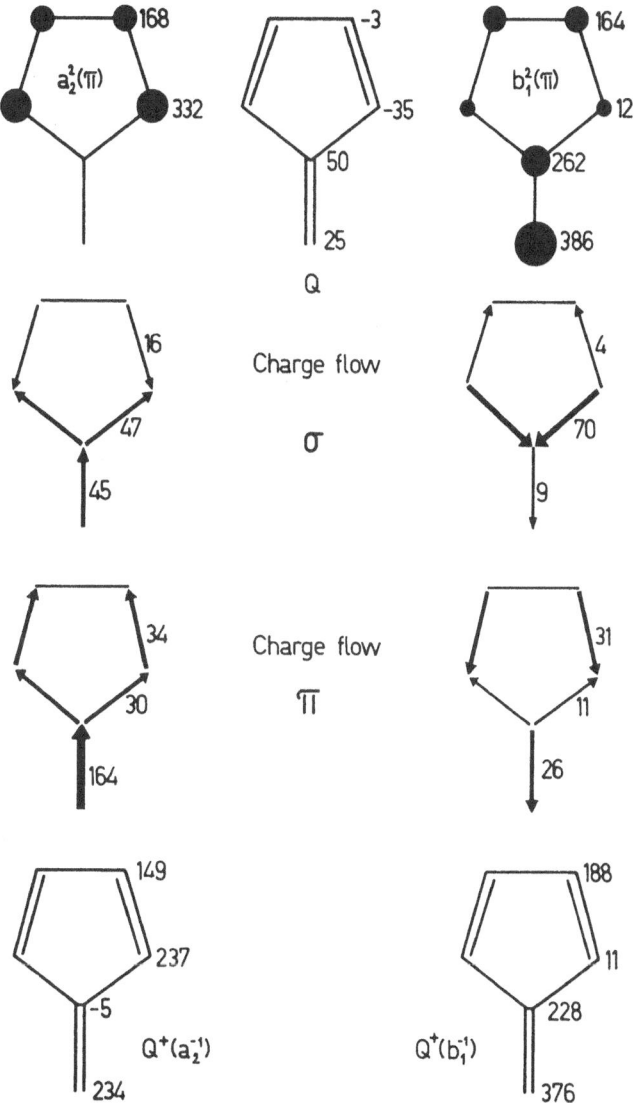

Fig. 8.   Electron drifts (in millielectrons) in the radical cation of fulvene **13** (H, H) following the Koopmans' processes leading to $^2\Psi_0 \equiv {}^2A_2$ and $^2\Psi_1 \equiv {}^2B_1$.

trend of the π-ionization potentials of cycloalkenes $14\,(n)$ [27] or of the benzocyclo-alkenes [28] $15\,(n)$ shown in Figure 9, let alone their absolute values.

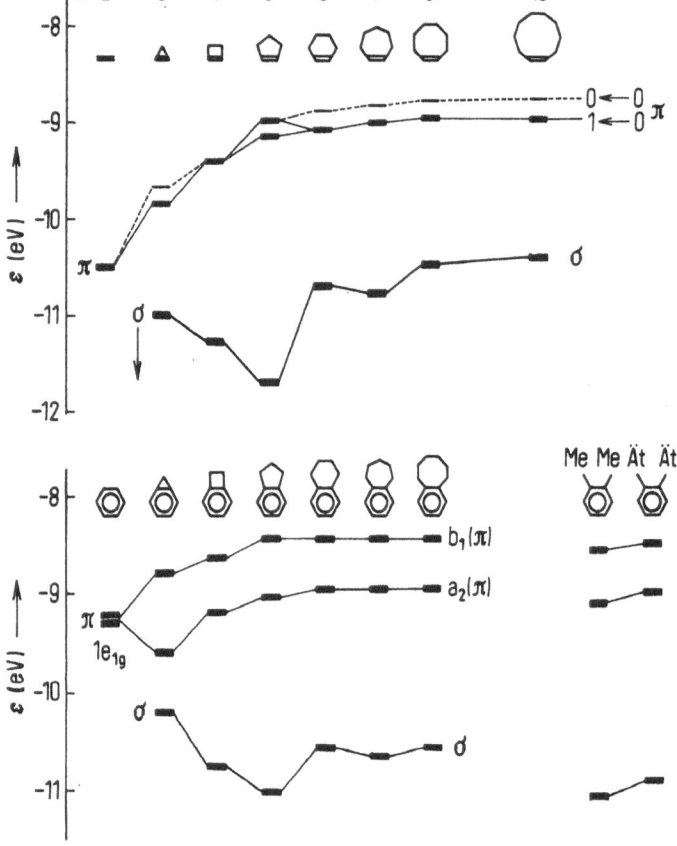

$$14\ (n)\qquad\qquad\qquad 15\ (n)$$

For example: in contrast to the experiment (see Figure 9) cyclopropene $14\,(3)$ is always calculated to be easier to ionize than cyclobutene $14\,(4)$.

On the other hand, such parametrizations have been quite successful for smaller classes of related molecules and the models so obtained may serve as useful starting points for treatments based on (9), i.e. which include the calculation of relaxation correction terms R_j. Presumably a heuristically useful and promising approach con-

Fig. 9.   Orbital correlation diagram for cycloalkenes **14** (n) and benzocycloalkenes **15** (n). The orbital energies given are those corresponding to the band positions in the PE spectra, according to (5) [27, 28].

sists in the transformation of the canonical SCF orbitals ψ_j, calculated by one or the other of these methods (e.g. MINDO/2 [29] or SPINDO [30]) into localized orbitals [31] and to determine their interaction matrix elements. From diagrams such as the one shown in Figure 10 for propene $H_2C=CH-CH_3$ it becomes possible to assess almost at a glance the essential differences between the various semiempirical models in terms of chemically appealing concepts. This in turn makes it much easier to gain insight into their relative limitations and advantages. For instance, from Figure 10 we see that the $\sigma(CH)$ basis orbitals for the out-of-plane CH bonds of the methyl group in propene are at lower energy in MINDO/2 (-17.63 eV) than in SPINDO

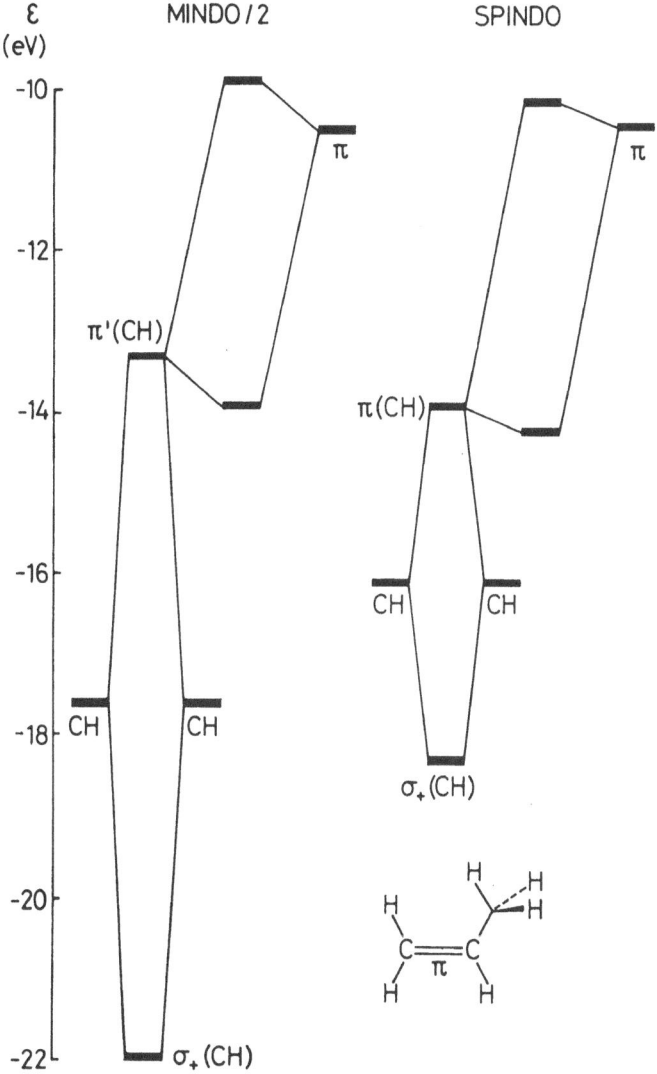

Fig. 10. Partial results of an Edmiston-Ruedenberg localization procedure applied to propene. CH = localized CH-orbitals, π = localized double-bond π-orbital. $\pi'(CH)$ and $\sigma_+(CH)$ denote the minus and plus combinations of the localized CH-orbitals.

(-16.11 eV). On the other hand, the interaction between these geminal CH-orbitals is almost twice as large in MINDO/2 (-4.32 eV) than in SPINDO (-2.22 eV) which places the pseudo-π-combination π' (CH) at -13.31 eV or -13.89 eV, respectively. Note that these latter energies differ only by about 0.5 eV, although their origin is quite different. The interaction matrix element between π' (CH) and the π-orbital of the double bond is larger in MINDO/2 (-1.43 eV) than in SPINDO (-1.09 eV). It is an empirical fact that MINDO/2 tends to over-emphasize σ, π-mixing whereas SPINDO definitely underestimates it.

From a purely pragmatic point of view the safest procedure that can be used for assigning bands in the PE spectra of large organic molecules, consists in constructing correlation diagrams on the basis of the PE spectroscopic results obtained from a set of closely related compounds [33]. A typical example is given in Figure 11, which leads to an unequivocal assignment of the PE spectrum of cyclooctatetraene [34].

Correlations obtained in this fashion, i.e. replacing theoretical calculations by extensive synthetic work, lead often to very simple empirical relationships which one would hardly have expected on the basis of the rather involved theoretical models available. As an example we show in Figure 13 the orbital correlation diagram derived

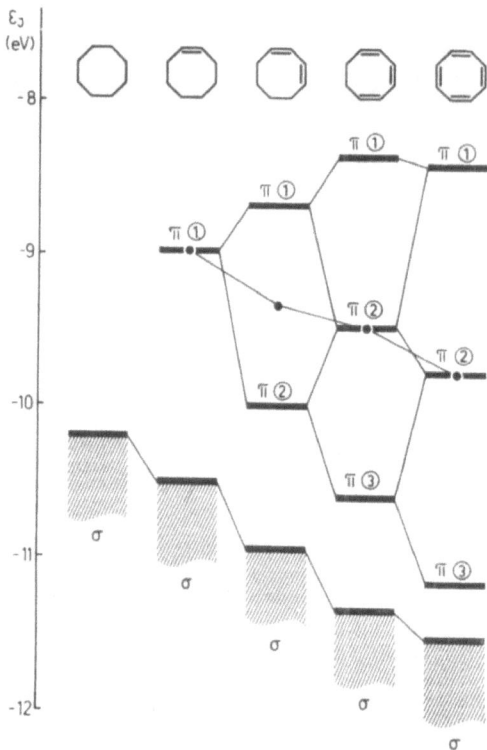

Fig. 11. Correlation diagram for the PE data of cyclooctatetraene and its di, tetra, hexa and octahydroderivatives [34]. ①, ②, ③ refer to the first, second and third PE band. The shaded area (σ) corresponds to the onset of the σ-system.

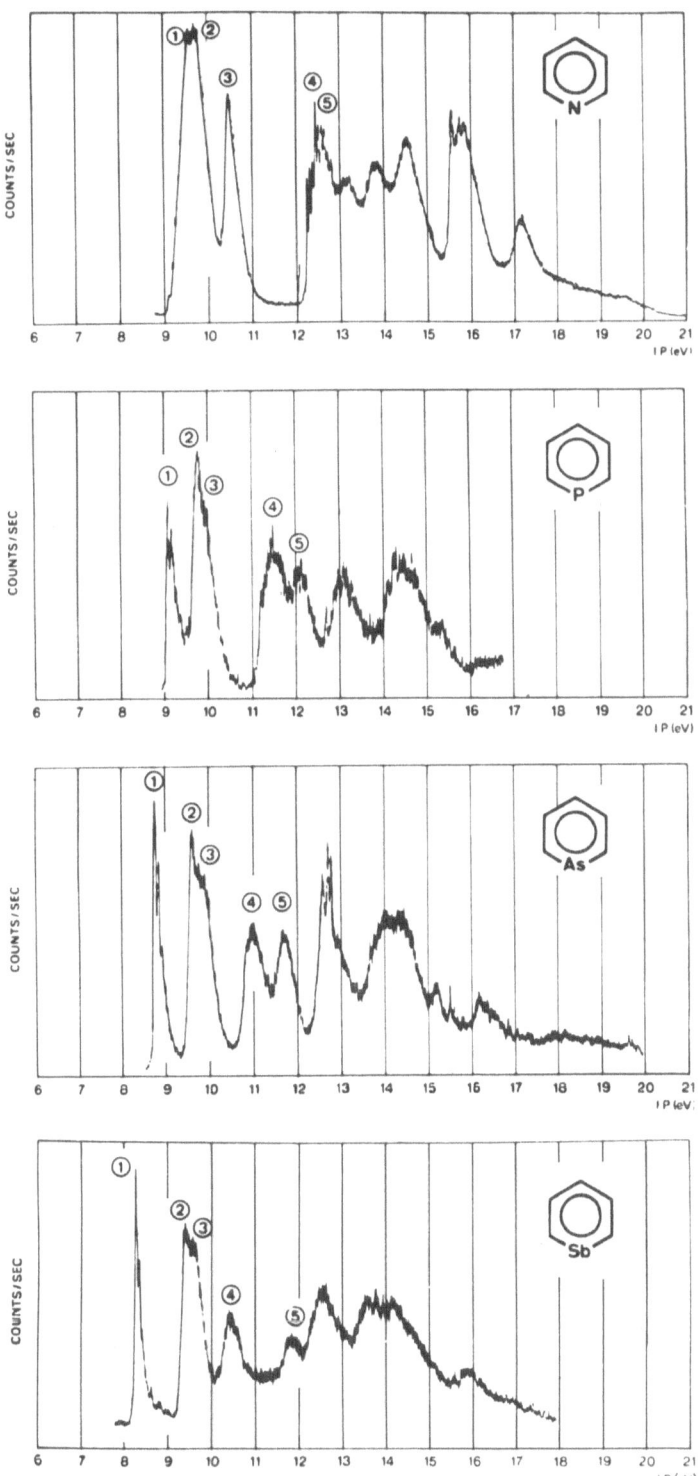

Fig. 12. PE spectra of pyridine (**12**), phosphabenzene (**16**), arsabenzene (**17**) and stibabenzene (**18**) [35].

from the PE spectra of benzene **5** pyridine **12**, phosphabenzene **16**, arsabenzene **17** and stibabenzene **18** (Figure 12) [35]. If the vertical ionization potentials corresponding to electron ejection from the π-orbitals belonging to the irreducible representation B_1 (i.e. $2b_1$, $3b_1$, $5b_1$, $7b_1$ and $1b_1$, $2b_1$, $4b_1$, $6b_1$ in the series **12, 16, 17, 18**) are plotted

16   **17**   **18**

vs. the ionization potentials $I(X)$ of the free atoms $X =$ N, P, As, Sb [36] the almost perfect linear regressions shown in Figure 14 are obtained. It is amusing to note, that the slopes 0.36 and 0.16 are practically identical to the squared orbital coefficients of the $2p$-atomic orbitals in position 1 of the parent π-orbitals in benzene, i.e. $(1/\sqrt{3})^2 = 0.33$ and $(1/\sqrt{6})^2 = 0.17$. In addition for $I(\text{C}) = 11.26$ eV [36] we obtain $I_v = 9.32$ eV and $I_v = 12.09$ eV to be compared to $I_v(e_{1g}) = 9.24$ eV and $I_v(a_{2u}) = 12.2$ eV found experimentally for benzene [1, 37]. Such simple, empirical relationships raise the hope that simpler theoretical treatments than those available at the moment can be developed for the interpretation of PE spectra of complex molecules.

The following section is devoted to the question whether the results obtained from

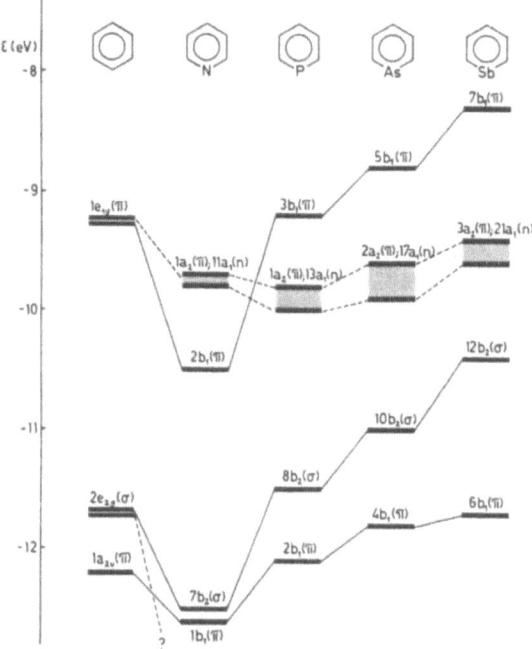

Fig. 13.   Orbital correlation diagram derived from the spectra shown in Figure 12.

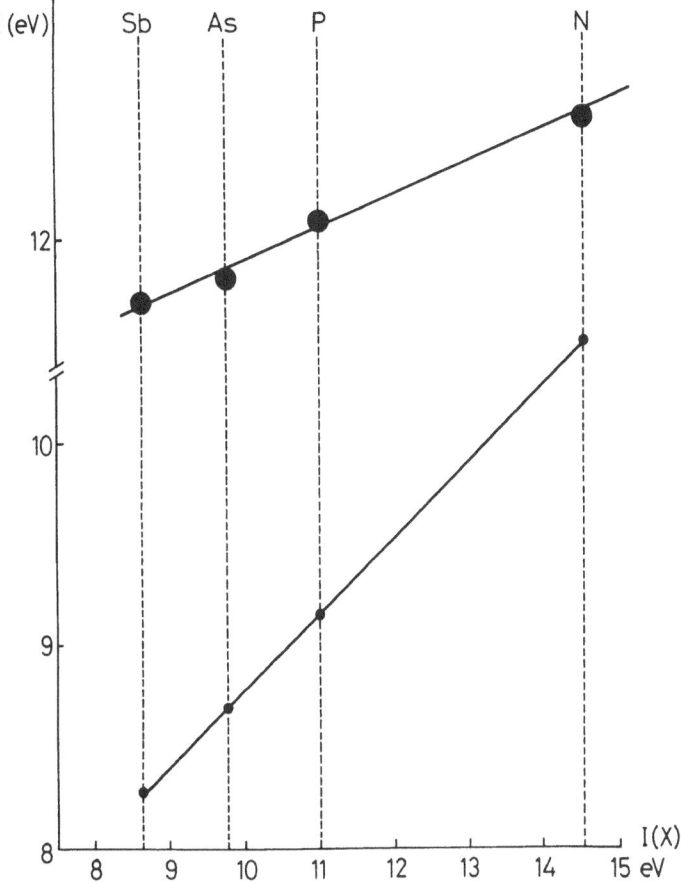

Fig. 14. Regression of I on $I(X)$ for pyridine (12), phosphabenzene (16), arsabenzene (17) and stibabenzene (18). I = ionization potentials of the bands corresponding to the ejection of an electron from π-orbitals of B_1 symmetry. $I(X)$ = ionization potential of the free atoms X = N, P, As, Sb.

PE spectroscopy can lead to a better understanding of electronically excited states of the parent molecule M.

For the description of electronically excited states of M one usually has to rely on a configuration interaction treatment. For this reason a direct correlation of ionization potentials with electronic excitation energies is impossible except in cases in which low lying excited states can reasonably well be described by single configurations. Sometimes this condition is met if (a) the molecule has high symmetry, leading to factorization of the CI-matrix into submatrices belonging to different irreducible representations, and (b) if the individual configurations in a given submatrix have widely different energies and thus interact only weakly.

Simple orbital diagrams, such as the one given in Figure 15 seem to suggest that

$$\Delta I(j, i) = \Delta E(j, i),\tag{14}$$

where

$$\Delta I(j, i) = I(j) - I(i)\tag{15}$$

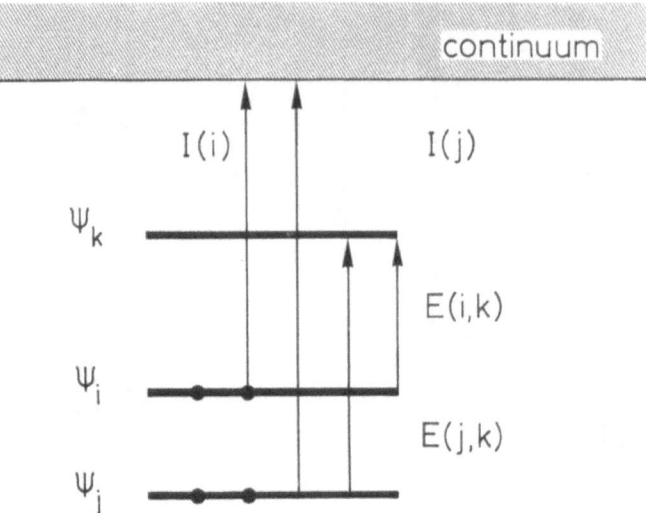

Fig. 15. Orbital diagram for the comparison of the differences $I(j) - I(i)$ in ionization potentials with those $E(j, k) - E(i, k)$ of the corresponding electronic excitations.

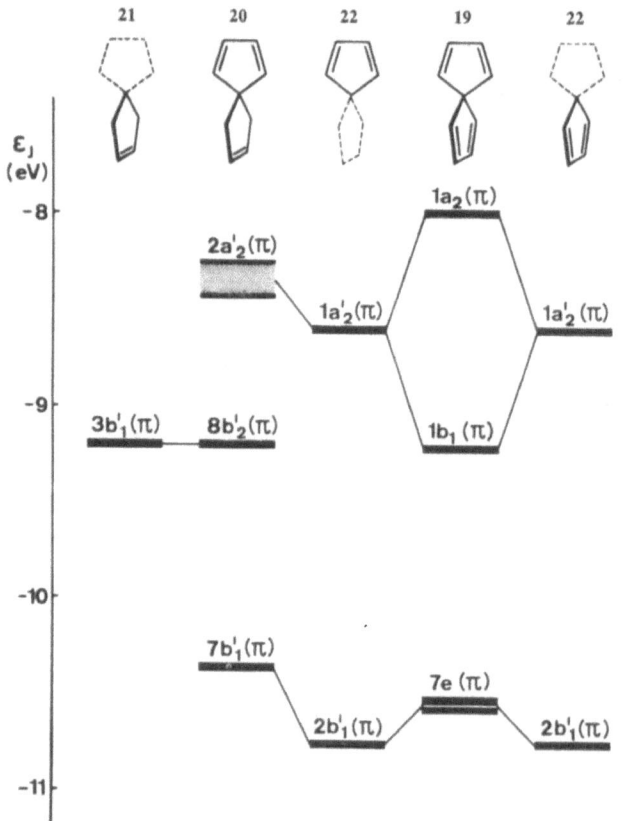

Fig. 16. Orbital correlation diagram for spiro [4, 4] nonatetraene (**19**), spiro [4, 4] nona-1, 3, 7-triene (**20**), cyclopentene (**21**) and cyclopentadiene (**22**).

$$\Delta E(j, i) = E(j, k) - E(i, k).$$

However, from the well-known matrix elements of the Hamiltonian for the ground and singly excited states of a closed-shell molecule M [38] it follows that [39]

$$^1\Delta E(j, i) = \Delta I(j, i) + (J_{ik} - J_{jk}) + 2(K_{jk} - K_{ik})$$ (16)
$$^3\Delta E(j, i) = \Delta I(j, i) + (J_{ik} - J_{jk}),$$

if use is made of Koopmans' theorem (5) [3]. Thus for singlet–singlet transitions (14) is true only if $(J_{ik} - J_{jk}) + 2(K_{jk} - K_{ik}) = 0$, which is generally not the case.

A remarkable exception is provided by spiro [4, 4] nonatetraene (19) [40], a member of the class of spiroconjugated compounds [41].

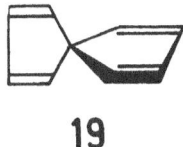

19

It can be shown, that due to its high symmetry (D_{2d}) and the alternancy of the subsystems the naive expectation (14) should be exceptionally fulfilled. The analysis of

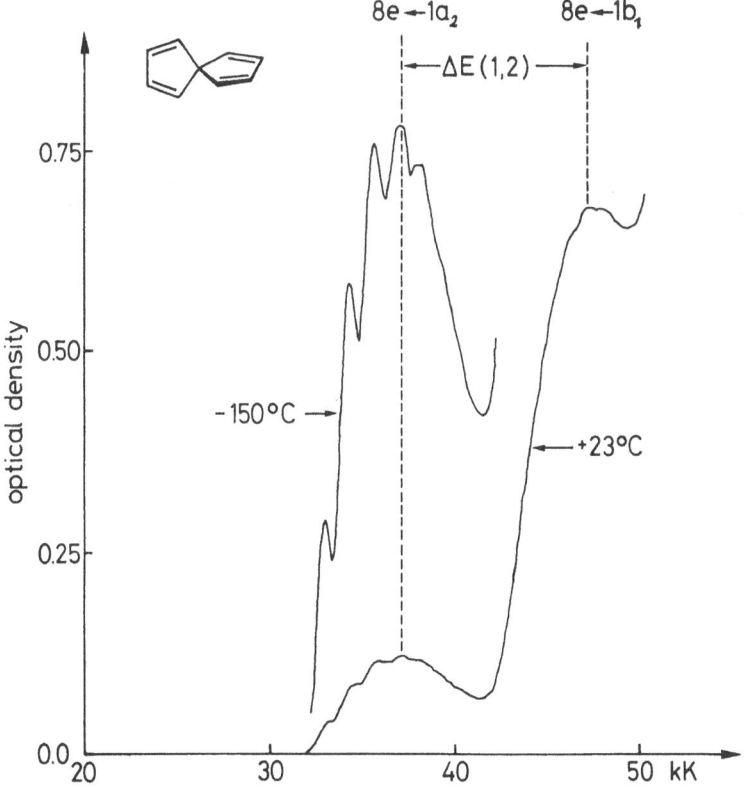

Fig. 17. Electronic spectrum of spiro [4, 4] nonatetraene (20). Solvent: rigisolve.

the PE spectra of **19**, of spiro [4, 4] nona 1, 3, 7-triene **20**, of cyclopentene **21** and cyclopentadiene **22** (see Figure 16) [42] yields $\Delta I = I(1b_1(\pi)) - I(1a_2(\pi)) = 1.23$ eV. The lowest antibonding orbital is $8e(\pi^*)$. It is easy to convince oneself that for $\psi_j \equiv 1b_1(\pi)$, $\psi_i \equiv 1a_2(\pi)$ and $\psi_k \equiv 8e(\pi^*)$ one has (in ZDO approximation) $J_{ik} - J_{jk} \approx 0$ and $K_{jk} - K_{ik} \approx 0$. In agreement with this result, one observes $^1\Delta E = 1.2$ to 1.3 eV, as shown in Figure 17.

Formula (16) can be used to obtain satisfactory correlations between lone-pair ionization potentials and $n \to \pi^*$ excitation energies. This has been done for *trans*-azomethane [43] and *p*-benzoquinone [44]. The correlation procedure will be illustrated, using pyrazine (**23**) and 2, 6-dimethylpyrazine (**24**) as an example [45].

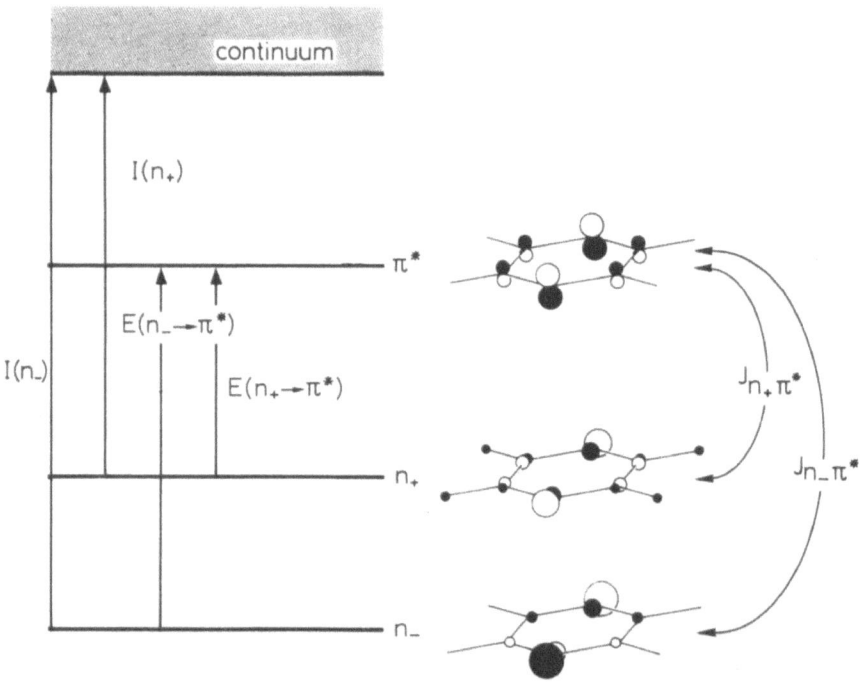

In **23** the splitting between the lone-pair PE bands (see Figure 18) is

$$\Delta I = I(n_-) - I(n_+) = \varepsilon(\psi(n_+)) - \varepsilon(\psi(n_-)) = 1.72 \text{ eV}. \tag{17}$$

Fig. 18.   Orbital diagram for the $\pi^* \leftarrow n$ bands of pyrazine (**23**) [45].

On the other hand, several experimental studies indicate that ΔE for **23** and **24** is rather small having a value not larger than 0.2 eV [46].

This discrepancy can be partly resolved by considering expression (16). For $n \to \pi^*$ transitions K_{jk} and K_{ik} turn out to be very small and can be neglected. Using the abbreviations defined in Figure 18 the splitting ΔE between the $n \to \pi^*$ states in **23** is given by

$$^1\Delta E \approx \Delta I + J_{n_+\pi^*} - J_{n_-\pi^*} = \Delta I + \Delta J. \tag{18}$$

From the diagrammatic representations of these molecular orbitals given in Figure 18 it is possible to derive at least an estimate of the sign of ΔJ. The n_+-MO has less space in common with the π^*-MO than the n_-MO, so that $J_{n_+\pi^*} < J_{n_-\pi^*}$ or $\Delta J < 0$. According to (18) this yields $\Delta E < \Delta I$, in qualitative agreement with the experimental evidence. Explicit calculation of ΔJ using the CNDO/2-method yields $\Delta J = -1.06$ eV for **23** and $\Delta J = -1.27$ eV for **24**. From PE spectroscopy, $\Delta I = 1.72$ eV for **23** and $\Delta I = 1.54$ eV for **24**. This leads according to (18) to $\Delta E = 0.66$ eV for **23** and $\Delta E = 0.27$ eV for **24**. Although there is not perfect numerical agreement with the experimental evidence for ΔE, the above results definitely support the idea that the splitting ΔE between the two $n \to \pi^*$ states in **23** and **24** is much smaller than the splitting ΔI. The reason for this discrepancy is that the two lone-pair orbitals $\psi(n_-)$ and $\psi(n_+)$ have different shapes, a necessary consequence of the 'through bond' interaction [47] between the semilocalized n_+-orbital and the CC-σ-orbitals of appropriate symmetry [48].

Acknowledgements

This work is part 57 of project No. 2.477.71 of the Schweizerischer Nationalfonds. Support by Ciba-Geigy S.A., F. Hoffmann-La Roche & Cie S. A. and Sandoz S. A. is gratefully acknowledged.

I thank the editors of the following journals for the permission to reproduce figures from previous publications: *Helvetica Chimica Acta* (Figures 2, 3, 4, 6, 18), *Journal of the American Chemical Society* (Figures 12, 13), *Chemische Berichte* (Figure 9) and *Journal of Electron Spectroscopy and Related Phenomena* (Figure 11).

References

[1] Al-Joboury, M. I. and Turner, D. W.: 1963, *J. Chem. Soc.*, 5141; Vilesov, F. I., Kurbatov, B. L., and Terenin, A. N.: 1961, *Dokl. Acad. Nauk USSR* **138**, 1329. Important reviews are: Turner, D. W., Baker, C., Baker, A. D. and Brundle, C. R.: 1970, *Molecular Spectroscopy*, Wiley-Interscience, London; Turner, D. W.: 1968, 'Molecular Photoelectron Spectroscopy' in *Molecular Spectroscopy* (ed. by P. Hepple), The Institute of Petroleum, London; Brundle, C. R. and Robin, M. B.: 1971, 'Photoelectron Spectroscopy' in *Determination of Organic Structures by Physical Methods*, Vol. 3 (ed. by F. C. Nachod and G. Zuckermann), Academic Press, New York, p. 1; Baker, A. D. and Betteridge D.: 1972, *Photoelectron Spectroscopy. Chemical and Analytical Aspects*, International Series of Monographs in Analytical Chemistry, Vol. 53 (ed. by R. Belcher and H. Freiser); Shirley, D. A. (ed.): 1972, *Electron Spectroscopy*, North-Holland, Amsterdam; Albridge, R. G.: 1972, 'Photoelectron Spectroscopy', in *Physical Methods in Chemistry* (ed. by A. Weissberger), Vol. 1, Wiley-Interscience, New York, Part 1.

[2] Siegbahn, K., Nordling, C., Fahlman, A., Nordberg, R., Hamrin, K., Hedman, J., Johansson G.,

Bergmark, T., Karlsson, S. E., Lindgren, I., and Lindberg, B.: 1967, *Atomic, Molecular and Solid State Structure by Means of Electron Spectroscopy*, Almqvist & Wiksell, Uppsala, Sweden; Siegbahn, K., Nordling, C., Johansson, G., Hedman, H., Heden, P. F., Hamrin, K., Gelius, U., Bergmark, T., Werme, L. O., Manne, R., and Baer, Y.: 1969, *ESCA, Applied to Free Molecules*, North-Holland, Amsterdam.

[3] Koopmans, T.: 1934, *Physica* 1, 104; Richards, W. G.: 1969, *Int. J. Mass Spectrom. Ion Physics* 2, 419.

[4] Koopmans, T.: 1970, *Scientific Papers of Tjalling C. Koopmans*, Springer Verlag, Berlin-Heidelberg-New York.

[5] Woodward, R. B. and Hoffmann, R.: 1965, *J. Am. Chem. Soc.* 87, 395, 2046, 2511; Woodward, R. B.: 1967, *Aromaticity*, Special Publication No. 21, The Chemical Society London, p. 217.

[6] Mulliken, R. S.: 1970, *Phil. Trans. Roy. Soc. London* 268, 3.

[7] Baker, C. and Turner, D. W.: 1969, *Chem. Comm.* 480.

[8] Brogli, F., Crandall, J. K., Heilbronner, E., Kloster-Jensen, E., and Sojka, S. A.: 1973, *J. Electron Spectrosc.* 2, 455 (1973/74).

[9] Brogli, F., Heilbronner, E., Kloster-Jensen, E., Schmelzer, A., Manocha, A. S., Pople, J. A., and Radom, L.: *Chem Physics*, in print.

[10] Pople, J. A. and Beveridge, D. L.: 1970, *Approximate Molecular Orbital Theory*, McGraw-Hill, New York.

[11] Radom, L. and Pople, J. A.: 1970, *J. Am. Chem. Soc.* 92, 4786; private communication.

[12] Murrell, J. N. and Harget, A. J.: 1972, *Semiempirical Self-Consistent-Field Molecular Orbital Theory of Molecules*, Wiley-Interscience, London-New York-Sydney-Toronto, 1972.

[13] Heilbronner, E. and Bock, H.: 1968, *Das HMO-Modell und seine Anwendung*, Verlag Chemie, Weinheim, Germany; Hoffmann, R.: 1963, *J. Chem. Phys.* 39, 1397; Hoffmann, R. and Lipscomb, W. N.: 1962, *J. Chem. Phys.* 36, 2179, 3489; 1962, 37, 2872.

[14] Clark, P. A., Brogli, F., and Heilbronner, E.: 1972, *Helv. Chim. Acta* 55, 1415.

[15] Boschi, R., Murrell, J. N., and Schmidt, W.: 1972, *Faraday Soc. Disc.* 54, 116.

[16] Coulson, C. A.: 1948, *Proc. Phys. Soc.* 60, 257.

[17] Brogli, F. and Heilbronner, E.: 1972, *Theor. Chim. Acta* 26, 289.

[18] Brogli, F. and Heilbronner, E.: 1972, *Angew. Chem.* 84, 551; 1972, *Angew. Chem.*, Int. Ed. 11, 538.

[19] Veillard, A.: 1969, *Chem. Comm.* 1022, 1427.

[20] Lloyd, D. R. and Lynaugh, N.: 1972, in *Electron Spectroscopy* (ed. by D. E. Shirley), North-Holland, Amsterdam.

[21] Rohmer, M. M. and Veillard, A.: 1973, *Chem. Comm.* 250.

[22] Veillard, A.: private communication.

[23] Heilbronner, E., Gleiter, R., Hopf, H., Hornung, V., and de Meijere, A.: 1971, *Helv. Chim. Acta* 54, 783.

[24] Brogli, F., Clark, P. A., Heilbronner, E., and Neuenschwander, M.: 1973, *Angew. Chem.* 85, 414.

[25] Brogli, F., Clark, P. A., Heilbronner, E., and Neuenschwander, M.: unpublished results.

[26] Pople, J. A., Beveridge, D. L., and Dobosh, P. A.: 1967, *J. Chem. Phys.* 47, 2026.

[27] Bischof, P. and Heilbronner, E.: 1970, *Helv. Chim. Acta* 53, 1677.

[28] Brogli, F., Giovannini, E., Heilbronner, E., and Schurter, R.: 1973, *Chem. Ber.* 106, 961.

[29] Dewar, M. J. S. and Haselbach, E.: 1970, *J. Ann. Chem. Soc.* 92, 590; Bodor, N., Dewar, M. J. S., Harget, A., and Haselbach, E.: 1970, *J. Am. Chem. Soc.* 92, 3854; Dewar, M. J. S., Haselbach, E., and Worley, S. D.: 1970, *Proc. Roy. Soc. London* A315, 431.

[30] Fridh, C., Åsbrink, L., Lindholm, E.: 1972, *Chem. Phys. Letters* 15, 282; Åsbrink., Fridh, C., and Lindholm, E.: 1972, *Chem. Phys. Letters* 15, 567; *J. Am. Chem. Soc.* 94, 5501.

[31] England, W., Salmon, L. S., and Ruedenberg, K.: 1971, *Fortschr. Chem. Forsch.* 23, 31; Edmiston, V., and Ruedenberg, K.: 1963, *Rev. Mod. Phys.* 34, 457.

[32] Heilbronner, E. and Schmelzer, A.: unpublished results.

[33] Bischof, P., Hashmall, J. A., Heilbronner, E., and Hornung, V.: 1969, *Helv. Chim. Acta* 52, 1745; Bischof, P., Gleiter, R., Heilbronner, E., Hornung, V., and Schröder, G.: 1970, *Helv. Chim. Acta* 53, 1645; Haselbach, E., Heilbronner, E., and Schröder, G.: 1971, *Helv. Chim. Acta* 54, 153; Goldstein, M. J., Natowsky, S., Heilbronner, E., and Hornung, V.: 1973, *Helv. Chim. Acta* 56, 294.

[34] Batich, C., Bischof, P., and Heilbronner, E.: 1972–73, *J. Electron Spectrosc.* 1, 333.

[35] Batich, C., Heilbronner, E., Hornung, V., Ashe III, A. J., Clark, D. T., Cobley, U. T., Kilcast, D., and Scanlan, I.: 1973, *J. Am. Chem. Soc.* **95**, 928.

[36] Moore(-Sitterley), C. E.: *Nat. Bur. Stand. (US), Circ. 467* **1** (1949), **2** (1952), **3** (1958).

[37] Åsbrink, L., Lindholm, E., and Edqvist, O.: 1970, *Chem. Phys. Letters* **5**, 609.

[38] Roothaan, C. C. J.: 1951, *Rev. Mod. Phys.* **23**, 69; Pople, J. A.: 1955, *Proc. Phys. Soc.* **A68**, 81; cf. Murrell, J. N.: 1963, *The Theory of the Electronic Spectra of Organic Molecules*, Methuen, London.

[39] Haselbach, E. and Schmelzer, A.: 1971, *Helv. Chim. Acta* **54**, 1575.

[40] Semmelhack, M. F., Foos, J. S., and Katz, S.: 1972, *J. Am. Chem. Soc.* **94**, 8637.

[41] Simmons, H. E. and Fukunaga, T.: 1967, *J. Am. Chem. Soc.* **89**, 5208; Hoffmann, R., Imamura, A., and Zeiss, G.: 1967, *J. Am. Chem. Soc.* **89**, 215.

[42] Batich, C., Heilbronner, E., and Semmelhack, M. F.: 1973, *Helv. Chim. Acta* **56**, 2110.

[43] Haselbach, E. and Schmelzer, A.: 1971, *Helv. Chim. Acta* **54**, 1575.

[44] Haselbach, E. and Schmelzer, A.: 1972, *Helv. Chim. Acta* **55**, 1745, 3130.

[45] Haselbach, E., Lanyiova, Z., and Rossi, M.: 1973, *Helv. Chim. Acta* **56**, 2889.

[46] Robinson, G. W. and El-Sayed, M. A.: 1961, *Mol. Physics* **4**, 273; Hochstrasser, R. and Marzzacco, C.: 1968, *J. Chem. Phys.* **49**, 971; Li, Y. H. and Lim, E. C.: 1971, *Chem. Phys. Letters* **9**, 514; Marzzacco, C. J. and Zalewski, E. F.: 1972, *J. Mol. Spectrosc.* **43**, 239; Innes, K. K., Kalantar, A. H., Khan, A. Y., and Durnick, T. J.: 1972, *J. Mol. Spectrosc.* **43**, 477; Moomaw, W. R., Decamp, M. R., and Podore, P. C.: 1972, *Chem. Phys. Letters* **14**, 255.

[47] Hoffmann, R., Imamura, A., and Hehre, W. J.: 1968, *J. Am. Chem. Soc.* **90**, 1499; Hoffmann, R.: 1971, *Accounts Chem. Res.* **4**, 1.

[48] Gleiter, R., Heilbronner, E., and Hornung, V.: 1972, *Helv. Chim. Acta* **55**, 255.

PART V

EFFECTS OF ENVIRONMENT ON THE
BEHAVIOUR OF MOLECULES

Chairman: Alberte Pullman

A working session.

INTRODUCTORY REMARKS

ALBERTE PULLMAN

*Institut de Biologie Physico-Chimique, Laboratoire de Biochimie Théorique
associé au C.N.R.S., 13, rue P. et M. Curie, Paris 5*

Even without any statistical search it may safely be stated that over 95 per cent of the past work in quantum chemistry has been devoted either to general theory or to the study of the isolated molecules. On the other hand, unless we consider the outer space, no molecule exists in isolation. Clearly this is a perplexing situation, the awareness of which has been spreading rapidly in the recent years. Indeed, most of us nowadays strongly feel that a serious effort should be undertaken for evaluating the effect of the environment so as to assess and possibly improve the significance of the majority of the quantum-chemical results for the description of the real physical and biological world. It is true that quantum chemistry, working in the isolated molecule approximation, has met with a very large number of remarkable successes over the years but then the reasons for this success itself become the problem. In fact, are not the very achievements of quantum chemistry more surprising than its failures?

I would like to stress the fact that the problem is in no way specific for the theoreticians but that it also faces the experimentalists in which case, truly, it is not merely the problem of the isolated versus medium-surrounded molecule but rather the problem of the differential effects of different surroundings. In important cases it may have an acute aspect: an outstanding example is the long-lasting controversy about the significance of X-ray crystal structure analysis of biomolecules and biopolymers for the understanding of their biological behavior which refers of course, to solution. As stated by Phillips (1970):

Crystals, for all their beauty, seem essentially dead and the least likely objects of study by biochemists, yet they have proved most important to the developments of biochemistry on more than one occasion.

Again the question arises, of course, about the reasons of this lucky situation.

But the best known aspect of the problem probably concerns the differential influence of different solvents on molecular structure, properties and modes of interaction. A considerable mass of information is accumulating rapidly in this field owing to the remarkable developments of various experimental techniques in the past few years. A classical example in this area taken in the field of fundamental molecular biology is the observation that the bases of the nucleic acids and their analogs when dissolved in nonpolar solvents form coplanar hydrogen-bonded pairs similar to those occurring in DNA, but that the same compounds dissolved in water form vertical stacks instead, with no detectable in-plane binding. (For a review, see Ts'o, 1970.)

Another spectacular instance is provided by the fascinating set of observations on differential solvent effects which have been accumulated in the last years by combining the use of various spectroscopic techniques (IR, NMR, ORD, CD) on the structure of

R. Daudel and B. Pullman (eds.), The World of Quantum Chemistry, 239–251. All Rights Reserved
Copyright © 1974 by D. Reidel Publishing Company, Dordrecht-Holland

a series of cyclic peptide and depsipeptide antibiotics which share the property to increase selectively the alkali-ion permeability of artificial and biological membranes. (For a recent review, see Ovchinnikov *et al.*, 1972.) For example, valinomycin which is a 36-membered cyclic compound made of alternating peptide and ester residues (Figure 1), has been shown to assume in nonpolar solvents a bracelet-like configuration where all the six available NH groups are hydrogen-bonded to the carbonyl oxygen of the next peptide unit (I in Figure 2) forming a relatively rigid framework of fixed ten-membered rings as schematized in the lower part of the figure. In solvents of medium polarity three of these hydrogen bonds are disrupted as in structure II, Figure 3 (development in Figure 4), whereas a further increase in the medium polarity yields a mixture of type-III structures where "all the NH group, particularly at high temperature form hydrogen bonds to the solvent" (Ovchinnikov *et al.*). On the other hand, the crystal structure analysis of the potassium-ion complex of the same molecule indicates (Pinkerton *et al.*, 1969) a bracelet-like configuration with six free carbonyls turned inwards and engulfing the ion in its center; while the crystal structure of the uncomplexed valimycin very recently obtained (Duax *et al.*, 1972) is claimed

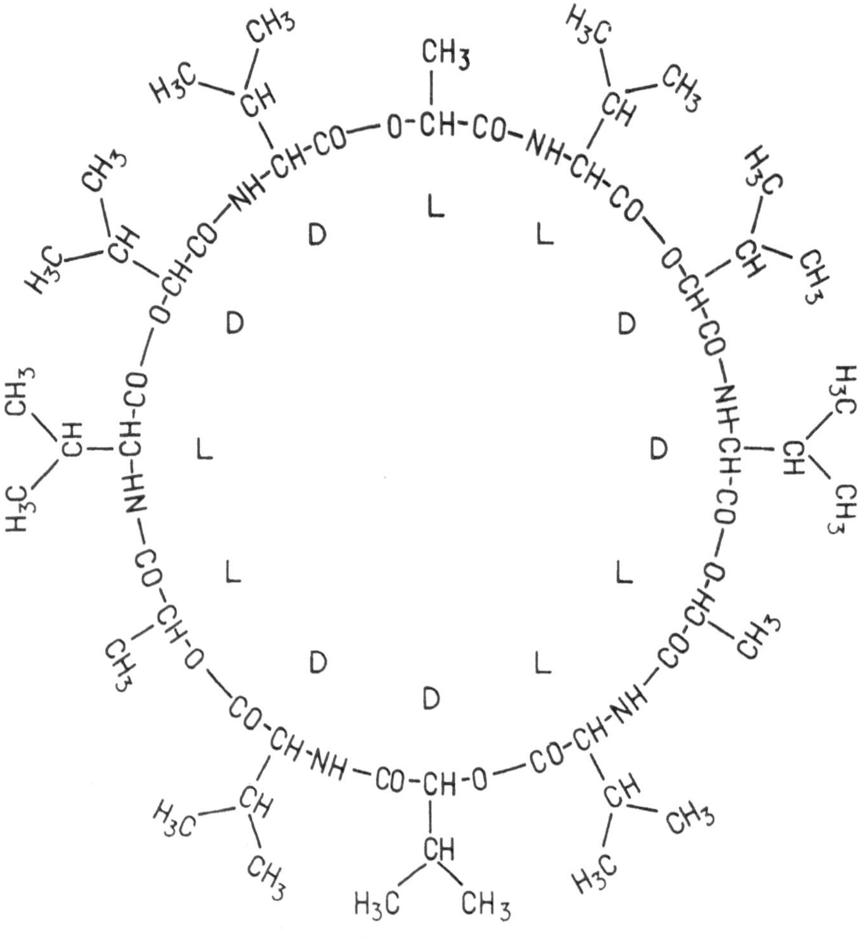

Fig. 1. The valinomycine molecule.

I

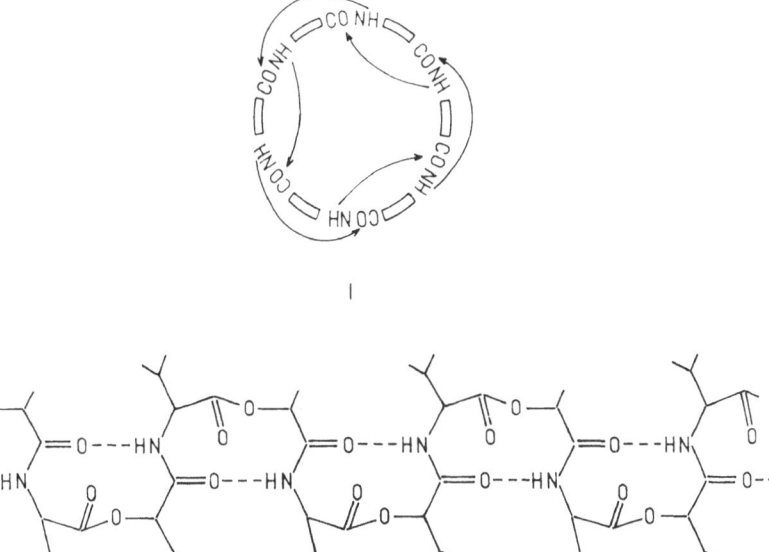

Fig. 2. Valinomycine in nonpolar solvents.

II III

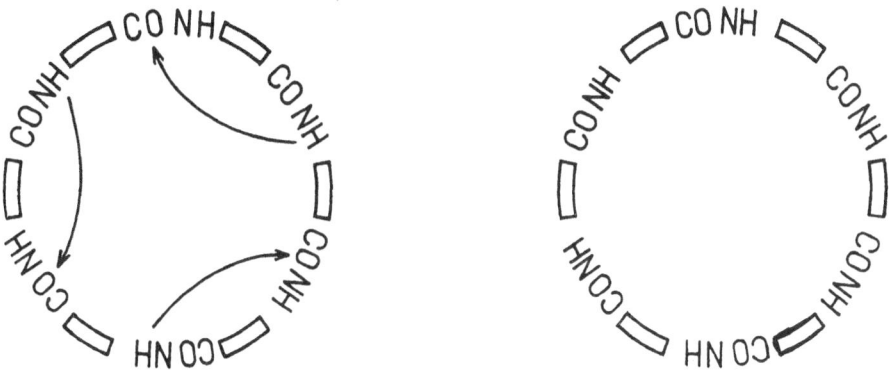

Fig. 3. Valinomycine in medium-polar (II) and highly polar solvents (III).

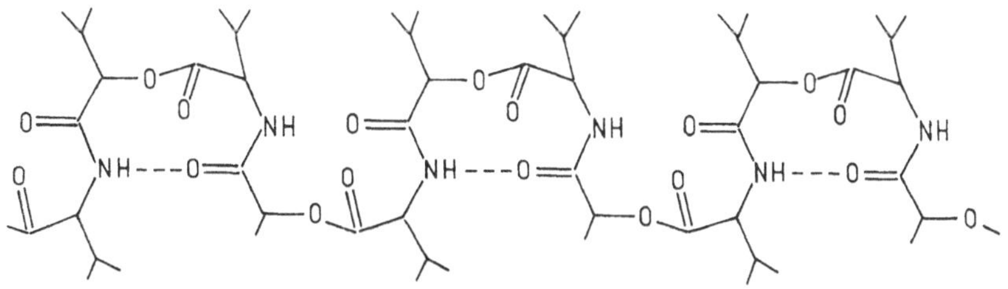

Fig. 4. The breaking of H-bonds in medium-polar solvents.

to be of a lower symmetry, with only four 10-membered hydrogen-bonded rings, the remaining two NH groups apparently forming two 3-membered rings by hydrogen-bonding to two ester carbonyls (see Figure 5). I only show you the pieces of the puzzle which have already been fitted into place but there are many more, the fitting-in of which could be highly speeded up if we had solid answers to questions like: what

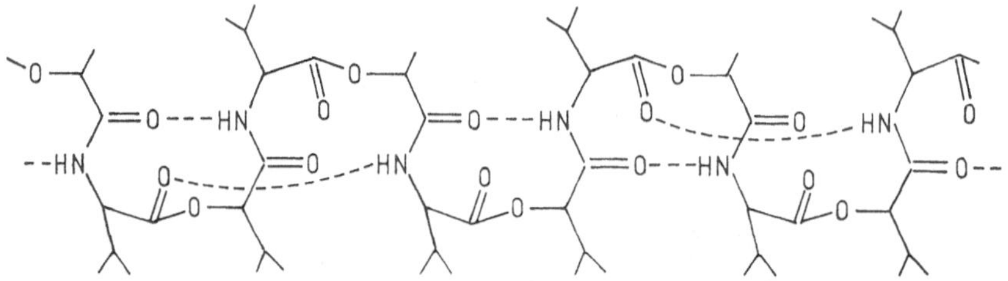

Fig. 5.   Proposed H-bonding in free valinomycin crystal.

does water precisely do to a depsipeptide molecule which induces the subtle changes observed? For instance, what kind of hydrogen bond is more likely to be disrupted? Why is the complexing with cations not observed in water? Does the solvation shell simply protect the cation? Or does the polar solvent destabilize the complex? Or is it that the form of the molecule which exists in highly polar solvents precludes the entrance or the binding of the ion? What are the nature and the strength of the cation binding?

Clearly, all these question-marks challenge us theoreticians to undertake a serious attack on environmental effects: the time has come when referring loosely to dipole–dipole interaction, hydrophobic forces, charge-transfer effects and the like, without the backing of a solid computation on a relevant case is dramatically insufficient.

Can we do more today? I think that we do because, in fact, the methodology largely exists and also the computing facilities. Indeed, broadly speaking, there are two principal methodologies for considering and studying environmental effects and I believe that we have among our distinguished speakers today the representatives of both these tendencies. First, the 'traditional' way of dealing with the problem through the use of a 'continuum' model which tries in one way or another to account for the bulk effect of the surrounding medium. Second, the more recent approach in the form of a 'discrete' treatment, in which one tries to establish the individual effect of the medium molecule upon the system studied.

The continuum model has proven in many cases its validity and we have among us to-day eminent specialists, Prof. Buckingham and Prof. Sinanoglu, who have been instrumental in the development of this field of research and will certainly teach us a great deal about their most recent achievements. On the other hand, the global approach may have striking failures linked to its very way of dealing with the problem. Let me give you an example. You remember perhaps from Prof. Pullman's lecture that the most stable forms of dipeptides are folded, assuming a characteristic seven-membered rings (called C_7) stabilized by a hydrogen bond (Figure 6). A group of Russian workers who use empirical partitioned potential functions (Lipkind *et al.*,

1970), and not quantum mechanics, have attempted to introduce the effect of the solvent essentially by varying the dielectric constant in the expression of the electrostatic component of the interaction energy: the conformational map corresponding to the nonpolar situation in the alanyl dipeptide is reproduced in Figure 7a, where

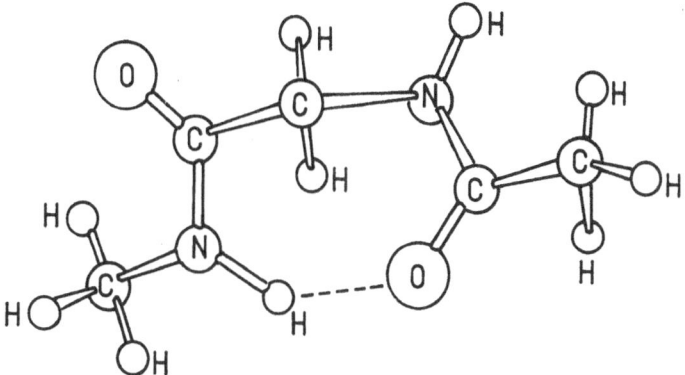

Fig. 6.   The C_7-form of a dipeptide.

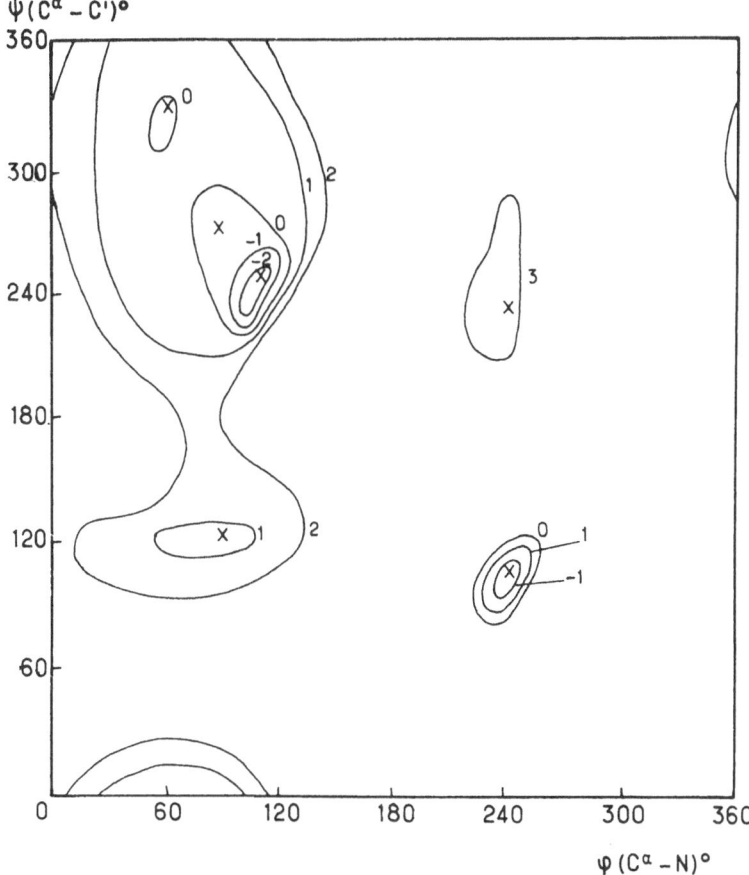

Fig. 7a.   Conformational map for the alanyl dipeptide in nonpolar solvents.

two deep minima appear for the two possible heptacycles corresponding to the axial and equatorial position of the methyl group on the α carbon. The corresponding map recomputed for a polar situation is shown in Figure 7b: the two C_7 minima have disappeared and a widening of the other previously secondary minimal regions has occurred. According to these results the C_7 structures should be destabilized by water. This conclusion is in conformity to the commonly accepted view on the destabilizing effect of polar solvents on hydrogen-bonded structures. However a series of recent experimental studies (Avignon, 1972) of dipeptides in various solutions by different spectroscopic techniques indicates that the C_7-form not only persists in water solution but in fact is stabilized by a molecule of water forming a bridge as shown in Figure 8 between the free carbonyl and NH groups. This could of course not be obtained in the computations of Lipkind *et al.* and shows the possible pitfalls of a global approach and the necessity of investigating more closely the interactions between the solute and the individual solvent molecules. The discrete model on the contrary can teach us a lot on the nature of the interactions at the local level and on the different possibilities of their occurrence and could possibly avoid such a failure.

I believe that a most exciting future for the study of environmental effects on molecular structure and properties resides in the 'discrete' approach, in particular in

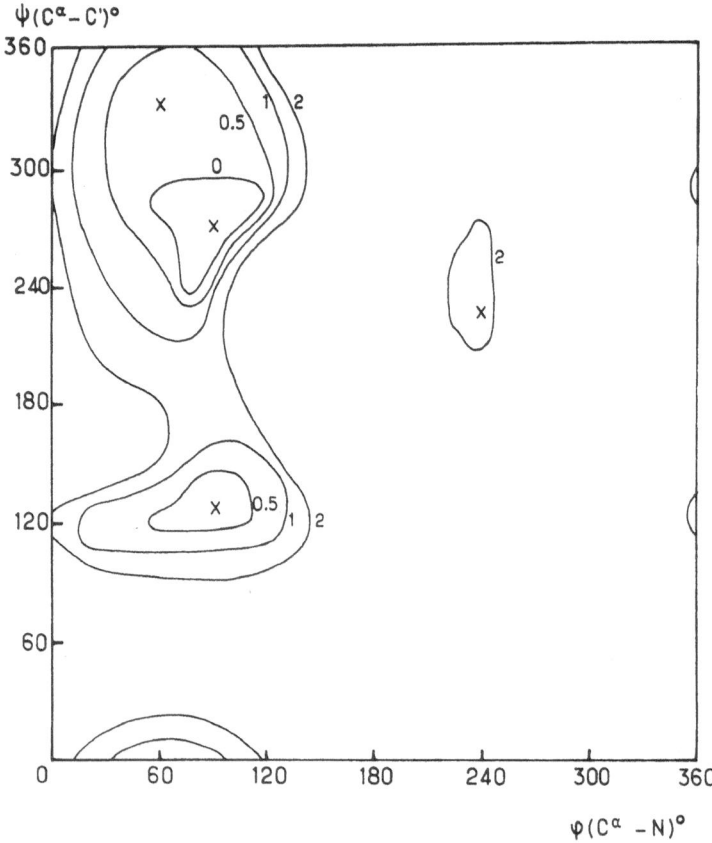

Fig. 7b.   Conformational map for the alanyl dipeptide in water.

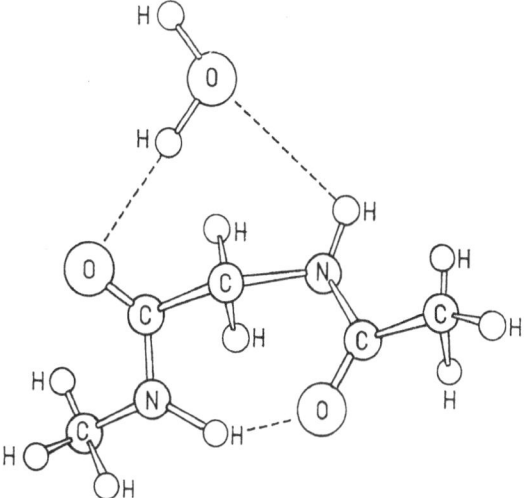

Fig. 8.   Stabilization of the C₇-form by water.

the 'supermolecule' discrete ·approach in which one treats the central system under study and the nearest-neighbour medium molecules as one supermolecular system dealt correctly with by the existing methods. Of course, if the principle of this approach is simple, the practical difficulties of such a treatment are enormous: the dimensions of the systems, the relative positioning of the molecules, the limits of the active surrounding, the modality of the displacements etc., without mentioning the theoretical inconveniences inherent to the supermolecule approach itself. Certainly we are far away yet from a coherent, well-defined technique. But the attempts which are carried out in this direction are extremely promising. For instance, the supermolecule approach has already led to a remarkable progress in the understanding of the hydrogen bond and of the effect of hydrogen bonding (one among the most important environmental effects) on molecular properties. Prof. Morokuma, who has himself been a pioneer in this field, will certainly tell us more about it here.

A recent development along these lines has led to the determination of the principal hydration sites of a number of fundamental molecules, among which formamide, model of the peptide bond, (Figure 9) (Alagona *et al.* 1973) and the nucleic acid bases (Figure 10) (Port and Pullman, 1973). This opens the possibility of studying the effect of the solvent (water) on the properties of such molecules and on their interactions. In fact, an explicit study of the effect of water binding on the conformational properties of an important biological molecule has been carried out recently in our laboratory with most fruitful results. The molecule is histamine (Figure 11), which, according to the isolated-molecule approach, should exist in a folded form (the side chain folded towards the ring) in its monocation but in the fully extended form in its dication. While this molecule exists probably in these different forms in its corresponding crystals, it exists as a mixture of both the extended and the folded forms in solution for both mono- and di-cation. After a preliminary determination of the principal hydration sites for the first hydration shell of the two cations, the construction of the conformational energy maps for the hydrated molecules accounts perfectly for this situation (Pullman and Port, 1973). This is, I believe, the first case in which the explicit introduction of the solvent at the discrete level of study, has led to such precise results and it certainly calls for further attemps.

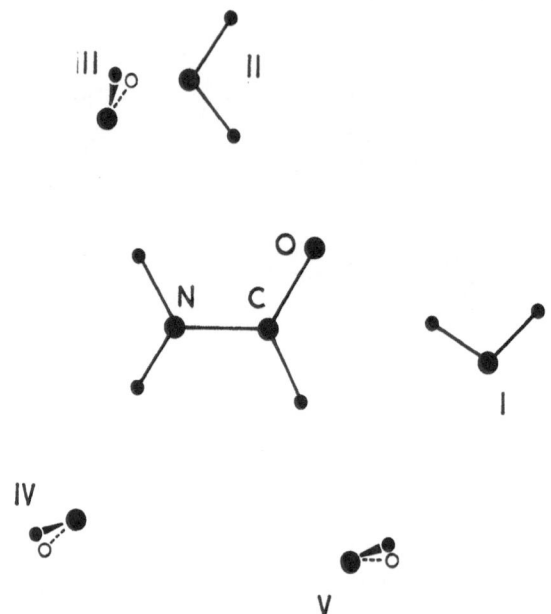

Fig. 9.   Hydration sites in formamide.

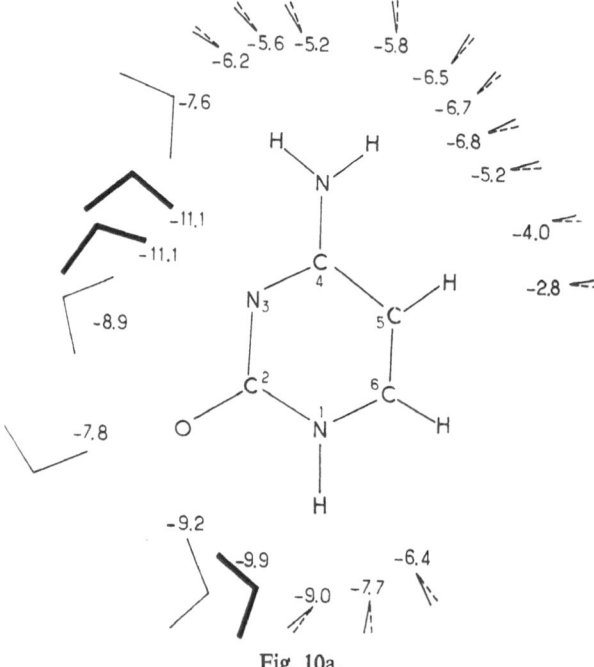

Fig. 10a.

Fig. 10.   Hydration sites in the nucleic bases; (a) cytosine, (b) thymine, (c) adenine, (d) guanine.

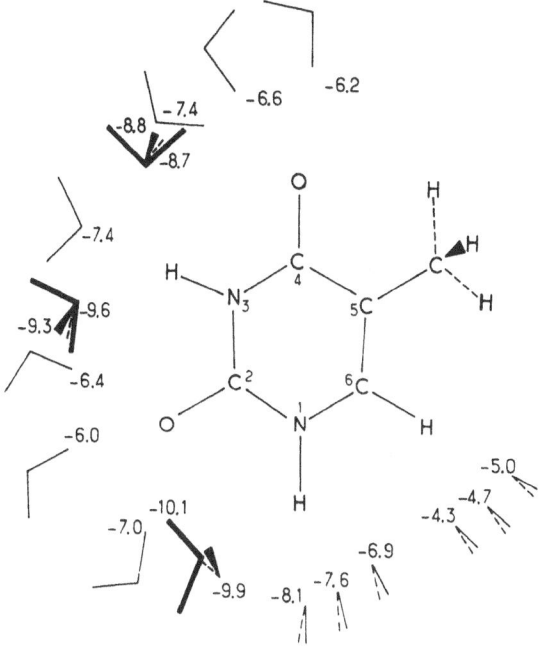

Fig. 10b.

(c)   (c)

Fig. 10c.

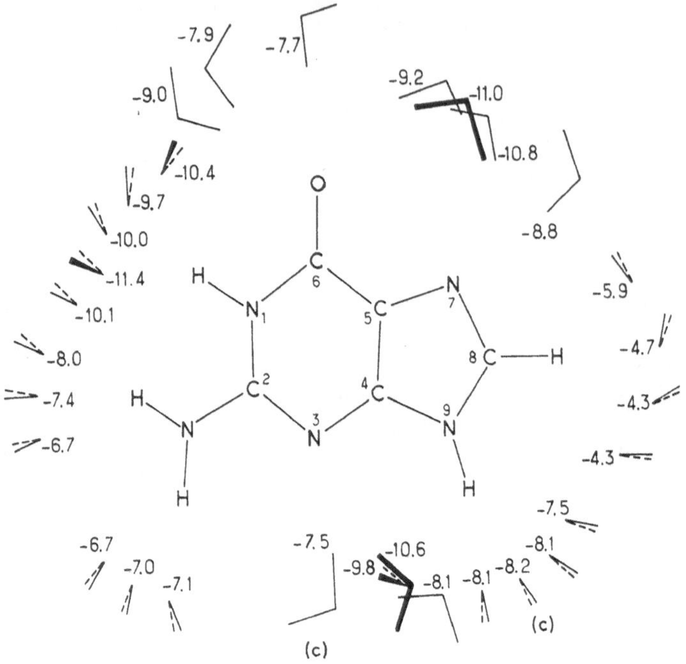

Fig. 10d.

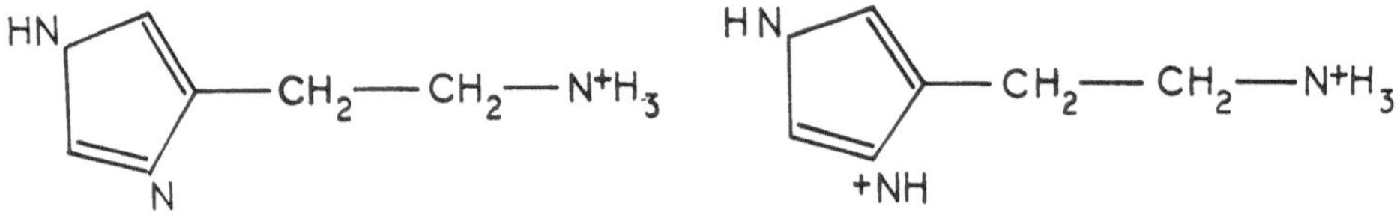

Fig. 11.   Mono- and di-cation of histamine.

Let me finish this introduction by recalling one case of an environmental effect which, I believe, is particularly extraordinary and exciting. It occurs in the field of spectroscopy in which many environmental studies have been carried out, but in which this one, although the most outstanding, somewhat escaped the attention of quantum-theoreticians. It concerns the 'bathochromic shift' in the visual pigments (Mantione and Pullman, 1971). The problem is the following. While retinal, the chromophore of the visual pigments absorbs around 380 nm and the protein opsin around 280 nm; rhodopsin, the combination of the two, absorbs at much longer wavelength, generally between 480 and 560 nm, depending on the animal species involved. This large bathochromic shift, so drastically responsible for our perception of the outside world, was considered to be one of the most important unexplained problems in the physical chemistry of the visual pigments.

In fact, a part of this shift is easily accounted for. Thus, it has been shown that the chromophore is most probably bound to the protein in the form of a *protonated Schiff base* which absorbs around 440 nm. The bathochromic shift, unaccounted for, attributable to residual interactions between this protonated base and the apoprotein would thus have to be measured from this limit. It still remains an unusually large shift of about 40–120 nm.

Numerous theories have been proposed in order to explain the origin of this spectral displacement. (For a review see Mantione and Pullman, 1971.) The most plausible among them is the *point-charge perturbation theory*, which postulates that the residual bathochromic shift is brought about by interaction of the protonated Schiff base of 11-*cis* retinal with some reactive groups on opsin placed in its vicinity.

In this particular case, the environment being relatively undefined, no supermolecular model could be built. The problem was treated in our laboratory in a simplified way. The procedure utilized consists of the evaluation of the differential stabilization of the ground state and of the first excited state of the protonated retinylidene iminium ion through interactions with an external point charge located at different positions along the periphery of the cationic chromophore. Both the electrostatic and the polarization effects are being considered. The procedure thus involves three steps:

(a) evaluation of the charge distribution in the ground and in the first excited state of the 11-*cis* retinylidene iminium ion (I).

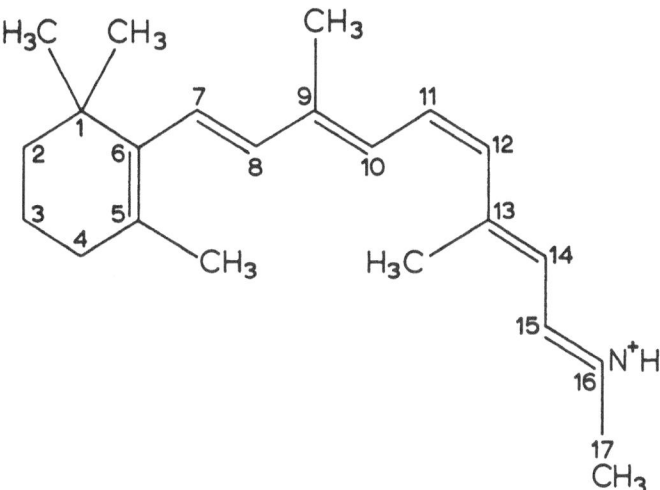

I. 11 — *cis* retinylidene iminium ion

(b) Calculation of the electrostatic interaction energy between these charge distributions in the cationic chromophore and an ion $X^{\delta-}$, placed at different positions around its periphery or in its vicinity.

(c) Calculation of the increment of interaction energy due to the polarization of retinylidene iminium ion by the $X^{\delta-}$ ion.

The results of the overall treatment are represented schematically in Figure 12 which indicates the difference in the electrostatic stabilization energy between the ground and the excited states. (The contribution of the polarization effect is negligible.) It is observed that this difference is in favor of the ground state when $X^{\delta-}$ is in the vicinity of the N^+ terminal of the chromophore (between C_{13} and N_{16}), and in favor

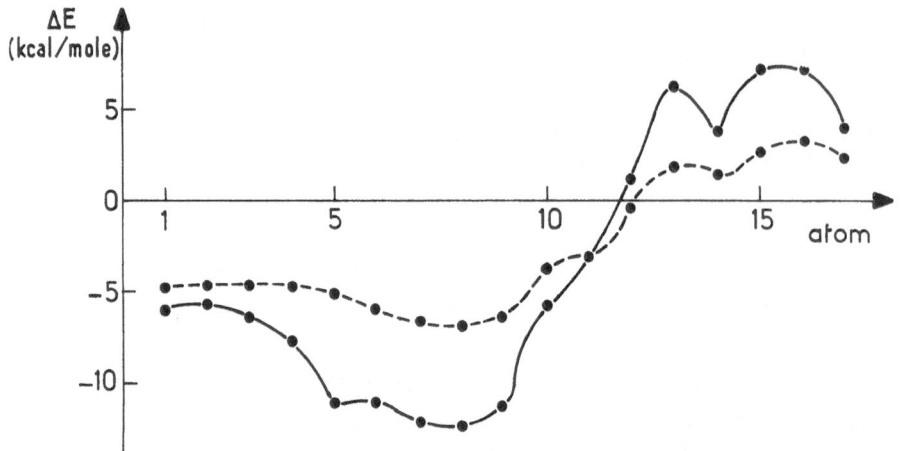

Fig. 12. Differential stabilization of the ground and excited state of the retinilydene iminium ion by a negative charge facing the vaɪious atoms of the chain (full line: close contact; dashed line: loose contact).

of the excited state for positions of $X^{\delta-}$ between C_1 and C_{12}, in particular for position of $X^{\delta-}$ in the vicinity of the region C_5–C_9. Thus, a bathochromic shift of the spectrum of retinylidene iminium ion may be expected to occur, as a result of the electrostatic factor, when the $X^{\delta-}$ group of opsin is located in the vicinity of these atoms. A quantitative evaluation shows that the appropriate magnitude of the shift is obtained provided that reasonable assumptions are adopted for the distance between an appropriate charge on the opsin and the chromophore.

No one can deny that the studies on molecules in the 'isolated system' approximation have been fruitful and useful. But nobody can deny either the necessity of going beyond this 'proud tower' attitude into the study of the environmental effects. Those may be of more or less importance but their investigation represents a desirable and to some extent inevitable extension of the work of the theoretician who wishes to maintain contact with physical reality and with... experimentalists.

References

Alagona, G., Pullman, A., Scrocco, E., and Tomasi, J.: 1973, *Int. J. Peptide Protein Res.* **5**, 251–59.
Avignon, M.: 1972, Thèse de doctorat d'État, Université de Bordeaux.
Dobler, M., Dunitz, J. D., and Krajewski, J.: 1969, *J. Mol. Biol.* **42**, 603–06.
Duax, W. L., Hauptman, H., Weeks, C. M., Worton, D. A.: 1972, *Science* **176**, 911–14.
Lipkind, G. M., Arkhipova, C. F., and Popov, E. M.: 1970, *Strukt. Khim.* **11**, 121.
Mantione, M. J. and Pullman, B.: 1971, *Int. J. Quant. Chem.* **5**, 349.
Ovchinnikov, Yu. A., Ivanov, V. T., Evstratov, A. V., Bystrov, V. F., Abdullaev, N. D., Popov, E. M., Lipkind, G. M., Arkhipova, S. F., Efremov, E. S., and Shemyakin, M. M.: 1969, *Biochem. Biophys. Res. Comm.* **37**, 668.
Ovchinnikov, Yu. A., Ivanov, V. T., and Shkrob, A. M.: 1972, in *Molecular Mechanisms of Antibiotic Action on Protein Biosynthesis and Membranes*, Proceedings of a Symposium in Granada (ed. by E. Muñoz, F. García-Ferrandiz, and E. Vazquez), Elsevier, Amsterdam.
Phillips, D. C.: 1970, *Biochem. Soc. Symp.* **30**, 11.

Pinkerton, M., Steinrauf, L. K., and (in part) Dawkins, P.: 1969, *Biochem. Biophys. Res. Comm.* **35**, 512–18.

Port, G. N. J. and Pullman, A.: 1973, *FEBS Letters* **31**, 70.

Pullman, B. and Port, G. N. J.: 1973, *Molecular Pharmacology*, in press.

Ts'o, P. O. P.: 1970, in *Fine Structure of Proteins and Nucleic Acid*, Vol. 4 of *Biological Macromolecules Series* (ed. by G. D. Fasman and S. N. Timasheff), M. Dekker, New York.

INTRODUCTION TO ...

...

INTERMOLECULAR FORCES AND THE ELECTRIC
AND MAGNETIC PROPERTIES OF MOLECULES

A. D. BUCKINGHAM

Dept. of Theoretical Chemistry, University of Cambridge, Cambridge, England

Contents

1. Introduction

Molecules attract one another when they are far apart – since liquids and solids exist – and repel one another when close together – since densities are finite. This well-known truth is one that could be revealed with pleasure to kind aunts and others; it is illustrated in Figure 1a. The study of the origin and details of potential energy surfaces is relevant to many branches of science, including physics, chemistry, crystallography, molecular biology, polymer science and even astronomy. In addition to the *forces* between molecules, that is, the change of energy with position, there are changes in *properties* such as dipole moment, polarizability, nuclear magnetic shielding, hyperfine constants, etc. It is with some of these changes in properties when molecules interact that this paper is concerned (Figure 1b).

2. Polarizability

The polarizability of a molecule determines the response of the charge distribution to an external field $\mathbf{E}$. The dipole polarizability α is defined in terms of the induced dipole moment μ:

$$\mu = \alpha \cdot \mathbf{E}. \tag{1}$$

The mean polarizability of a molecule in a bulk sample is related to the refractive index n by the Lorentz–Lorenz equation

$$\frac{n^2 - 1}{n^2 + 2} \varrho^{-1} = \frac{N\alpha}{3\varepsilon_0}, \tag{2}$$

R. Daudel and B. Pullman (eds.), The World of Quantum Chemistry, 253–264. All Rights Reserved
Copyright © 1974 by D. Reidel Publishing Company, Dordrecht-Holland

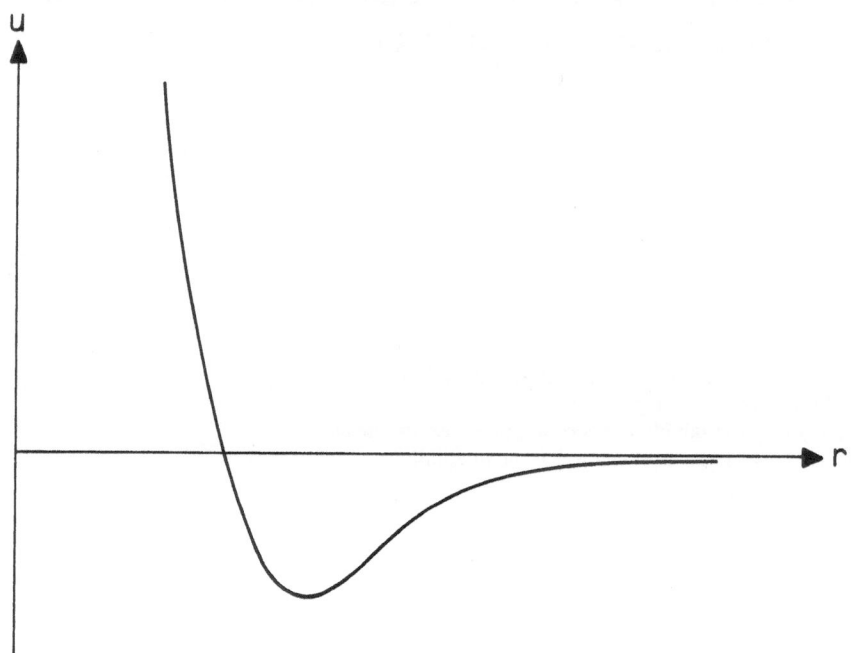

Fig. 1a. Typical dependence of the interaction energy u on separation r.

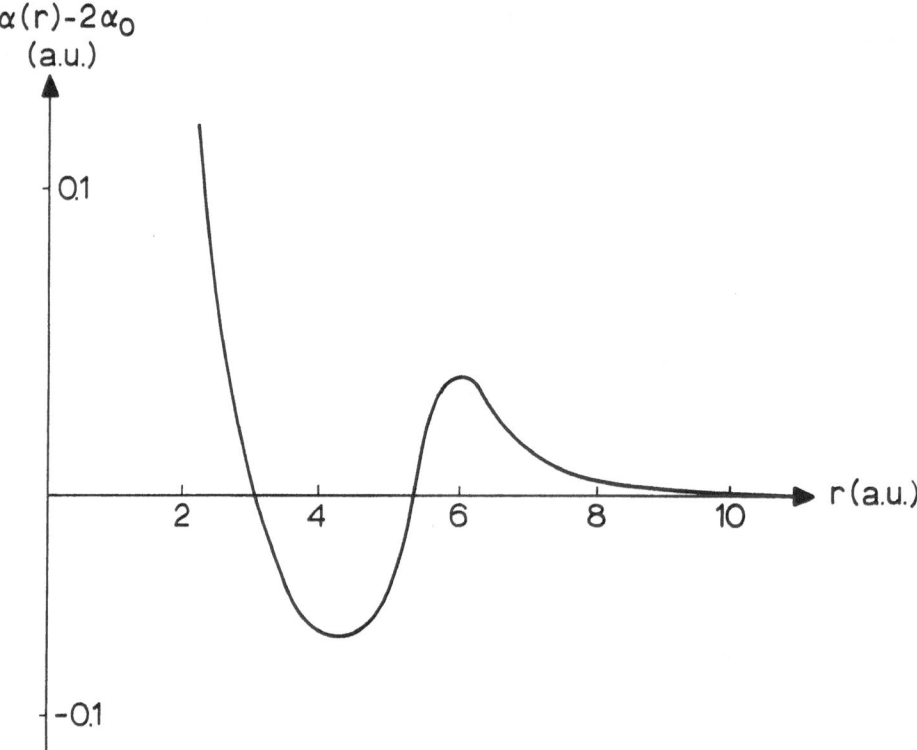

Fig. 1b. Schematic representation of the change in polarizability of two helium atoms as a function of the internuclear distance r. The accurate value of α_0 is 1.38319 a.u.

where ε_0 is the permittivity of free space ($4\pi\varepsilon_0 = 1$ e.s.u.). If the density ϱ is in moles per unit volume, the left-hand side of Equation (2) is the molar refraction R_m and N is Avogadro's number. The polarizability is a function of temperature T, density ϱ and the fequency v of the electromagnetic wave used to measure n. For a gas at low density $n \approx 1$ and (2) reduces to

$$\tfrac{2}{3}(n-1)\,\varrho^{-1} = N\alpha_0/3\varepsilon_0, \tag{3}$$

where α_0 is the polarizability of an isolated molecule. At higher densites α may differ from α_0 because of molecular interactions, and for imperfect gases it is convenient to expand the molar refraction as a power series in ϱ (Buckingham and Pople, 1956)

$$R_m = \frac{n^2 - 1}{n^2 + 2}\,\varrho^{-1} = A_R + B_R\varrho + C_R\varrho^2 + \cdots. \tag{4}$$

The coefficients $A_R, B_R, C_R, \ldots$ are the first, second, third, ... *refractivity virial coefficients* and are functions of T and v but not of ϱ. Thus

$$A_R = [R_m]_{\varrho=0} = N\alpha_0/3\varepsilon_0. \tag{5}$$

B_R is determined by the change in polarizability resulting from pairwise interactions and, using classical statistical mechanics,

$$B_R = [(R_m - A_R)\,\varrho^{-1}]_{\varrho=0} = \frac{N}{3\varepsilon_0}\,[\langle\alpha - \alpha_0\rangle\,\varrho^{-1}]_{\varrho=0} =$$

$$= \frac{N^2}{3\varepsilon_0}\int [\tfrac{1}{2}\alpha_{12}(\tau) - \alpha_0]\exp[-u_{12}(\tau)/kT]\,\mathrm{d}\tau, \tag{6}$$

where $\alpha_{12}(\tau)$ is the polarizability and $u_{12}(\tau)$ the potential energy of the pair of molecules 1, 2 when in the relative configuration τ; the integration extends over a large spherical region. For a monatomic gas the relative configuration τ can be expressed in terms of the internuclear distance r alone, and

$$B_R = \frac{4\pi N^2}{3\varepsilon_0}\int_0^\infty [\tfrac{1}{2}\alpha_{12}(r) - \alpha_0]\exp[-u_{12}(r)/kT]\,r^2\,\mathrm{d}r. \tag{7}$$

Minor quantum corrections to Equation (7) have been evaluated for He and Ne (Ely and McQuarrie, 1971), using a partition-function approach due to Kirkwood (1933). If $u_{12}(r)$ is known, measurements of B_R over a range of T provide information about the variation of α_{12} with r.

There are four more-or-less distinct contributions, A, B, C and D, to $\tfrac{1}{2}\alpha_{12}(\tau) - \alpha_0$.

A. CLASSICAL DIPOLAR INTERACTIONS

The classical dipole interaction (Silberstein, 1917) leads to an extra field at one molcule due to the dipole of the other, and for two spheres the model gives, for the change in polarizability of the pair,

$$\alpha_{12}(r) - 2\alpha_0 = \tfrac{2}{3}\alpha_0 \left[\left(1 - \frac{2\alpha_0 r^{-3}}{4\pi\varepsilon_0} \right)^{-1} + 2 \left(1 + \frac{\alpha_0 r^{-3}}{4\pi\varepsilon_0} \right)^{-1} - 3 \right]. \tag{8}$$

This expression diverges at $r = [2\alpha_0 (4\pi\varepsilon_0)^{-1}]^{1/3}$ but this is not serious since it occurs at a separation (1.404 a.u. for He$_2$) at which there is an extensive overlap where the model is inappropriate. Equation (8) may be expanded as a power series in r^{-3}:

$$\alpha_{12}(r) - 2\alpha_0 = 4\alpha_0^3 r^{-6} (4\pi\varepsilon_0)^{-2} + 4\alpha_0^4 r^{-9} (4\pi\varepsilon_0)^{-3} + \dots. \tag{9}$$

When allowance is made for the field of the atomic quadrupole CE' induced by the field gradient $\mathbf{E}'$ at a nucleus an additional term $20\alpha_0^2 Cr^{-8}(4\pi\varepsilon_0)^{-2}$ is added to (9) (Buckingham et al., 1973). The classical interaction has been much used (Kirkwood, 1936; Yvon, 1936; Brown, 1956); it produces a positive contribution to B_R but is inaccurate and does not even give the correct behaviour at long range (see B below).

B. DISTORTION DUE TO DISPERSION-TYPE INTERACTIONS

The intrinsic molecular polarizability changes as a result of dispersion-type interactions at long range (Jansen and Mazur, 1955; Buckingham, 1956; Certain and Fortune, 1971; Buckingham et al., 1973). The polarizability change is

$$\alpha_{12}(r) - 2\alpha_0 = A_6 r^{-6} + A_8 r^{-8} + A_9 r^{-9} + \dots. \tag{10}$$

For He$_2$ $A_6 = 38.98$ a.u. $= 14.86\ \alpha_0^3 (4\pi\varepsilon_0)^{-2}$ (Certain and Fortune, 1971), and for two H atoms, $A_6 = 1698.37$ a.u. $= 18.6378\ \alpha_0{}^3 (4\pi\varepsilon_0)^{-2}$, $A_8 = 44886.7$ a.u. $= 295.550$ $\alpha_0^2 C (4\pi\varepsilon_0)^{-2}$ (Buckingham et al., 1973). In comparing these results with the clasical values (see A above) of $4\alpha_0^3 (4\pi\varepsilon_0)^{-2}$ and $20\alpha_0^2 C (4\pi\varepsilon_0)^{-2}$, the deficiencies of the simple model are obvious.

C. SHORT-RANGE 'OVERLAP' EFFECTS

The effects of electron overlap and exchange interactions at short range lead to repulsive intermolecular forces and to changes in α_{12} (Michels et al., 1937; ten Seldam and de Groot, 1952; Lim et al., 1970; O'Brien et al., 1973; Buckingham and Watts, 1973). This effect reduces α_{12} and therefore contributes negatively to B_R. By means of SCF calculations on He$_2$ in the presence of various finite electric fields $\big(0.05, 0.075$ and 0.10 a.u., 1 a.u. $= ea^{-2}(4\pi\varepsilon_0)^{-1} = 17.15 \times 10^6$ e.s.u. $= 0.5142 \times 10^{12}$ V m$^{-1}\big)$, it has been shown that $\alpha_{12}(r) - 2\alpha_0$ changes from positive values at large r to negative values at intermediate r (Buckingham and Watts, 1973). The basis set comprised 16 functions on a total of 14 centres – two 1s Slater functions at each nucleus and six spherical Gaussians arranged octahedrally around each nucleus at a distance of 0.2 a.u., four being on the internuclear axis. In a similar calculation, O'Brien et al. (1973) used a basis set comprising five s, four p and one d Gaussian orbitals on each nucleus, and although the total energy is higher, the polarizability is more accurate and actually leads to a negative B_R at 300 K. The calculated polarizability components $\alpha_\parallel (r)$ and $\alpha_\perp (r)$ together with $\alpha(r) = \tfrac{1}{3}[\alpha_\parallel (r) + 2\alpha_\perp (r)]$ are shown in Figure 2. The qualitative

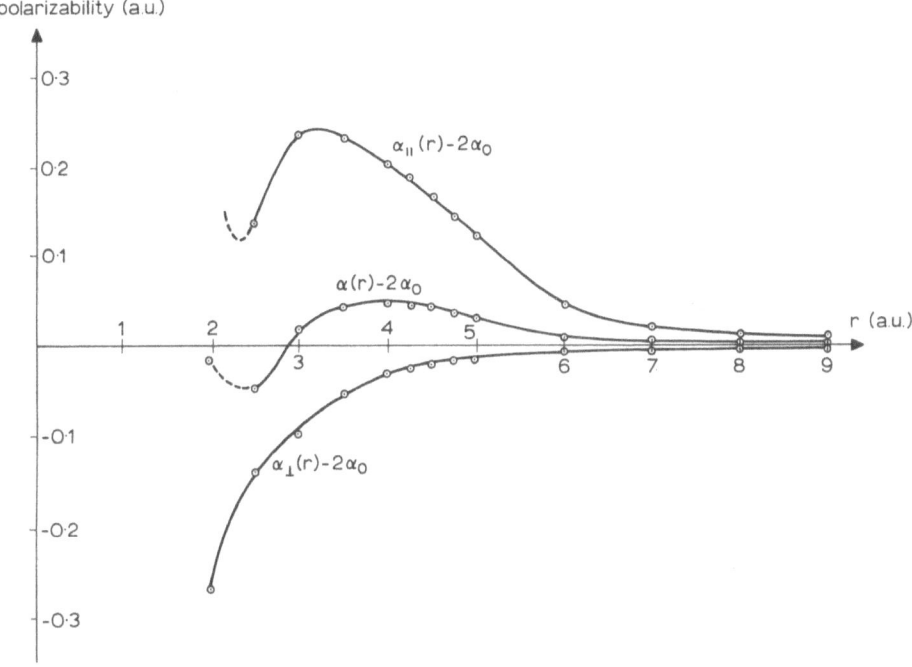

Fig. 2.   The change in the polarizabilities of He₂ with internuclear distance r, as calculated using a 16-function basis set.

features of the schematic curve in Figure 1b are reproduced, but the results are not definitive since the basis set is limited and because of electron-correlation effects excluded by the SCF technique. The physical basis for the reduction in $\alpha(r)$ due to overlap and exchange might be attributed to the increased effective charge experienced by the electrons when the nuclei are close together, producing a contraction of the electron cloud and a consequent reduction in polarizability.

At very small separations, the polarizability of He₂ rises steeply to reach the value for the beryllium atom in its ground $1s^2 2s^2$ state which is approximately 16 times $2\alpha_0$ (Langhoff *et al.*, 1966).

Calculations on pairs of larger atoms would be of interest but none has been reported.

D. HYPERPOLARIZABILITY AND STRONG INTERMOLECULAR FIELDS

The strong fields arising from the permanent multipole moments of a molecule lead to nonlinear polarization of a neighbouring molecule by the external field. These effects may be related to the higher polarizabilities of an isolated molecule. Thus if a single molecule is in a strong field **E** its dipole moment may be written as (Buckingham and Pople, 1955a)

$$\boldsymbol{\mu} = \boldsymbol{\mu}_0 + (4\pi\varepsilon_0)\,\boldsymbol{\alpha}_0 \cdot \mathbf{E} + \tfrac{1}{2}(4\pi\varepsilon_0)^2\,\boldsymbol{\beta} : \mathbf{E}^2 + \tfrac{1}{6}(4\pi\varepsilon_0)^3\boldsymbol{\gamma} \vdots \mathbf{E}^3 + \dots, \qquad (11)$$

where $\boldsymbol{\mu}_0$ is the permanent moment and $\boldsymbol{\beta}$ and $\boldsymbol{\gamma}$ are the first and second hyperpolarizabilities. There may be additional terms involving the field gradient at the molecule. If **E** in (11) is the sum of the external field $\mathbf{E}^{(0)}$ and the field **F** of another molecule, then the effective polarizability is

$$\alpha = \left(\frac{\partial\mu}{\partial\mathbf{E}^{(0)}}\right)_{E^{(0)}=0} = \alpha_0 + 4\pi\varepsilon_0\,\boldsymbol{\beta}\cdot\mathbf{F} + \tfrac{1}{2}(4\pi\varepsilon_0)^2\,\boldsymbol{\gamma}:\mathbf{F}^2 + \cdots. \tag{12}$$

If **F** is the field of a polar molecule, the contribution of $\boldsymbol{\beta}\cdot\mathbf{F}$ to B_R may be considerable (Buckingham, 1956).

E. THE MEASUREMENT OF CHANGES IN POLARIZABILITY

The success (see Van Vleck, 1932; Born and Wolf, 1959) of the Lorentz–Lorenz Equation (2) in yielding polarizabilities α which are nearly independent of density and state implies that α is changed very little by intermolecular forces. Consequently, accurate measurements of these changes have been difficult and are few in number; published values of second refractivity virial coefficients B_R are not very reliable (Graham, 1972; Sutter, 1972). A new *differential* method of determining B_R has been developed in our laboratory; it involves changing the density but not the amount of gas in the optical path (see Figure 3). New values of B_R have been obtained for Ne, Ar, N_2, CO_2, CH_4, SF_6 and CHF_3 at room temperature (Graham, 1972) and are in Table I.

3. Dipole Moments

The effect of interactions on the electric dipole moment μ of a molecule may be mea-

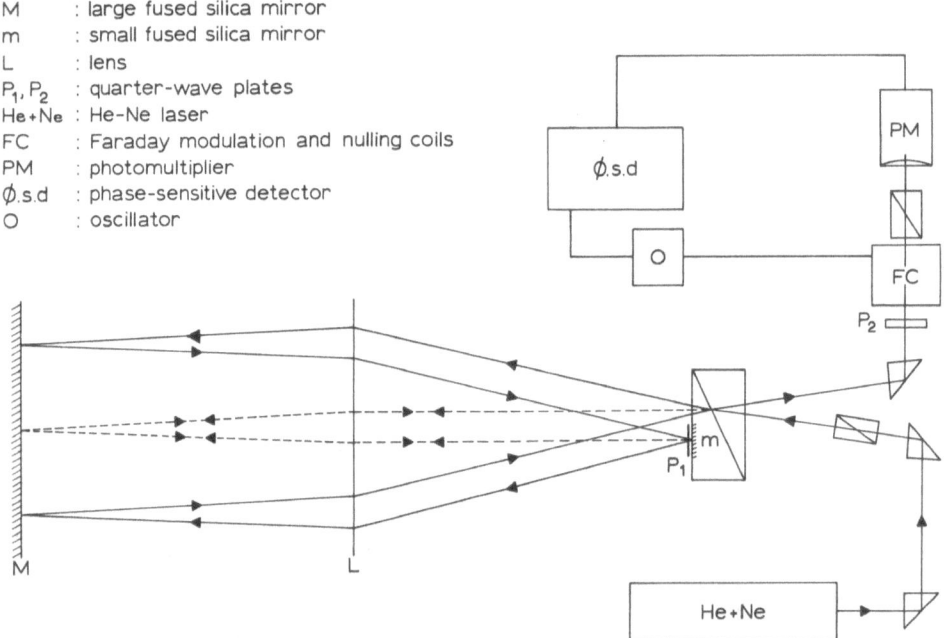

M	: large fused silica mirror
m	: small fused silica mirror
L	: lens
P_1, P_2	: quarter-wave plates
He+Ne	: He-Ne laser
FC	: Faraday modulation and nulling coils
PM	: photomultiplier
$\emptyset$.s.d	: phase-sensitive detector
O	: oscillator

Fig. 3.   A stable interferometer for determining differentially the effect of intermolecular forces on refractivities of gases (Graham, 1972).

TABLE I

New experimental values of refractivity
virial coefficients at 299 K

Gas	A_R	B_R
	$10^{-6} \text{ m}^3 \text{ mole}^{-1}$	$10^{-12} \text{ m}^6 \text{ mole}^{-2}$
Ne	1.003	-0.06 ± 0.09
Ar	4.207	2.16 ± 0.10
N_2	4.460	1.0 ± 0.15
CO_2	6.650	3.2 ± 0.97
CH_4	6.600	7.15 ± 0.04
SF_6	11.34	29 ± 3
CHF_3	7.052	3.4 ± 0.4

sured through studies of the density dependence of the permittivity (Harris and Alder, 1953; Buckingham and Pople, 1955b, 1956):

$$\frac{\varepsilon - 1}{\varepsilon + 2} \varrho^{-1} = A_\varepsilon + B_\varepsilon \varrho + C_\varepsilon \varrho^2 + \cdots. \tag{13}$$

The first and second dielectric virial coefficients are given by

$$A_\varepsilon = \tfrac{1}{3} N \varepsilon_0^{-1} (\alpha_0 + \mu_0^2/3kT), \tag{14}$$

$$B_\varepsilon = \tfrac{1}{3} N^2 \varepsilon_0^{-1} \int \left[(\tfrac{1}{2}\alpha_{12}(\tau) - \alpha_0) + (3kT)^{-1} \{\tfrac{1}{2}(\mu_{12}(\tau))^2 - \mu_0^2\} \right] \times$$
$$\times \exp\left(- \mu_{12}(\tau)/kT\right) d\tau, \tag{15}$$

where $\mu_{12}(\tau)$ is the total dipole moment of the pair of molecules 1, 2 in the relative configuration τ as in (6).

For a pair of nonpolarizable dipoles, $\tfrac{1}{2}(\mu_{12}(\tau))^2 - \mu_0^2 = \mu_1 \cdot \mu_2$ and B_ε is determined by the angular correlation $\langle \cos\theta_{12}\rangle$, where $\cos\theta_{12} = (\mu_1 \cdot \mu_2)\mu_0^{-2} = \cos\theta_1 \cos\theta_2 + \sin\theta_1 \sin\theta_2 \cos\varphi$ and the angles θ_1 and θ_2 are the polar angles between the permanent molecular dipoles and the internuclear axis and $\varphi = \varphi_1 - \varphi_2$ is the difference in the azimuthal angles about this axis. In Stockmayer's model (1941) for the intermolecular potential energy of a pair of polar molecules

$$u_{12}(\tau) = u_{L-J} + \mu_0^2 r^{-3} (4\pi\varepsilon_0)^{-1} \left[- 2\cos\theta_1\cos\theta_2 + \sin\theta_1\sin\theta_2\cos\varphi\right], \tag{16}$$

where u_{L-J} is a Lennard-Jones angle-independent potential. For this model B_ε is positive since $\theta_1 \approx \theta_2 \approx 0$ is favoured by the potential.

A more realistic model is the polarizable dipole model in which the linear polarization of each molecule by the other is incorporated in both $\mu_{12}(\tau)$ and $u_{12}(\tau)$. This extension of the Stockmayer model makes little difference to computed second virial coefficients B in the equation of state, $p\varrho^{-1}(RT)^{-1} = 1 + B\varrho + C\varrho^2 + ...$, but increases B_ε by a factor of 4.4 for parameters representing methyl fluoride CHF_3 (Buckingham and Pople, 1955b); the Lennard-Jones parameters in the two potentials were adjusted to fit observed values of B and the factor would have been even larger had this not been done. The importance to B_ε of distortion is therefore very large indeed; in this simple polarizable dipole model, the total dipole moment μ_{12} for the configuration $\rightarrow \rightarrow$ is increased by 25% over $2\mu_0$ if $(4\pi\varepsilon_0)^{-1} \alpha r^{-3} = 0.1$. Nonlinear polarization and the dipole induced by the field-gradients or by overlap effects must also have their effects on B_ε.

In practice it is very difficult to predict values for B_ε, for so little is known about the short-range forces between simple polar molecules (Sutter and Cole, 1971; Sutter, 1972). The angle-dependence of these interactions is of crucial importance in determining the sign and magnitude of B_ε since they may favour either of the configurations $\rightarrow \rightarrow$ or $\leftrightarrows$ in which $\frac{1}{2}\mu_{12}^2(\tau) - \mu_0^2$ is positive or negative, respectively (Buckingham and Pople, 1956).

Recently experimental (Dyke et al., 1972) and ab initio molecular-orbital theoretical (Kollman and Allen, 1970) studies of the dimer $(HF)_2$ have been made; the electric dipole moment was determined, the distortion effects being quite small, though significant.

Pairs of dissimilar inert-gas atoms, e.g. He–Ar, possess dipole moments $\mu_{12}(r)$ along the internuclear axis. This dipole gives rise to far-infrared 'translational' spectra (Kiss and Welsh, 1959; Bosomworth and Gush, 1965; Marteau et al., 1970; Bar-Ziv and Weiss, 1972). The dipole consists of a long-range 'dispersion' contribution varying as r^{-7} (Buckingham, 1957; Brown and Whisnant, 1970) and a short-range 'overlap' part generally of opposite sign, which has been evaluated for He–Ne, He–Ar and Ne–Ar by accurate SCF calculations (Matcha and Nesbet, 1967).

4. Magnetic Susceptibility

Intermolecular forces affect the dia- and paramagnetism of molecules; the change in the former is generally small, but for the latter large changes can be anticipated when the unpaired electrons overlap. In the case of gaseous oxygen, the interacting pair $(O_2)_2$ could be a singlet, triplet or quintuplet; when the singlet or triplet is favoured energetically, the magnetic susceptibility χ decreases with density, and in the unlikely event of the quintuplet having the lowest potential χ would rise with ϱ (Buckingham and Pople, 1956). If the singlet, triplet and quintuplet potential curves are the same, χ is unaffected by pairwise interactions. Recent infrared and visible spectra of gaseous O_2 at 90 K (Long and Ewing, 1973) provide evidence for bound van der Waals molecules $(O_2)_2$ but there is apparently no significant pairing of the electrons at this temperature, implying that energy differences between the singlet, triplet and quintuplet are small.

5. Nuclear Magnetic Shielding

NMR spectroscopy is a useful probe of molecular interactions (Raynes et al., 1962; Laszlo, 1967). In the gas phase it is convenient to express the nuclear shielding constant of nucleus X, $\sigma(X)$, as

$$\sigma(X) = \sigma_0(X) + \sigma_1(X-A)\varrho_A + \sigma_1(X-X)\varrho_X + \dots, \tag{17}$$

where $\sigma_0(X)$ is the shielding at zero density, and ϱ_X and ϱ_A are the densities of the molecular species containing nucleus X and of the foreign gas; $\sigma_1(X–A)$ and $\sigma_1(X–X)$ result from pair interactions and the former is given by

$$\sigma_1(X–A) = N \int \sigma(\tau) \exp\left[-u_{12}(\tau)/kT\right] d\tau, \tag{18}$$

where $\sigma(\tau)$ is the shift of nucleus X for the particular XA pair configuration τ; N is Avogadro's number if ϱ_A is in moles per unit volume.

A. DIAMAGNETIC SAMPLES

It is usual to consider four distinct contributions to $\sigma(\tau)$ in diamagnetic gases:

(i) $\sigma^{(bulk)}$ is the bulk susceptibility contribution to the shielding. It arises from the magnetization of the gas at large distances from the nucleus under consideration and is zero for a spherical sample. We shall therefore suppose that the integration in (18) is over a sphere.

(ii) If the perturbing molecule A is anisotropically magnetizable, there is a magnetic field at nucleus X resulting from the moment induced in A. If A is axially symmetric with molecular susceptibilities $\chi_\parallel$ and $\chi_\perp$, the contribution to the shift of X at r, θ is

$$\sigma^{(anis.)} = -\frac{1}{3}\frac{\mu_0}{4\pi}(\chi_\parallel - \chi_\perp)(3\cos^2\theta - 1)r^{-3}, \tag{19}$$

where μ_0 is the permeability of free space ($\mu_0/4\pi = 1$ e.m.u.). This contribution to $\sigma_1(X–A)$ is generally quite small.

(iii) The electric field $\mathbf{E}$ at nucleus X due to the charge distribution of neighbours distorts the electronic structure and induces a shift $\sigma^{(E)}$ (Buckingham, 1960). For an axially symmetric molecule

$$\sigma^{(E)} = -aE_z - bE^2 - \dots, \tag{20}$$

where E_z is the field along the axis and a and b depend on the nature of the chemical bonds embracing X.

(iv) The dispersion and overlap interactions may cause significant shifts and these are generally to low field. These 'van der Waals' contributions to the shlieding are not well understood. The long-range deshielding varies as r^{-6} and is sometimes expressed approximately in terms of b in (20).

Values of σ_1 for the protons in HCl and HBr at various temperatures have been reported (Raynes and Chadburn, 1973); the results clearly indicate the need for a more refined theory.

B. PARAMAGNETIC GASES

It has been observed (Jameson and Jameson, 1971) that ^{129}Xe nuclear magnetic resonance is shifted substantially to lower field by the paramagnetic gases O_2 and NO. The results are summarized in Table II. The shifts in the paramagnetic gases are exceptional, particularly when $\sigma_1(Xe–A)$ is plotted against the polarizability of A

TABLE II

Foreign gas contributions to nuclear magnetic shielding of Xe at 298 K (Jameson *et al.*, 1970; Jameson and Jameson, 1971). The shifts are in ppm mole^{-1} cm^3 for a spherical sample

σ_1(Xe–Ar)	σ_1(Xe–CO$_2$)	σ_1(Xe–Xe)	σ_1(Xe–NO)	σ_1(Xe–O$_2$)
−3080	−3779	−12188	−20736	−28015

(Jameson and Jameson, 1971), and can be successfully related to the 'contact' shift arising from overlap of the unpaired electrons with the Xe 5s orbital (Buckingham and Kollman, 1972). The magnetic field favours β electron spin (in O$_2$, $\overline{S}_z = -4\mu_B B_z/3kT$, where $\mu_B = e\hbar/2m_e$ is the Bohr magneton), and overlap imparts a nonvanishing spin density at the Xe nucleus proportional to the square of the overlap integral and to $(5s(0))^2$, where $5s(0)$ is the value of the 5s Xe orbital at the nucleus. The Fermi contact Hamiltonian then leads to a negative temperature-dependent chemical shift. The best calculated values of σ_1(Xe–O$_2$) are about -10^5 ppm mole^{-1} cm^3, but are very sensitive to the Xe 5s orbital at both large and small distances from the nucleus, as well as to the outer reaches of the O$_2$ π_g^* orbital. The effect of 'exchange polarization' tends to draw the Xe 5$s\beta$ electron further from the nucleus, and this effect must reduce σ_1(Xe–O$_2$).

The 'pseudo-contact' shift, due to anisotropy in the susceptibility of the paramagnetic gas (see (19)), is present in NO but it contributes only about 100 ppm mole^{-1} cm^3 to σ_1(Xe–NO) and is sensitive to the angle-dependent potential (Buckingham and Kollman, 1972). This effect is absent in O$_2$.

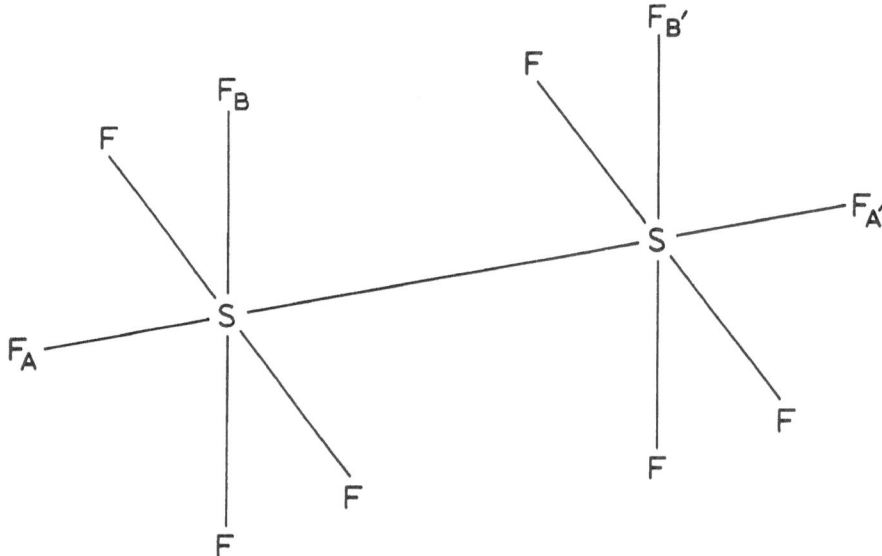

Fig. 4. The fluorine spin–spin coupling constants and chemical shift in S$_2$F$_{10}$ (Finer and Harris, 1968). $J_{AB} = \pm 137.9 \pm 0.5$ Hz; $J_{AB'} = \mp 5.1 \pm 0.3$ Hz; $J_{AA'} = \mp 0.5 \pm 0.3$ Hz; $J_{BB'} = \mp 51.3 \pm 0.5$ Hz; $\delta_{AB} = (\nu_B - \nu_A)/\nu = (3.72 \pm 0.01) \times 10^{-6}$.

It is clear that much can be learnt about wave functions and about molecular inter-actions through careful experimental and theoretical studies of gaseous chemical shifts in O_2 and NO.

Nuclear spin–spin coupling of nonbonded nuclei can proceed through a similar mechanism. The electron spin arising from the interaction of nucleus 1 with the elec-trons in its vicinity is transferred through the overlap to the electrons in the vicinity of nucleus 2, yielding a coupling constant J_{12} proportional to the square of the over-lap. This is presumably the mechanism responsible for the large F–F coupling constant $J_{BB'}$ relative to $J_{AA'}$ and $J_{AB'}$ in the molecule S_2F_{10} (see Figure 4), and for the very large ^{203}Tl–^{205}Tl coupling (2.56 kHz) in the thallous ethoxide tetramer, $(TlOC_2H_5)_4$ (Schneider and Buckingham, 1962).

References

Bar-Ziv, E. and Weiss, S.: 1972, *J. Chem. Phys.* **57**, 29.
Born, M. and Wolf, E.: 1959, *Principles of Optics*, Pergamon Press, p. 86.
Bosomworth, D. R. and Gush, H. P.: 1965, *Can. J. Phys.* **43**, 751.
Brown, W. B. and Whisnant, D. M.: 1970, *Chem. Phys. Letters* **7**, 329.
Brown, W. F.: 1956, *Handbuch der Physik* (ed. by S. Flügge), **17**, 1.
Buckingham, A. D.: 1956, *Trans. Faraday Soc.* **52**, 747, 1035.
Buckingham, A. D.: 1957, CNRS, Colloque Propriétés optiques et acoustiques des fluides comprimés et actions intermoléculaires **77**, 57.
Buckingham, A. D.: 1960, *Can. J. Chem.* **38**, 300.
Buckingham, A. D. and Kollman, P. A.: 1972, *Mol. Phys.* **23**, 65.
Buckingham, A. D. and Pople, J. A.: 1955a, *Proc. Phys. Soc.* **A68**, 905.
Buckingham, A. D. and Pople, J. A.: 1955b, *Trans. Faraday Soc.* **51**, 1029, 1173, 1179.
Buckingham, A. D. and Pople, J. A.: 1956, *Disc. Faraday Soc.* **17**, 22.
Buckingham, A. D. and Watts, R. S.: 1973, *Mol. Phys.* **26**, 7.
Buckingham, A. D., Martin, P. H., and Watts, R. S.: 1973, *Chem. Phys. Letters* **21**, 186.
Certain, P. R. and Fortune, P. J.: 1971, *J. Chem. Phys.* **55**, 5818.
Dyke, T. R., Howard, B. J., and Klemperer, W.: 1972, *J. Chem. Phys.* **56**, 2442.
Ely, J. F., and McQuarrie, D. A.: 1971, *J. Chem. Phys.* **54**, 2885.
Finer, E. G. and Harris, R. K.: 1968, *Spectrochim. Acta* **24A**, 1939.
Graham, C.: 1972, Ph.D. Thesis, University of Cambridge.
Harris, F. E. and Alder, B. J.: 1953, *J. Chem. Phys.* **21**, 1351.
Jameson, A. K., Jameson, C. J., and Gutowsky, H. S.: 1970, *J. Chem. Phys.* **53**, 2310.
Jameson, C. J. and Jameson, A. K.: 1971, *Mol. Phys.* **20**, 957.
Jansen, L. and Mazur, P.: 1955, *Physica* **21**, 193.
Kirkwood, J. G.: 1933, *Phys. Rev.* **44**, 31.
Kirkwood, J. G.: 1936, *J. Chem. Phys.* **4**, 592.
Kiss, Z. J. and Welsh, H. L.: 1959, *Phys. Rev. Letters* **2**, 166.
Kollman, P. A. and Allen, L. C.: 1970, *J. Chem. Phys.* **52**, 5085.
Langhoff, P. W., Karplus, M., and Hurst, R. P.: 1966, *J. Chem. Phys.* **44**, 505.
Laszlo, P.: 1967, *Progress in Nuclear Magnetic Resonance Spectroscopy* **3**, 231–402.
Lim, T.-K., Linder, B., and Kromhout, R. A.: 1970, *J. Chem. Phys.* **52**, 3831.
Long, C. A. and Ewing, G. E.: 1973, *J. Chem. Phys.* **58**, 4824.
Marteau, P., Vu, H., and Vodar, B.: 1970, *J. Quant. Spectrosc. Rad. Transfer* **10**, 283.
Matcha, R. L. and Nesbet, R. K.: 1967, *Phys. Rev.* **160**, 72.
Michels, A., de Boer, J., and Bijl, A.: 1937, *Physica* **4**, 981.
O'Brien, E. F., Gutschick, V. P., McKoy, V., and McTague, J. P.: 1973, *Phys. Rev. A* **8**, 690.
Phillipson, P. E.: 1962, *Phys. Rev.* **125**, 1981.
Raynes, W. T. and Chadburn, B. P.: 1973, *J. Magnetic Resonance* **10**, 218.
Raynes, W. T., Buckingham, A. D., and Bernstein, H. J.: 1962, *J. Chem. Phys.* **36**, 3481.

Schneider, W. G. and Buckingham, A. D.: 1962, *Disc. Faraday Soc.* **34**, 147.

Silberstein, L.: 1917, *Phil. Mag.* **33**, 521.

Stockmayer, W. H.: 1941, *J. Chem. Phys.* **9**, 398.

Sutter, H.: 1972, 'Dielectric and Related Molecular Processes', Specialist Periodical Report, Chem. Soc. London **1**, 65.

Sutter, H. and Cole, R. H.: 1971, *J. Chem. Phys.* **54**, 4988.

Ten Seldam, C. A. and de Groot, S. R.: 1952, *Physica* **18**, 905, 910.

Van Vleck, J. H.: 1932, *The Theory of Electric and Magnetic Susceptibilities*, Oxford University Press, p. 15.

Yvon, J.: 1936, *C.R. Acad. Sci.* **202**, 35.

THREE TYPES OF POTENTIAL NEEDED
IN PREDICTING CONFORMATIONS OF MOLECULES
IN SOLUTION AND THEIR USE

OKTAY SINANOĞLU*

Sterling Chemistry Laboratory, Yale University, New Haven, Conn. 06520, U.S.A.

1. Molecule in Vacuo – the U-Potential

Quantum-mechanically one calculates an adiabatic [1] potential energy (PE) surface $U(R_M)$ in studying the shape of a molecule M by itself (*in vacuo*). To distinguish this from the others below, we shall refer to it as the *U-potential*.

As atom positions $\{R_1, R_2, ... R_n\} \equiv R_M$ vary, the surface covers from short-range repulsions, to directional chemical forces restraining bond bends and stretches, to long-range (intramolecular) van der Waals (vdW) attractions. It used to be thought necessary to treat each of these regions with a different theory with the difficulty of patching up the regions in between. But now, both chemical and van der Waals forces are obtainable from the same theory covering the entire $U(R_M)$ from short to inter-mediate to long ranges. For this theory we refer the reader to reference [2].

Prediction of a molecular shape is a problem not just of $U(R_M)$, but of statistical mechanics. With the $U(R_M)$ one looks for minima, barrier tops, etc. If such prominent features exist, one breaks the phase space (P_M, R_M) into 'significant structures' as Eyring calls them. For each *significant structure*, A, its $U(R_M)$ region is used to cal-culate a partition function $(pf)_A$. Then we have equilibrium, or quasi-equilibrium in activated complex rate treatment, between such regions, A, B, $C^{\neq}$,...

$$A \rightleftharpoons B; \quad \Delta\mu = 0 \tag{1}$$
$$x_A \quad x_B;$$

the chemical potential $\mu_i = \mu_i^0 + RT \ln x_i$.

If no such prominent features are notable on $U(R_M)$, then the molecule simply flops around a good many spatial configurations $\{R_M\}$. The pf method is then not of much use. One needs to integrate over the wide $U(R_M)$ surface [3].

The U-potential maps into the Helmholtz free energy A of a molecule *in vacuo*, via the configuration integral [4] Q of the canonical partition function (PF),

$$Q = \int_{\{R_M\}} e^{-U(R_M)/kT} \, d\tau_M \tag{2}$$

$$\boxed{A(T, V) = -kT \ln(\text{PF})} \tag{3a}$$

* Also: Institut für Organische Chemie, Technische Universität München, W. Germany, and Boğaziçi Evrenkenti (Üniversitesi), Istanbul, Türkiye.

R. Daudel and B. Pullman (eds.), The World of Quantum Chemistry, 265–276. All Rights Reserved
Copyright © 1974 by D. Reidel Publishing Company, Dordrecht-Holland

and for a mole of perfect gas of M,

$$PF = \frac{(pf)^{N_0}}{N_0!}; \qquad pf = \Lambda Q. \tag{3b}$$

Here N_0 is Avogadro's number and for single molecule in gas, M,

$$\Lambda = \Lambda_M = \prod_i^n \lambda_i; \qquad \lambda_i = (2\pi m_i kT/h^2)^{3/2},$$

with m_i = the mass of atom i of molecule M.

We shall return to these matters below, but now what happens to the above, if the molecule M whose shape we wish to study happens to be immersed in a solution?

2. Molecule in Solution – the C-Potential

The analog of the $U(R_M)$ to examine, when the molecule M is in solution, is easily derived from the theory of solvent effects on *inter-* or *intra*-molecular associations we have given before [5]. We shall call it the *C-potential*.*

One still calculates quantum mechanically the *in vacuo* $U(R_M)$ for the molecule, but adds to each point the potential of the *solvophobic force* [5], the force due to solvent which tends to drive molecular groups together even against their mutual intrinsic repulsions which they may have had. Is is clear therefore, the solvophobic force may change the likely conformation of M, contort the $U(R_M)$, and dislocate its prominent features wiping out some minima, accentuating others, etc.

We shall state the $C(R_M)$, derive it from a molecular point of view, then show how it can be used in statistical mechanics (in PF, in Q, etc.) as if it were a true potential energy function even though it is part thermodynamic.

Placing the molecule M in a fixed spatial configuration R_M with solvent, packing, distributing around this constraint, at temperature T, the C-potential is given by

$$C(R_M) = U(R_M) + A^0_{\text{solv. eff.}}(R_M), \tag{4}$$

with $U(R_M)$ = the quantum-mechanical potential energy surface point of molecule at shape R_M; $A^0_{\text{solv. eff}}(R_M)$ = Hemholtz free energy of placing molecule M (rigidified at shape R_M) into solvent at T, V; that is

$$M(\text{g.}; R_M) + \text{solvent (1.)} \rightleftharpoons M(\text{in soln.}; R_M); \qquad \Delta A^0 = A^0_{\text{solv. eff}}(R_M). \tag{5}$$

The magnitude of $A^0_{\text{solv. eff}}$ depends parametrically on the molecular configuration (kept rigid) R_M. The solution is dilute, i.e. in the Henry's law region (no M–M inter-action). The A^0 refers to the hypothetical Henry's law standard state $x^0_M \equiv 1$ (the so-called 'unitary' state).

As the Gibbs free energy $F = A + PV$, and as experiments are carried out at constant

---

* As S for solvo-potential would get mixed with entropy, we use C for 'çözgen'.

T, P, for convenience we can also use, approximately, Equation (4) with

$$C(R_M) \cong U(R_M) + F_{\text{solv. eff}}^0(R_M). \tag{6}$$

The difference added, $PV_{\text{solv.eff}} \approx 0$ at $P=1$ atm usually, so there is no practical consequence.

From the solvent effect theory [5], the $F_{\text{solv. eff}}^0(R_M)$ is given by:

$$F_{\text{solv. eff}}^0(R_M) \cong F_c(R_M) + F_{\text{int}}(R_M) + kT \ln \frac{kT/P_0}{v_1}. \tag{7}$$

The $F_c(R_M)$ is the free energy required to make a cavity with walls appropriate to the immersed $M(R_M)$, molecule at shape R_M. The $F_{\text{int}}(R_M)$ is the *free* energy of interaction of $M(R_M)$ with the solvent around. The third term is an approximate 'free volume' entropy effect of $M(R_M)$ in liquid ① whose average molecular volume is $[(V_1 \text{ cm}^3/\text{mole})/N_0] = v_1$ with N_0 Avogadro's number. $P_0 = 1$ atm.

Although the $F_{\text{solv. eff}}^0 \cong A_{\text{solv. eff}}^0$, Equations (6) and (7) can be visualized physically as taking place in two steps: first making the cavity, then placing the M in it; this is really a mathematical separation. For instance, the walls of the cavity, though the cavity is empty, have the solvent molecules (even in water) oriented in anticipation of the M to come. The detailed expression for $F_c(R_M)$ includes such 'solvent structuring' effects in the equation [5]

$$F_c(R_M) = \kappa_1^e(\varphi_{1\,RM}^{-1/3}) \times \sigma(R_M)\gamma_1(1 - W_{1M}), \tag{8a}$$

$$W_{1M} = (1 - \eta_{1\,RM}) \frac{\partial \ln \gamma_1}{\partial \ln T} + \tfrac{2}{3}\mathscr{A}_M T, \tag{8b}$$

where

$$\eta_{1\,RM} = \frac{\kappa_1^s(\varphi_{1\,RM}^{-1/3})}{\kappa_1^e(\varphi_{1\,RM}^{-1/3})}. \tag{8c}$$

Equation (8) includes some geometric properties of $M(R_M)$ and some properties of the pure solvent ①, the latter all obtained from simple macroscopic experimental quantities pertinent to the pure liquids. The $\sigma(R_M)$ = geometric surface area of the molecule M at its fixed conformation R_M, and $v(R_M)$ = geometric volume of $M(R_M)$ are easily obtained, for example from space-filling models.

For the solvent: γ_1 = pure solvent ordinary surface tension at T, P, $\varphi_{1\,RM}$ = volume ratio $v_1/v(R_M)$; v_1 is obtained from density of pure solvent at T, P and its molecular weight.

Further, $\kappa_1^e(\varphi_{1\,RM}^{-1/3})$ is a dimensionless function giving the deviation of the energy ($\approx$ enthalpy) part of the microscopic surface tension at the molecular dimensions R_M from the macroscopic surface tension part [5]. The $\kappa_1^s(\varphi)$ is a similar function in Equation (8c), for the entropy part of γ. This author has evaluated $\kappa_1^e(\varphi)$ and $\kappa_1^s(\varphi)$ from theory he developed on pure liquids (where $\varphi = 1$) and dilute solutions which allows reliable curves to be obtained from thermodynamic data on pure liquids and simple solutions of varying φ. For each $\kappa(\varphi)$ a separate curve is obtained, one for nonpolar solvents, one for polar solvents which include water.

The $\kappa_1^e(\varphi=1)$ and $\kappa_1^s(\varphi=1)$ values are tabulated for many common solvents in reference [6]. These can be used wherever $M(R_M)$ and solvent molecule sizes are roughly comparable. The full $\kappa(\varphi)$ curves will also be published. Note also that for all $\kappa(\varphi)$,

$$\kappa(\varphi) \to 1 \tag{9}$$

as $\varphi \to 0$, i.e. if $V(R_M) \gg V_1$.

For polar and structuring liquids, $\kappa_1^s \neq \kappa_1^e$ in general, however, if for some liquids one takes roughly $\kappa_1^s \approx \kappa_1^e$, the $W_{1M} \gg 0$.

If configuration R_M of M is quite similar to the ordinary average shape of molecules M in a pure liquid made of M at the same reduced temperature $T^r = T/T_M^c$ as that of solvent $T^r = T/T_1^c$, then molecular volume v_M can be estimated from the density of pure liquid M at T^r. Then also

$$\sigma(R_M) \approx v_M^{2/3} \times 4.836. \tag{10}$$

If M is not available in pure liquid form one can use a model compound (of similar shape and size, but say, nonpolar) to estimate v_M, and also $\mathscr{A}_M$ of Equation (8b). Then $\mathscr{A}_M =$ coefficient of thermal expansion of pure liquid of M or its model compound at T^r.

We retain the same $\mathscr{A}_M$ in Equation (8b) even for other hypothetical geometries R_M of $M(R_M)$.

This leaves the interaction term $F_{int}(R_M)$ in Equation (7). The $F_{int}(R_M)$ with inert solvents has in general van der Waals and electrostatic (e.s.) parts. These are given in some detail in [5] and [6]. The e.s. part for a polar or charged molecule M for each R_M, can also be obtained by quantum chemical MO-calculations involving either $M(R_M)$ alone (to get its spatial charge distribution), or on $M(R_M)$ together with a number of discrete solvent molecules packed around (to get electrostatic and induction forces).

3. The Non Polar Solute–Solvent Interaction Part of the C-Potential

We have

$$F_{int}(R_M) \cong A_{int}(R_M) = A_{e.s.}(R_M) + A_{vdW}(R_M), \tag{11}$$

where we used the term vdW to mean the entire nonpolar interaction.

In conformation studies, the $M(R_M)$ usually has some molecular side groups like phenyl, alkyl, etc. which can orient in different ways. In the azobenzene example we showed in detail [6], the two rings yield the $U(R_M)$ features: cis, trans, and the presumably flat transition state. With such side groups R_M^g, the molecule is then like:

$$M(R_M, \approx \bigcup_g M_g(R_M^g), \tag{12}$$

(with $\bigcup_g$ the 'union' sign over groups g), as far as its exposure to solvent is concerned. Then

$$F_{\mathrm{vdW}}(R_M) \approx \sum_g F^g_{\mathrm{vdW}}(R^g_M), \tag{13}$$

where $F^g_{\mathrm{vdW}}(R^g_M)$ is the nonpolar interaction of the gth molecular group with solvent surrounding the entire R_M cavity, while R^g_M orients the g and makes up part of R_M.
We have [5–6]

$$F^g_{\mathrm{vdW}}(R^g_M) = -f(\varphi_{1g}, l_{1g})\,\Delta_{1g}D_1D_gB_{1g}, \tag{14}$$

with

$$\Delta_{1g} = \mu\,\frac{I_g I_1}{I_g + I_1}\,; \qquad \mu \cong 1.35. \tag{15}$$

I_g resp. I_1 = ionization potential of side group g resp. solvent molecule ①.

$$D_i = \frac{n_i^2 - 1}{n_i^2 + 2} = \frac{4\pi}{3v_i}N_0\bar\alpha_i. \tag{16}$$

n_i, v_i, $\bar\alpha_i$ = refractive index, molecular volume (V_i/N_0), average polarizability of $i \in \{$side group g or solvent ①$\}$; properties derived from pure liquids (solvent and a liquid of models of g).
The $f(\varphi_{1g}, l_{1g})$ is a dimensionless function of relative solvent–solute size and molecular 'core sizes' are given [5–7] in terms of simple macroscopic handbook properties of the pure liquids involved. The $f(\varphi, l)$ results from integration of the (g-solvent molecule) interaction over the first discrete solvation layer, then over the rest of the solvent.
In Equation (14) we have

$$-f(\varphi_{1g}, l_{1g})B_{1g} = \frac{27}{8\pi}(1 - x)(Q' + Q''). \tag{17}$$

$x \approx 0.436$ for most solvents, polar, nonpolar [6].
Q' is obtained analytically,

$$Q' = \frac{\pi}{6y^3(1-t)^3}\left[\frac{1}{2}\left(\frac{\mu}{\tau}\right)^6\left(\frac{1}{9} + \frac{1}{5\tau} + \frac{1}{11\tau^2}\right) - \left(\frac{1}{3} + \frac{1}{2\tau} + \frac{1}{5\tau^2}\right)\right], \tag{18}$$

where

$$\mu = \frac{3}{7\varpi + 0.24}, \tag{19a}$$

$$\tau = \frac{1}{t} - 1\,; \qquad t = \frac{l}{R_s}\,; \qquad l = \left(\frac{7\varpi + 0.24}{7\varpi + 3.24}\right)\frac{2}{\beta}\left(\frac{3v_1}{4\pi}\right)^{1/3}\,;$$

$$R_s = \left(\frac{3}{4\pi}\right)^{1/3}(v_g^{1/3} + \chi v_1^{1/3}), \tag{19b}$$

with $\beta \cong 1.15$; $y = \frac{1}{2}(1+\chi)$; $\chi \cong 0.85$; v_g, v^1 = average molecular volumes of liquid of g and of solvent ① as in Equation (16); ϖ = accentric factor (a macroscopic property of a pure liquid). (The t is with l average between that of liquid of g and of solvent.) All these quantities (β, χ, ϖ) are used in [7].

Q'' is an integral which is evaluated numerically. But it is a smaller part. For most liquids we have found $(Q''/Q') \cong 0.1$.

The above $F_{vdW}^g(R_M^g)$ has resulted essentially from an integral of the type (the form of $U^e(r)$ given in [7] and approximations for $g^{(2)}$ also used in [7] gave rise to Q' and Q'' above)

$$e_{vdW}^g \equiv \varrho \int_0^\infty U_{1g}^e(r)\, g^{(2)}(r; R_M)\, 4\pi r^2\, \mathrm{d}r. \tag{20}$$

Here $\varrho =$ number density of solvent ①; $g^{(2)}(r; R_M) =$ radial (r) distribution function for the solvent molecules, around the molecular group g as center, with parametric dependence however on the placements of all other parts of the molecule M at $R_M = \bigcup g R_M^g$; $U_{1g}^e(r) =$ nonpolar intermolecular 'effective pair potential' $U^e(r)$ between g and a solvent molecule ① at distance r from g. The $U^e(r)$ is given in detail in a separate work by Sinanoğlu [7]. We refer to this third kind of potential (as distinct from the U- and the C-potentials already seen) as the U^e-potential.

We have now completely specified the C-potential for $M(R_M)$ in a solvent ① with Equations (4), (6)–(8), (11), (13), and (17) and no adjustable parameters. All quantities in these equations are either geometric or come from handbook properties of some pure liquids.

We show now why the C-potential can be used just as though it were an ordinary potential like $U(R_M)$, in all the statistical mechanical, thermodynamic calculations. This will also give further molecular insight into the $C(R_M)$ as if one were obtaining it from a quantum calculation on M plus a number of solvent molecules around it.

4. Theorem on the C-Potential for Its Use in Thermodynamics

THEOREM. The configurational motions, the conformations of a molecule M in solution are determined by its C-potential, Equation (4), $C(R_M)$, a potential surface on the spatial configuration space $\{R_M\}$. The C-potential can be used in statistical mechanical equations as if it were an ordinary molecular potential energy surface though it includes the solvent effects and the solvophobic force.

Proof. We first define what is meant by a function of R_M to behave as if it were the *in vacuo* $U(R_M)$,

DEFINITION. A function $Y(R_M)$ can be used as a potential function, a function of $\{R_M\}$ alone, if it yields the unitary free energy A_M^0 of the molecule* (say in the solvent medium) from a configurational integral Q which is over the molecular coordinates R_M only; i.e.

$$A_M^0 \text{ (in soln.)} = -kT \ln \Lambda_M Q, \tag{21}$$

---

* Note again that $A_M{}^0$ (in soln.) is the unitary quantity, i.e. per molecule of M and its surroundings ($x_M{}^0 \equiv 1$). Hence we use its pf only, and $A_M{}^0 = -kT \ln(\text{pf})$, not $\text{PF} = (\text{pf})\, N_0/N_0!$.

$$Q = \int_{\{R_M\}} e^{-Y(R_M)/kT} \, d\tau_M, \tag{22}$$

$(d\tau_M = d^3 R_1 d^3 R_2 ... d^3 R_n).$

The *in vacuo* $U(R_M)$ trivially satisfies the definition yielding for M (in dilute gas $(g.)$).

$$A_M (g.) = - kT \ln (PF); \qquad PF = (pf)^{N_0}/N_0!; \qquad pf = \Lambda_M Q (g.)$$

$$Q (g.) = \int_{\{R_m\}} e^{-U(R_M)/kT} \, d\tau_M. \tag{23}$$

For the C-potential for M (in soln.; $x_M^0 \equiv 1$), we wish to show from Equation (4) put into Equation (22) that

$$C(R_M) = U(R_M) + A_{\text{solv. eff}}^0 (R_M) \tag{24}$$

gives

$$A_M^0 (T, V, \text{in soln.}) = -kT \ln \Lambda_M Q,$$

$$Q = \int_{\{R_M\}} e^{-C(R_M)/kT} \, d\tau_M. \tag{25}$$

The theorem clearly holds for the special case of M being in a 'significant structure' R_M^a, a being a minimum on $C(R_M)$ such that the well $C(R_M^a) \ll kT$. For, then, M has definite conformation, all the play in R_M being molecular vibrations, etc. as in the usual (pf) problem. But then

$$A_{\text{solv. eff}}^0 (R_M) \approx A_{\text{solv. eff}}^0 (R_M^a) = \text{const}. \tag{26}$$

and we get from Equations (25) and (26),

$$A_M^0 a (T, V; \text{in soln.}) = - kT \ln \Lambda_M \left[\int_{\{R_M\}} e^{-U(R_M)/kT} \, d\tau_M \right] \times$$

$$\times \, e^{-A^0 (R_M^a, \text{solv. eff})/kT}, \tag{27}$$

$$\therefore A_M^0 (T, V: x_i^0 = 1; R_M^a) = A_M^0 (T, V; g.) + A_{\text{solv. eff}}^0, \tag{28}$$

(The $A_M^a (T, V; g.)$ too is for a single molecule, now of gaseous M.) This is the equation used whenever there are definite R_M^i, significant structures to calculate equilibrium constants and concentrations of each 'conformer' i at T, V (better T, P with F_M^0's) as in reference [6] and Equation (1).

Now most generally, to prove Equation (25), consider a collection of some solvent molecules Ⓢ with the $M(R_M)$ sitting in their midst as in Figure 1 $(R_M = R_M^p \cup R_M^q)$.

The ordinary interaction potential of the collection is

$$U_t = U(R_M) + U_{MS}(\{R_{MS}\}) + U_{SS}(\{R_{SS}\}), \tag{28}$$

where U_{MS} is PE of interaction of M with molecules $\{s\}$, and U_{SS} is the PE of all $(N_S$ in number) solvent molecules with each other. The free energy of the system:

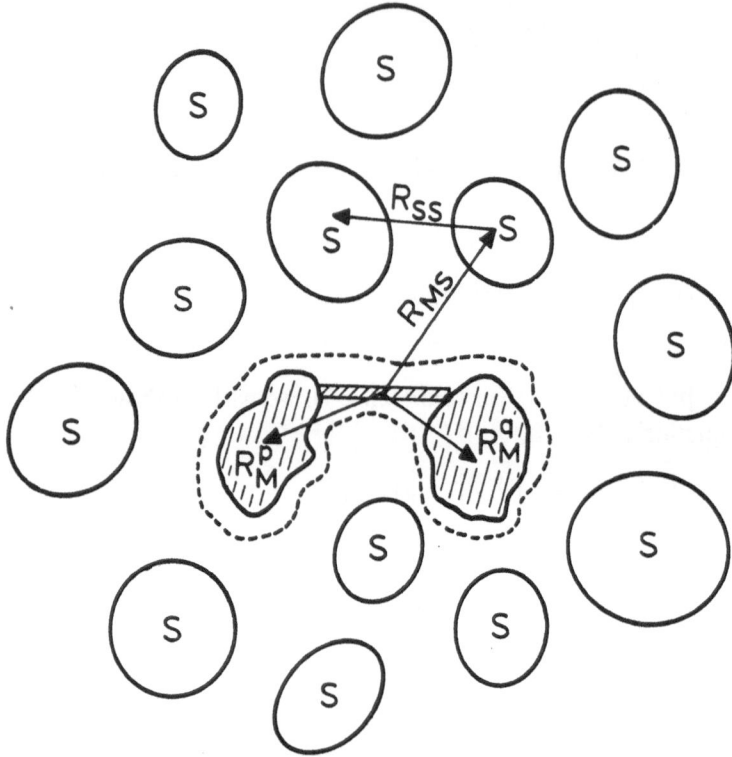

Fig. 1.   Molecule M in configuration R_M (with side groups oriented $R_M{}^p$, $R_M{}^q$) in the midst of solvent molecules Ⓢ.

M plus solvent

$$A^0(T, V) = -kT\ln\Lambda Q,$$

$$Q = \int\limits_{R_M} \int\limits_{\{S\}} e^{-U_t/kT}\, \mathrm{d}\tau_M\, \mathrm{d}\tau_S, \tag{29}$$

with $\mathrm{d}T_S = (\mathrm{d}T_{S_1}\mathrm{d}T_{S_2}...\mathrm{d}T_{SN_S})$, and where, now having a factor in Λ for each molecule, $\Lambda = \Lambda_M\Lambda_{\{S\}}$ and $\Lambda_{\{S\}} = (1/N_{S!})(2\pi m_S kT/h^2)^{3\,N_S/2}$. The integral is over all the internal motions of M, i.e. over R_M, and all motions and wanderings of the solvent molecules $\{S\}$ around M.

Then,

$$Q = \int\limits_{R_M} \mathrm{d}\tau_M\, e^{-U(R_M)/kT} \left[\int\limits_{\{S\}} e^{-(U_{MS}+U_{SS})/kT}\, \mathrm{d}\tau_S \right]. \tag{30}$$

The partial integral in [] over $\{S\}$ is a number parametrically dependent on R_M. The rest of Equation (30) involves integration over R_M only. Hence, clearly, Equation (30) gives us the precise (and molecular) definition of $A^0_{\text{solv. eff}}$

$$A^0_{\text{solv. eff}}(R_M) \equiv -kT \ln \Lambda_{\{S\}} \left\{ \int\limits_{\{S\}} e^{-(U_{MS}+U_{SS})/kT}\, \mathrm{d}\tau_S \right\}, \tag{31a}$$

or

$$\left\{ \int_{\{S\}} e^{-(U_{MS}+U_{SS})/kT} \, d\tau_S \right\} = \frac{1}{\Lambda_{\{S\}}} e^{-A^0 \text{solv. eff.} (R_M)/RT} \tag{31b}$$

Putting Equation (31b) into (30), then (29),

$$A^0(T, V) = -RT \ln \Lambda_M Q,$$

with

$$Q = \int_{R_M} d\tau_M \, e^{-C(R_M)/kT}, \tag{25'}$$

$$C(R_M) = U(R_M) + A^0_{\text{solv. eff}}(R_M) \qquad \text{Q.E.D.} \tag{24'}$$

Remark 1. The molecular definition of $A_{\text{solv.eff}}(R_M)$ (equation (30)) has the molecule M–solvent interaction part coming from $e^{-U_{MS}/kT}$ showing how the $F^0_{\text{int}}(R_M) = A^0_{\text{int}}(R_M)$ arises in Equation (7).

Remark 2. In the $\{R_S\}$-integrations in Equation (31a), as an s-molecule comes within the hard-core-like repulsions of the $M(R_M)$, $U_{MS} \to \infty$; then $U_{MS} \gg U_{SS}$, and integrand vanishes. This effect is reproduced by an R_M-cavity. It also allows approximate factorization of the integral

$$\int_{\{S\}} e^{-(U_{MS}+U_{SS})/kT} \, d\tau_{MS} \, d\tau_{SS} \cong \left(\int_C^\infty e^{-U_{MS}/kT} \, d\tau_{MS} \right) \left(\int_C^\infty e^{-U_{SS}/kT} \, d\tau_{SS} \right),$$

$$\tag{32}$$

with $d\tau_{\{S\}} \approx d\tau_{\{MS\}} \, d\tau_{\{SS\}}$ for the definite integrals. We get the $F^0_{\text{int}}(R_M)$ and the $F^0_C(R_M)$ terms of Equation (7) from Equation (31a).

Remark 3. The U_{SS} in Equations (28–32) can be taken relative to the free energy $A^0_S(1.)$ of the same amount of pure liquid (with no (R_M) in it). We can assume this to be the reference point of U_{SS} in these equations, i.e. $U_{SS} = \bar{U}_{SS} - A^0_S(1.)$ with $\bar{U}_{SS}$ the mutual potential energy of solvent molecules measured from separated (gaseous) solvent molecules as zero. As $A^0_S(1.)$ is a constant within Q, integrals are unaffected, but $A_{\text{solv·eff}}$ becomes consistent with Equation (5). Then also the U_{SS} terms in Equation (32) become identified with Equation (8a), the cavity surface tension effect, the solvophobic force.

In quantum, physical organic, or physical biochemistry applications where molecular associations or conformations in solution are involved, the strong driving force coming from solvent's interaction with itself in general was apparently overlooked until the present theory [5] which started with a physical explanation and prediction of relative stabilities of DNA-bases associating in different solvents including water (references in [5]). 'Hydrophobic bonding' itself was shown to be due [8] to the U_{SS}, the surface tension effect, rather than to the 'water structuring' effect [9] which turns out to play a smaller role. The $F_C(R_M)$ formula gives the U_{SS} effect most directly and simply which otherwise would be most difficult to calculate molecularly.

5. The Third Potential, the U^e-Potential in Liquids
as One Input to the C-Potential

For ease of presentation we omitted up to now another effect that goes into the C-potential of $M(R_M)$.

In the C-potential, Equation (24), we have the $U(R_M)$, the bare PE surface of M, but also the U_{MS} and the U_{SS} through Equation (31). Each of these will have nonpolar (vdW) parts, and sometimes electrostatic (e.s.; polar) ones. It is customary to take such interactions (between the molecular groups M_g of M, or between different, e.g. M and ⓢ molecules) in pairwise form:

$$U = \sum_{i>j} U_{ij}(R_{ij}). \tag{33}$$

Again it turned out, that a bare $U_{ij}(R_{ij})$ itself is strongly affected due to many-atom (three-body, etc.) forces in a liquid by as much as 30–40% in, e.g. liquid benzene, CCl_4, etc., an effect also overlooked in the field previously. Adding such forces to the U's would complicate the statistical mechanics. The effect was observed in [7] (and our earlier work quoted therein). It is sensitive to the solvation structure of ⓢ's around a given pair (i, j) in interaction [7]. But both the solvation aspects and the magnitude could be incorporated into $\sum_{i>j} U_{ij}$ analytically obtaining an 'effective pair potential' $U_{ij}^{\text{eff}}(R_{ij}) \equiv U_{ij}^e(R_{ij})$. This is the U^e-potential. Both its form and all its parameters (none of which are adjustable, all come from liquid properties) are given in [7]. It will not be repeated here.

For the electrostatic parts of U's, too, there are reductions, some familiar, discussed in [5].

The conclusion of this section is that the C-potential for R_M in solution involves not

$$U \neq \sum_{i>j} U_{ij} \quad \text{(for } M, \text{ forms, for } SS),$$

but

$$U = \sum_{i>j} U_{ij}^e(R_{ij}; \varrho; \varphi^{-1/3}), \tag{34}$$

ϱ being density of solvent, φ relative solute (i or j)-solvent size. Thus *the complete C-potential is*:

$$\boxed{C(R_M) = U^e(R_M) + A_{\text{solv. eff}}^0(R_M)} \tag{35}$$

with U_{MS}^e and U_{SS}^e in Equation (31a) and (28). The expressions for $F_{\text{int}}^0(R_M)$ and $F_C^0(R_M)$, Equations (14) and (8), are already with the U^e's as noted in Equation (20) also.

The difference $[U^e(R_M) - U(R_M)]$ of the $C(R_M)$ vs. $U(R_M)$ gives rise to the 'reduction' term [5], $F_{\text{red}}^0(R_M)$, of the solvent effect theory, so that

$$F_M^0 \text{ (in soln.)} = F_M^0 \text{ (g.)} + F_{\text{solv. eff}}^0,$$

with

$$F_{\text{solv. eff}}^0 = F_{\text{int}} + F_C + F_{\text{red}} + RT \ln \frac{kT/P_0}{v_1}. \tag{7'}$$

The F_{red} is fairly constant in different solvents and not too large, but may be worth-while to ihclude on a $C(R_M)$ surface search.

6. Experimental Tests and Use of Theory

The C-potential, Equation (35), is tested most directly through its thermodynamic form, Equation (7'). As noted in Equation (1), when the $C(R_M)$ shows a number of prominent features (extrema) with their resulting 'significant structures', such as *cis–trans* forms, folded–unfolded forms, stacked–unstacked bases, helix–random coil forms, etc., one has a $F_M^0(k$th conformer; in solution), Equation (7'), for each 'species', 'transition state', etc., in quasi- or actual equilibrium. Then for the equilibrium, Equation (1), one has:

$$M_k \rightleftharpoons M_{k'}; \quad \Delta F_M^0,$$

with

$$\Delta F_M^0 (k \rightarrow k'; \text{in soln.}) = \Delta F_M^0 (k - k'; \text{g.}) + \Delta F_{\text{solv. eff}}^0 \tag{36}$$

with the equilibrium constant K given by

$$K = e^{-\Delta F_M^0 (\text{in soln.})/RT} = x_{k'}/x_k. \tag{37}$$

The K (or, for rates, $K^{\neq} \propto$ rate const.) are measured at T, P in different pure solvents, polar, nonpolar. From Equation (36)

$$\ln K = \ln K_g. - \frac{1}{RT} \Delta F_{\text{solv. eff}}^0. \tag{38}$$

A plot of $(\ln K)$ vs. the theoretical $(F_{\text{solv. eff}}^0)$ yields a straight line with intercept $(\ln K_g)$, the equilibrium constant for the isomerization of M in the gas phase.

This test has worked well in the past several years as carried out by various workers (cf. references in [5] and [6] and later ones in the literature; some from pharmaco-logical, molecular, biophysical examples) on systems like quinone-hydroquinone asso-ciation (then R_M refers to the composite bimolecular system), *cis-trans* azobenzene isomerization rate [6], a thymine dimer (T_pT), actinomycin, etc.

In turn, Equation (38) allows sequences of solvents [5] to be predicted with regard to their ability to drive certain isomerizations, conformation changes, etc.

In many organic and biochemical systems, it is not possible to carry out a reaction, or to observe conformations of a molecule M, in the gas phase. Then, if data are obtained in a number of solvents, from Equation (38) one can derive a ΔF_M^0 (in gas) value. (This could not be done prior to the present theory [5] as $\ln K$'s would not correlate with any solvent property for a range of different solvents, a situation well known in physical organic chemistry [10].)

This application would be particularly relevant to quantum chemistry, as *a priori* MO, etc. calculations on a molecule would yield only $U(R_M)$ (and after pf calculations ΔF_M^0 (in g.)). Experimental data for comparison can thus be provided from the $\ln K_g$ of Equation (38) using the solvent effect theory [5].

What of cases where no conformation knowledge in solution is available for a mole
cule? There may also not be any significant features and structures on $C(R_M)$. This
is not rare in biochemical, pharmacological and chemical kinetics contexts. Then
$C(R_M)$, the C-surface, has to be obtained and examined theoretically in one or more
solvents. The $U(R_M)$ is calculated either *a priori* (MO; NCMET) [1–2], or from 'experimental' force fields and side group pair potentials $[U^e(R_M)]$ [7], Then
$A^0_{\text{solv. eff}}(R_M)$, $F^0_{\text{solv. eff}}(R_M)$ is obtained at each R_M using the equations and data of
this work (here and in [5] and [6]). The $C(R_M)$ then is observed with respect to its
minima, etc. If no 'significant structures' turn up, the free energy of M per mole (in
soln.) can still be obtained from the C-surface and Equation (25). This gives the
thermodynamic average for properties of M covering all its rapidly interconverting
conformers in solution.

Acknowledgement

This work was supported by a grant from the U.S. National Science Foundation.

References

[1] For a critical survey of recent progress on calculating PE surfaces and their use in molecular
    beam kinetics and chemical kinetics see: Sinanoğlu, O.: 1971, *Comm. Atomic Mol. Phys.* III, 53.
[2] Sinanoğlu, O.: *J. Mol. Structure* (Special Proceedings Issue of International Conference on
    Molecular Spectroscopy, Wrøclaw, Poland, Sept. 15–19, 1972).
[3] See e.g. how this happens already in diatomic gas equilibria $A_2 \rightleftharpoons 2A$ at higher temperatures,
    necessitating full U(R) methods for thermodynamic properties: Sinanoğlu, O. and Pitzer, K. S.:
    1959, *J. Chem. Phys.* **31**, 960.
[4] See e.g. Hill, T. L.: 1956, *Statistical Mechanics*, McGraw-Hill, New York.
[5] Sinanoglu, O.: 1968, in *Molecular Associations in Biology* (ed. by B. Pullman), Academic Press,
    New York, p. 427.
[6] Halıcıoğlu, T. and Sinanoğen, O.: 1969, *Ann. N.Y. Acad. Sci.* **158**, 308.
[7] Sinanoğlu, O.: 1968, *Adv. Chem. Phys.* **12**, 283.
[8] Sinanoğlu, O. and Abdulnur, S.: 1965, *Federation Proc.* **24**, part III, S-2.
[9] See Frank, H. S. and Evans, M. W.: 1945, *J. Chem. Phys.* **13**, 507; and later workers, Kauzmann,
    W.: 1959, *Adv. Protein Chem.* **14**, 1–64; Nemethy, G. and Scheraga, H. A.: 1962, *J. Chem.
    Phys.* **36**, 3401; 1962, *J. Phys. Chem.* **66**, 1773.
[10] See e.g. Wiberg, K. B.: 1964, *Physical Organic Chemistry*, Wiley, New York, 1964.

MOLECULAR INTERACTIONS IN GROUND
AND EXCITED STATES

KEIJI MOROKUMA, SUEHIRO IWATA, and W. A. LATHAN

Dept. of Chemistry, The University of Rochester, Rochester, N.Y. 14627, U.S.A.

Contents

1. Introduction

The interaction between molecules is the central issue of chemistry. Molecules interact to break old chemical bonds and to form new bonds in chemical reactions. In weaker interactions molecules associate themselves to form molecular complexes or aggregates. In solutions interactions between solute and solvent molecules play the essential role. Much experimental evidence, both chemical and spectroscopic, demonstrates the existence and the nature of such molecular interactions.

Special attention has been directed to two specific molecular interactions: hydrogen bonds and charge transfer complexes; and for these a huge collection of experimental results are available (Pimentel and McClellan, 1960; Mulliken and Person, 1969; Foster, 1969).

In the past several years *ab initio* molecular orbital calculations, especially with the Roothaan–Hartree–Fock SCF methods (Roothaan, 1951), have made significant contributions to the prediction of the geometry and the understanding of the origin of the hydrogen bonds. For the water dimer, the first *ab initio* calculation with a small Gaussian basis set (Morokuma and Pedersen, 1968) predicted an open structure with a linear hydrogen bond in which one OH bond of an H_2O molecule is aligned collinear with the oxygen atom of the other H_2O molecule, contrary to an early IR experimental conclusion that the dimer has a cyclic structure (Van Thiel *et al.*, 1957). Later more

R. Daudel and B. Pullman (eds.), The World of Quantum Chemistry, 277–316. All Rights Reserved

accurate calculations (Kollman and Allen, 1969; Morokuma and Winick, 1970; Del Bene and Pople, 1970; Hankins *et al.*, 1970; Dierksen, 1970) all supported the linear hydrogen bonds. Recent IR experiments by Tusi and Nixon (1970) favor the linear structure. Most recently a molecular beam electric deflection study by Dyke and Muenter (1972) confirmed the polar nature of the dimer. A further microwave study of the dimer (Dyke and Muenter, 1973) established the structure of the dimer which is in good agreement with recent theoretical predictions. Other pioneering molecular orbital studies of hydrogen bonds include the strong symmetric hydrogen bond in bifluoride $(FHF)^-$ by McLean and Yoshimine (1968) and the strong interactions to form NH_4Cl from NH_3 and HCl by Clementi (1967). Many *ab initio* molecular orbital studies published recently are reviewed by Kollman and Allen (1972).

So far we have restricted our introduction to the interaction between ground state molecules. But there are many important problems associated with the electronic structure of excited states of complexes, or, interaction between an excited molecule and a ground state molecule or molecules. The appearance of the charge-transfer band in the spectra of the molecular complexes and the shifts of $n-\pi^*$ and $\pi-\pi^*$ transitions upon hydrogen bonding are among the most essential features of such interactions. The behavior of excited states of complexes has a direct significance in photochemistry and in the emission spectra of molecules. Such interactions in excited states have some biological implications as well (Löwdin, 1963). To our knowledge very little has been done with *ab initio* molecular orbital methods for excited states of hydrogen-bonded systems or any molecular complexes.

In the present paper we would like to present results of our recent effort to understand excited states as well as the ground state of complexes by carrying out *ab initio* molecular orbital calculations and analyzing energy components. In Section 2 we will discuss the methods, the electron-hole potential (EHP) method and two-configuration electron-hole potential (TCEHP) method, and the energy decomposition scheme. In Section 3 hydrogen bonds in the ground and excited states of formaldehyde–water complex will be discussed, and also calculation for the 1:2 formaldehyde–water complex and the acrolein–water complex will be presented to explain shifts of $n-\pi^*$ and $\pi-\pi^*$ transitions. In Section 4 the ground and excited states of the formic acid dimer will be studied, and in Section 5 the symmetric hydrogen bond in hydrogen maleate will be examined. Section 6 deals with the ground and charge-transfer states of the molecular complex between cyanocarbonyl and water. Section 7 summarizes the conclusions and gives a perspective view of the near future.

2. Method

A. ELECTRON-HOLE POTENTIAL (EHP) METHOD

Molecular orbital studies of interactions between ground state molecules have shown the Hartree–Fock method to be capable of making semiquantitative predictions concerning the geometry of such interacting systems, the interaction energy and other physical properties. However, as mentioned in Section 1, many interesting interactions occur between a molecule in an excited state and another molecule in the ground state. Thus, we seek theoretical methods which can describe such interactions or excited states of an interacting system as accurately as the Hartree-Fock method can do for the ground state. The spin unrestricted Hartree-Fock method would be able to describe lower triplet excited states quite well but would be useless for singlet states. The configuration interaction method would be rather tedious; inclusion of only all the singly excited configurations would be prohibitive for calculations involving large molecules. The spin-restricted Hartree–Fock method for open-shell molecules (Davidson, 1973; Iwata and Morokuma, 1973c) is the best candidate. This method gives the best single configuration wave function for excited states. Since the method is a variational one, it cannot be rigorously applied to excited states which have the same symmetry and multiplicity as the ground state. The molecular orbitals obtained in this method are different from the molecular orbitals of the ground state Hartree–Fock calculation, hence causing some inconvenience in calculating matrix elements between the two states and in interpreting the results within the single electron excitation framework.

The electron-hole potential (EHP) or the extended Hartree–Fock method proposed by us (Morokuma and Iwata, 1972) is an approximate Hartree–Fock method for excited states and has some convenient features, i.e., the method is well defined (no adjustable parameters or arbitrary choice of configurations), the method is applicable to an excited state of the same symmetry as the ground state and the molecular orbitals are common between the excited and ground states. We have used this EHP method for calculations of molecular interactions between an excited state molecule and a ground state molecule or molecules with success.

The starting point of the EHP method is the set of molecular orbitals which for a given basis set diagonalizes the Hartree–Fock operator of the (closed-shell) ground state of the system. We call this set $\{\psi_l\}$ the SCF molecular orbitals (SCF MO's). The total wave function of the ground state is given by a single Slater determinant as

$$\Psi_0 = \mathscr{A}\psi_1\bar{\psi}_1\psi_2\bar{\psi}_2\ldots\psi_N\bar{\psi}_N. \tag{2.1}$$

The Hartree–Fock operator F has the following matrix element.

$$F_{kl} = \langle\psi_k| H + \sum_{m}^{occ} (2J_{mm} - K_{mm}) |\psi_l\rangle = \delta_{kl}\varepsilon_k, \tag{2.2}$$

where J_{mm} and K_{mm} are Coulomb and exchange operators, respectively, which are in general defined as

$$J_{mn}\psi(1) = \int \psi_m^*(2)\,\psi_n(2)\,(1/r_{12})\,\psi(1)\,\mathrm{d}\tau_2,$$

$$K_{mn}\psi(1) = \int \psi_m^*(2)\,\psi(2)\,(1/r_{12})\,\psi_n(1)\,\mathrm{d}\tau_2. \tag{2.3}$$

We would like to determine a new set of MO's $\{\phi\}$ that gives the lowest energy for a single configuration $^{1,3}\Psi(\alpha \to \mu)$ corresponding to an electron excitation from MO ϕ_α

to ϕ_μ:

$$^{1,3}\Psi\,(\alpha\to\mu)=\mathscr{A}\phi_1\bar{\phi}_1\cdots(\phi_\alpha\bar{\phi}_\mu\pm\phi_\mu\bar{\phi}_\alpha)\,2^{-1/2}\cdots\phi_N\bar{\phi}_N,\tag{2.4}$$

where $\pm$ corresponds to the singlet and triplet, respectively.

The determination of $\{\phi\}$ without any restriction will lead to the spin-restricted Hartree–Fock method. Instead, in the EHP method, we require that the new MO's $\{\phi_i\}$, $i=1, 2, ..., N$ including α, be expanded within the occupied orbital space $\{\psi_j\}, j=1, 2, ..., N$ of the ground state SCF MO's as

$$\phi_i=\sum_j^{occ} a_{ij}\psi_j,\quad i=1, 2, ..., \alpha, ..., N,\tag{2.5}$$

and the new MO ϕ_μ is expanded within the vacant orbital space $\{\psi_v\}$, $v=N+1, ...,$ $N+M$ of the ground state SCF MO's as

$$\phi_\mu=\sum_v^{vac} b_{\mu v}\psi_v.\tag{2.6}$$

The coefficients $a_{\alpha j}$ and $b_{\mu v}$ are determined by the variation method so that the excitation energy $^{1,3}\Delta E(\phi_\alpha\to\phi_\mu)$ is minimized. After some manipulations, one can show that the optimized MO's satisfy the following equation:

$$[F+P^\dagger L_{\mu\mu}P-(1-P^\dagger)\,L_{\alpha\alpha}(1-P)]\,\phi=\lambda\phi,\tag{2.7}$$

where F is the standard Hartree–Fock operator of the ground state and P is the projection operator onto the occupied orbital space:

$$P=\sum_j^{occ}|\psi_j\rangle\langle\psi_j|,\tag{2.8}$$

and $L_{\mu\mu}$ and $L_{\alpha\alpha}$ are operators whose general definition is given by

$$L_{mn}=J_{mn}-K_{mn}\mp K_{mn},\tag{2.9}$$

where the upper and lower signs correspond to the singlet and triplet state, respectively. If the form of new orbitals, Equation (2.5) and (2.6), is explicitly specified, Equation (2.7) can be written as the following simultaneous pseudo-eigenvalue problems which are to be solved iteratively:

$$(F+L_{\mu\mu})\,\phi_i=\lambda_i\phi_i,\quad i=1, ..., \alpha, ..., N\tag{2.10}$$
$$(F-L_{\alpha\alpha})\,\phi_\mu=\lambda_\mu\phi_\mu.\tag{2.11}$$

The excitation energy is then simply given by

$$\Delta E(\phi_\alpha\to\phi_\mu)=\lambda_\mu-\lambda_\alpha+\langle\mu|L_{\alpha\alpha}|\mu\rangle.\tag{2.12}$$

Physically Equations (2.10) and (2.11) mean that the orbital of the α-hole feeling the potential field of the μ-electron is determined within the occupied orbital space of the ground state, while the orbital of the μ-electron moving in the field of the α-hole is determined within the vacant orbital space of the ground state. Therefore we call this method *the electron-hole potential method*.

The convergence of iteration of Equations (2.10) and (2.11) is very fast. The integral tape has to be read twice for each cycle, and usually in a few cycles one reaches the energy convergence of 10^{-5} Hartree or less.

By using the new MO's $\{\phi_j\}, j=1, 2, ..., N$ obtained by Equation (2.10) one can rewrite the SCF wave function Ψ_0, Equation (2.1), for the ground state as

$$\Psi_0 = \mathscr{A}\phi_1\bar{\phi}_1\phi_2\bar{\phi}_2 \cdots \phi_N\bar{\phi}_N. \tag{2.13}$$

That is to say, one can use the common set of MO's $\{\phi_i\}$ for both the ground and $\phi_\alpha \to \phi_\mu$ excited states.

A remarkable feature of the EHP theory is an extension of Brillouin's theorem. Assume that the EHP procedure has been completed for the excited state $^s(\phi_\alpha \to \phi_\mu)$, and that EHP MO's determined by Equations (2.10) and (2.11) are used also for all other states. Here s is the singlet (1) or triplet (3). Then Brillouin's theorem for the $^s(\alpha \to \mu)$ state is valid not only with the ground state $|0\rangle$

$$\langle^s(\phi_\alpha \to \phi_\mu)|\mathscr{H}|0\rangle = 0, \tag{2.14}$$

but also with some other singly excited states:

$$\langle^s(\phi_\alpha \to \phi_\mu)|\mathscr{H}|^s(\phi_\alpha \to \phi_\nu)\rangle = 0, \quad \nu \neq \mu \tag{2.15}$$
$$\langle^s(\phi_\alpha \to \phi_\mu)|\mathscr{H}|^s(\phi_\beta \to \phi_\mu)\rangle = 0, \quad \beta \neq \alpha.$$

The excited state wave function $^{1,3}\Psi(\alpha \to \mu)$, Equation (2.4), can be expanded as a linear combination of singly excited configurations based on SCF MO's of the ground state by using Equations (2.5) and (2.6), as:

$$\Psi(\alpha \to \mu) = \sum_j^{\text{occ}} \sum_k^{\text{vac}} a_{\alpha j}b_{\mu k}\Psi(\psi_j \to \psi_k). \tag{2.16}$$

Equation (2.16) is determined by $(n+m)$ coefficients, $a_{\alpha j}$ and $b_{\mu\nu}$, where n and m are the number of occupied and vacant MO's belonging to the symmetry of orbitals α and μ, respectively. One can compare EHP with the configuration interaction (CI) including only all the singly excited configurations (called complete SECI), in which the wave function of the excited state is written as

$$\Phi = \sum_j^{\text{occ}} \sum_k^{\text{vac}} C_{jk}\Psi(\psi_j \to \psi_k). \tag{2.17}$$

The comparison of Equations (2.16) and (2.17) reveals that $m \times n$ variation parameters in CI, C_{jk}, are decoupled in the EHP method into a product $a_{\alpha j}b_{\mu k}$ of new sets of $(m+n)$ variational coefficients, $a_{\alpha j}$ and $b_{\mu k}$.

The efficiency of the EHP method is demonstrated in Table I in comparison of the vertical excitation energies of formaldehyde (other examples in Morokuma and Iwata, 1972). The basis set used is the STO-3G set with scale factors of Hehre *et al.* (1969) augmented with a set of diffuse p orbitals (called the STO-3G$+p$ set). Column 2 is calculated by using the occupied and virtual SCF MO's of the ground state. Column 4 is obtained from the single configuration wave function of the EHP method. Column 6

TABLE I

Comparison of the vertical excitation energies (Hartree) of H_2CO calculated by various methods[a]

State	Ground state SCF MO's		EHP		Full single excitation CI(no of confs.)
	One conf.	Two conf.	One conf.	Two conf.	
$^3(n-\pi^*)$	0.21318	0.21318	0.13270	0.13261	0.13063 (9)
$^1(n-\pi^*)$	0.22952	0.22952	0.16243	0.16241	0.16085 (9)
$^3(\pi-\pi^*)$	0.28862	0.28862[b]	0.17931	0.17908[b]	0.17563 (29)
$^1(\pi-\pi^*)$	0.44421	0.39632[b]	0.44389	0.38981[b]	0.38260 (29)
$^1(\sigma-\pi^*)$	0.42328	0.42326	0.34390	0.34374	0.34253 (19)
$^1(n-\sigma^*)$	0.37710	0.37703	0.37419	0.37384	0.37273 (23)

[a] R(CH) = 1.120 Å, <HCH = 118°, R(CO) = 1.210 Å and planar, symmetric (Morokuma, 1971). Basis set: STO-3G + p (exponents 0.106 and 0.060 for oxygen and carbon, respectively).
[b] Approximated by $C_1(\pi-\pi^*) + C_2(n-n^*)$. When $C_1'(\pi-\pi^*) + C_2'(\sigma-\sigma^*)$ is used, 0.28862 (ground state SCF MO's) and 0.17680 (EHP) for the triplet state, and 0.44401 (ground state SCF MO's) and 0.40982 (EHP) for the singlet state.

gives the results of the complete SECI. One can see for all the states examined except for $^1(\pi-\pi^*)$ that the single configuration EHP calculation reproduces the SECI result within 0.004 Hartree $\sim$0.1 eV! The efficiency of the EHP method is also well represented in the EHP MO's. For instance, in the H_2CO-H_2O hydrogen bonded system, the highest occupied ψ_{HO} and lowest vacant ψ_{LV} SCF MO's of the ground state are delocalized throughout the system. After the EHP procedure is applied for the transition from ϕ_{HO} to ϕ_{LV}, the new MO's ϕ_{HO} and ϕ_{LV} are both almost completely localized on H_2CO and look like the lone pair and π^* orbitals of the isolated H_2CO (specific example in Iwata and Morokuma, 1973a, b). Of course this EHP picture is more realistic than the SCF MO picture when one is discussing weak molecular interactions.

B. TWO-CONFIGURATION ELECTRON-HOLE POTENTIAL (TCEHP) METHOD

The EHP method presented in the preceding section has been shown to produce results almost indistinguishable from the full SECI results for most of the states examined. However, there are cases in which the single configuration wave function, however good it may be, cannot predict characteristics of the correct wave function. One such example is the lowest singlet $\pi-\pi^*$ state of H_2CO, discussed in Section 3A. Here the mixing of the $\pi-\pi^*$ configuration with $\sigma-\sigma^*$ and/or $n-n^*$ configurations is essential. A similar situation has been found for the $B^3\Sigma_u^-$ state of the oxygen molecule (Morokuma and Konishi, 1971) and the V state of ethylene (Bender et al., 1972).

Another is the case of excited states of a dimer AA of identical molecules. Let us assume that in the dimer two molecules occupy the positions which are related by a symmetry operation to each other. In the MO theory the occupied MO's of the dimer will have the form:

$$\psi_{\pm}^{HO} \sim \psi_1^{HO} \pm \psi_2^{HO}, \qquad (2.18)$$

which are the plus and minus linear combinations of the occupied MO's of the monomer. Similarly the vacant orbitals of the dimer are

$$\psi_{\pm}^{LV} \sim \psi_1^{LV} \pm \psi_2^{LV}. \tag{2.19}$$

Any single configuration cannot describe the interaction between the two molecules
appropriately. For instance,

$$\Psi(\psi_+^{HO} \to \psi_+^{LV}) \sim \Psi(\psi_1^{HO} \to \psi_1^{LV}) + \Psi(\psi_2^{HO} \to \psi_2^{LV}) +$$
$$+ \Psi(\psi_1^{HO} \to \psi_2^{LV}) + \Psi(\psi_2^{HO} \to \psi_1^{LV}) \tag{2.20}$$

includes with an equal weight the intramolecular excitation (the first two terms of the
right-hand side) and the intermolecular charge-transfer excitation (the last two terms).
The energy of the intramolecular excited state is usually different from that of the
intermolecular state. In order to allow correct mixing of the two kinds of excitation,
one has to use wave functions of the form

$$C_1 \Psi(\psi_+^{HO} \to \psi_+^{LV}) + C_2 \Psi(\psi_-^{HO} \to \psi_-^{LV})$$

and $\qquad\qquad\qquad\qquad\qquad\qquad\qquad\qquad\qquad\qquad\qquad\qquad$ (2.21)

$$C_3 \Psi(\psi_+^{HO} \to \psi_-^{LV}) + C_4 \Psi(\psi_-^{HO} \to \psi_+^{LV}).$$

Multiconfiguration (MC) SCF methods have been successfully used for such a
case (Roothaan and Bagus, 1963; Das and Wahl, 1972). Considering the efficiency
of the EHP method for the single configuration wave function, we have developed a
multiconfiguration EHP method (Iwata and Morokuma, 1973d). In the present
paper we will restrict ourselves to the two-configuration (TC) EPH method. The
extension of the method to more than two configurations is straightforward, though
the computer program would require some modifications.

The starting point for the TCEHP is again the SCF MO's $\{\psi_i\}$ for the (closed-shell)
ground state. We would like to variationally determine the best two-configuration wave
function ψ^{TC} for the excited state

$$\Psi^{TC} = B_{\alpha\mu} \Psi(\phi_\alpha \to \phi_\mu) + B_{\beta\nu} \Psi(\phi_\beta \to \phi_\nu), \tag{2.22}$$

where $\Psi(\phi_\alpha \to \phi_\mu)$ and $\Psi(\phi_\beta \to \phi_\nu)$ are single-configuration wave functions as defined
by Equation (2.4). Instead of allowing complete freedom in the choice of MO's, in
the spirit of EHP we require that MO's for the hole, ϕ_α and ϕ_β, be expressed as linear
combinations of *occupied* SCF MO's of the ground state

$$\phi_\alpha = \sum_j^{occ} b_{\alpha j} \psi_j, \qquad \phi_\beta = \sum_j^{occ} b_{\beta j} \psi_j, \tag{2.23}$$

with the orthogonality requirement:

$$\langle \phi_\alpha | \phi_\beta \rangle = 0. \tag{2.24}$$

Also the MO's for the electrons, ϕ_μ and ϕ_ν, are restricted to be linear combinations of *vacant* SCF MO's of the ground state:

$$\phi_\mu = \sum_k^{\text{vac}} b_{\mu k} \psi_k, \qquad \phi_v = \sum_k^{\text{vac}} b_{vk} \psi_k, \tag{2.25}$$

with the requirement of orthogonality:

$$\langle \phi_\mu \mid \phi_v \rangle = 0. \tag{2.26}$$

Application of the variation method to the energy of Equation (2.22) with respect to MO coefficients b's of Equation (2.23) and Equation (2.25) and the CI coefficient B's of Equation (2.22) leads to the following set of simultaneous equations:

$$
\begin{aligned}
B_{\alpha\mu}^2 (F + L_{\mu\mu}) \phi_\alpha + B_{\alpha\mu} B_{\beta v} L_{\mu v} \phi_\beta &= \lambda_{\alpha\alpha} \phi_\alpha + \lambda_{\alpha\beta} \phi_\beta, \\
B_{\alpha\mu} B_{\beta v} L_{v\mu} \phi_\alpha + B_{\beta v}^2 (F + L_{vv}) \phi_\beta &= \lambda_{\beta\alpha} \phi_\alpha + \lambda_{\beta\beta} \phi_\beta, \\
B_{\alpha\mu}^2 (F - L_{\alpha\alpha}) \phi_\mu - B_{\alpha\mu} B_{\beta v} L_{\alpha\beta} \phi_v &= \lambda_{\mu\mu} \phi_\mu + \lambda_{\mu v} \phi_v, \\
- B_{\alpha\mu} B_{\beta v} L_{\beta\alpha} \phi_\mu + B_{\beta v}^2 (F - L_{\beta\beta}) \phi_v &= \lambda_{v\mu} \phi_\mu + \lambda_{vv} \phi_v.
\end{aligned} \tag{2.27}
$$

These equations are not pseudo-eigenvalue problems, but can be conveniently solved by using quadratic convergence method (Iwata and Morokuma, 1973c). The convergence of the calculation is usually very good: a few cycles of iteration give the convergence of the total energy of 10^{-5} Hartree or less.

An actual timing of calculations may be instructive. For formic acid with 26 basis functions, the ground state SCF calculation using Gaussian 70 (Hehre *et al.*, 1973) took 12 cycles and 2.3 CPU min on IBM 360/65. The (single-configuration) EHP method took 2 to 3 cycles and 1.0 to 1.3 min per state, and the TCEHP method required 2 to 4 cycles and 2.2 to 4.4 min per state.

The TCEHP wave function, Equation (2.22) can be expanded as a linear combination of singly excited configurations based on SCF MO's of the ground state as

$$\Psi^{TC} = \sum_j^{\text{occ}} \sum_k^{\text{vac}} (B_{\alpha\mu} b_{\alpha j} b_{\mu k} + B_{\beta v} b_{\beta j} b_{vk}) \Psi (\psi_j \to \psi_k). \tag{2.28}$$

The TCEHP wave function, Equation (2.28), is determined by $2(n+m)$ variational b parameters and two more B parameters, with a total of $2(n+m+1)$. This is compared to $(n+m)$ variational parameters for the EHP wavefunction of Equation (2.16) and $n \times m$ variational parameters for the complete SECI wavefunction of Equation (2.17).

As in EHP (Equations (2.14) and (2.15)), extensions of the Brillouin theorem can be proved for TCEHP. The TCEHP wave function, Equation (2.22), which is a linear combination of $|\phi_\alpha \to \phi_\mu\rangle$ and $|\phi_\beta \to \phi_v\rangle$, satisfies the following relationships.

$$
\begin{aligned}
\langle \Psi^{TC} | \mathcal{H} | 0 \rangle &= 0 \\
\langle \Psi^{TC} | \mathcal{H} | \phi_\alpha \to \phi_\lambda \rangle &= 0, \quad \lambda \neq \mu \quad \text{but including } \lambda = v \\
\langle \Psi^{TC} | \mathcal{H} | \phi_\beta \to \phi_\lambda \rangle &= 0, \quad \lambda \neq v \quad \text{but including } \lambda = \mu \\
\langle \Psi^{TC} | \mathcal{H} | \phi_\gamma \to \phi_\mu \rangle &= 0, \quad \gamma \neq \alpha \quad \text{but including } \gamma = \beta \\
\langle \Psi^{TC} | \mathcal{H} | \phi_\gamma \to \phi_v \rangle &= 0, \quad \gamma \neq \beta \quad \text{but including } \gamma = \alpha.
\end{aligned} \tag{2.29}
$$

Examples of calculations using the TCEHP method are shown in Column 5 of Table I. For $n-\pi^*$, $\sigma-\pi^*$ and $n-\sigma^*$ states, the second configurations to be mixed are, from symmetry considerations, $\pi-n^*$, $\pi-\sigma^*$ and $\sigma-n^*$ states, respectively. For $\pi-\pi^*$ states, $n-n^*$ or $\sigma-\sigma^*$ can be mixed. As discussed in Section 3A, for states other than $^1(\pi-\pi^*)$ the EHP results are already very close to the complete SECI results, and therefore the effect of the second configuration is small. For the singlet $(\pi-\pi^*)$ state the mixing of $(n-n^*)$ configuration gives an energy only 0.007 Hartree above the complete SECI result, illustrating the efficiency of the TCEHP method. Another example of EHP and TCEHP calculations is shown in Figure 1 for formic acid HCOOH with the same (STO-3$G+p$) set. Again a drastic effect of the TCEHP method on the singlet $(\pi-\pi^*)$ state is observed.

C. ENERGY DECOMPOSITION

In discussing weak interactions between atoms and molecules, energy terms such as electrostatic energy, polarization and dispersion energy, exchange repulsion energy and charge transfer energy have been traditionally used. Usually these terms are cal-

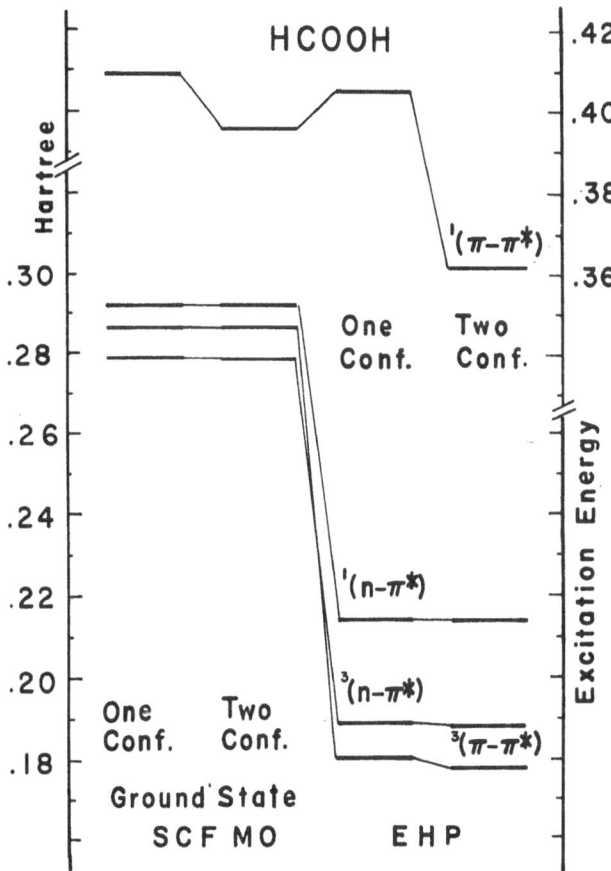

Fig. 1. Comparison of vertical excitation energies of HCOOH between virtual ground state SCF MO, EHP and TCEHP methods.

culated separately by perturbation methods and are added to make up the total inter-action energy (Margenau, 1949; Hirschfelder *et al.*, 1964; Hirschfelder, 1967; Margenau and Kestner, 1971).

In the present model, the entire interacting system is included in the molecular orbital calculation, and the total interaction energy is obtained directly as the difference of the total energy between the interacting system and the isolated molecules. Of course such a calculation is more reliable than the collection of individual perturbation terms. Nevertheless, various components of energy are very useful in order to examine the origin of molecular interactions.

Morokuma (1971) proposed an energy decomposition scheme for interactions between ground state molecules and applied it to the H_2CO-H_2O and H_2O-H_2O hydrogen bonding systems. The decomposition scheme is based on specific approximate forms of wavefunctions and does not invoke any perturbation expansion. Therefore it is anticipated to be valid even for rather strong molecular interactions. The method has been extended to interactions between a ground state molecule and an excited state molecule (Iwata and Morokuma, 1973b), and also to the analysis of the charge transfer state D^+A^- of molecular complexes (Lathan and Morokuma, 1973a).

In the scheme the total interaction energy is decomposed into the electrostatic energy E_{es}, the electron exchange repulsion E_{ex}, the polarization and resonance energy E_{pr} and the charge-transfer or delocalization energy E_{ct}. Figure 2 shows the relationship between various approximate wave functions in the ground and excited states and how they are calculated. Figure 3 schematically shows what components of energy are included in each of four energies calculated from approximate wave functions.

At infinite separation, the isolated molecules D and A are in the ground state, $\Phi(D_0^G)$ and $\Phi(A_0^G)$, in one of the excited states $\Phi(A_0^*)$ or in the ionic (charge-transfer) state, $\Phi(D_0^+)$ and $\Phi(A_0^-)$. The reference energies of the ground, intramolecular excited and charge-transfer states of the complex are, respectively,

$$E_0^G = E(D_0^G) + E(A_0^G),$$
$$E_0^* = E(D_0^G) + E(A_0^*), \qquad (2.30)$$
$$E_0^C = E(D_0^+) + E(A_0^-).$$

In the HF-EHP approximation, the ground state wave functions are determined by the Hartree–Fock method, and the excited state wave function $\Phi(A_0^*)$ is determined from A_0^G SCF MO's by the EHP method. The EHP method for the anion and the cation is equivalent to Koopmans' theorem: anion or cation MO's are equal to the ground state MO's of the neutral species. Therefore the difference between E_0^C and E_0^G, namely, the ionization potential of D_0^G minus the electron affinity of A_0^G is simply the difference of the SCF MO energy of the two levels involved in the charge transfer.

$$E_0^C - E_0^G = \varepsilon^{LV}(D_0^G) - \varepsilon^{HO}(A_0^G). \qquad (2.31)$$

For the complex of given geometry, one can define four types of wave functions and energies associated with them.

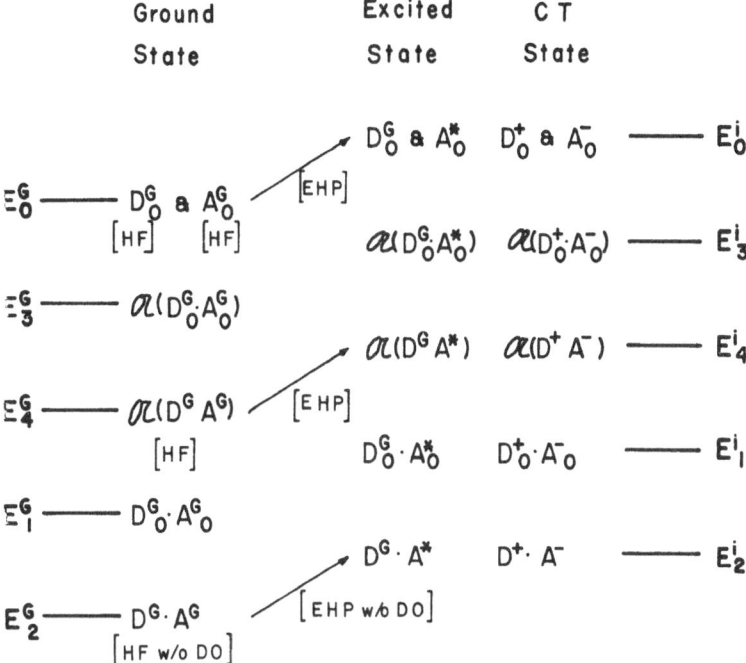

Fig. 2. Relationship between five wave functions in the ground, charge transfer (CT) and excited states. [HF] means that the Roothaan–Hartree–Fock calculation is to be carried out for this wave function. w/oDO means the differential overlap between AO's belonging to different molecules should be neglected. [EHP] and an arrow indicate that an EHP calculation is to be carried out starting from the SCF MO's of the arrow tail state to obtain EHP MO's for the arrow head state.

(1) The Hartree product of molecular wave functions determined at the infinite separation

$$\Phi(D_0^G)\cdot\Phi(A_0^G) \qquad \text{Energy } E_1^G$$
$$\Phi(D_0^G)\cdot\Phi(A_0^*) \qquad E_1^* \qquad\qquad (2.32)$$
$$\Phi(D_0^+)\cdot\Phi(A_0^-) \qquad E_1^C.$$

These wave functions correspond to the electrostatic interaction of two unpolarizable molecules; therefore, the electrostatic energy is

$$E_0^i - E_1^i = E_{es}^i \qquad i = G, *, \text{ and } C. \qquad\qquad (2.33)$$

The interaction energy is defined as positive when stabilization results.

The electrostatic interaction includes, in the perturbation's sense, all the interaction between net charge (in the charge transfer state only), permanent dipole, permanent quadrupole and all multipoles.

(2) The Hartree product of two molecular wave functions optimized by the HF or EHP procedure in the existence of the other molecule

$$\Phi(D^G)\cdot\Phi(A^G) \qquad \text{Energy } E_2^G$$
$$\Phi(D^G)\cdot\Phi(A^*) \qquad E_2^* \qquad\qquad (2.34)$$
$$\Phi(D^+)\cdot\Phi(A^-) \qquad E_2^C.$$

Energies

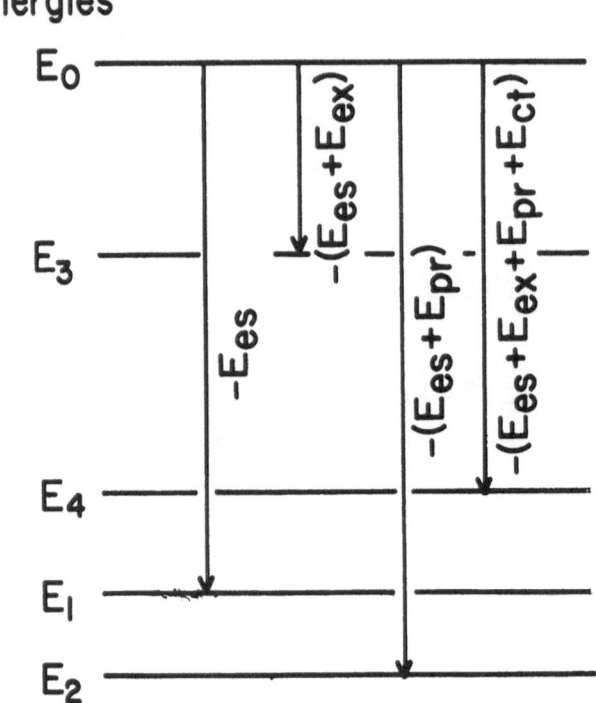

Fig. 3.   Five energies and energy components included in the difference.

The energy difference between E_2^i and E_0^i includes the electrostatic energy and the polarization-resonance energy E_{pr}

$$E_0^i - E_2^i = E_{es}^i + E_{pr}^i, \qquad i = G, \ast, \text{ and } C. \tag{2.35}$$

The polarization energy would include the interaction between the permanent electron distribution of D and the change in the distribution induced by D on A, and vice versa. In addition the interaction between the induced electron distributions is included. Such an interaction would, in the perturbation expansion, contain a term like

$$2 \sum_k \frac{|\langle \Phi(D^i) \cdot \Phi(A^i)| \, V \, |\Phi(D^f) \cdot \Phi(A^f)\rangle|^2}{E(D^i \cdot A^i) - E(D^f \cdot A^f)}, \tag{2.36}$$

where i denotes the state the system is in, and f denotes the states which can interact with the i state. The contribution becomes extremely large when the denominator becomes close to zero. This situation, called resonance, occurs when the molecule D is excited from D^i to D^f while the molecule A is de-excited from A^i to A^f and the energy absorbed by D is close to the energy emitted by A. The resonance is observed only when i and f belong to the same multiplicity, and therefore for the molecules whose ground state is the closed-shell singlet, the resonance is observed only for the singlet excited state.

(3) The antisymmetrized product of two molecular wave functions determined at

the infinite separation

$$\mathscr{A}\{\Phi(D_0^G)\cdot\Phi(A_0^G)\} \quad \text{Energy } E_3^G$$
$$\mathscr{A}\{\Phi(D_0^G)\cdot\Phi(A_0^*)\} \quad E_3^* \quad (2.37)$$
$$\mathscr{A}\{\Phi(D_0^+)\cdot\Phi(A_0^-)\} \quad E_3^C.$$

The antisymmetrizer $\mathscr{A}$ allows the exchange of electrons between the two molecules. E_3 therefore should include the electrostatic energy E_{es} plus the exchange repulsion E_{ex} (a negative quantity) between the electron clouds.

$$E_0^i - E_3^i = E_{es}^i + E_{ex}^i, \quad i = G, *, \text{ and } C. \quad (2.38)$$

(4) The HF or EHP wave function for the whole complex.

$$\Phi(D^G A^G) \quad \text{Energy } E_4^G$$
$$\Phi(D^G A^*) \quad E_4^* \quad (2.39)$$
$$\Phi(D^+ A^-) \quad E_4^C.$$

The total interaction energy E_T^i is of course defined as

$$E_T^i = E_0^i - E_4^i. \quad (2.40)$$

The energy E_T^i can be looked upon as the sum of all the contributions including the charge transfer or electron delocalization energy E_{ct}^i

$$E_T^i = E_{es} + E_{pr} + E_{ex} + E_{ct}. \quad (2.41)$$

From Equations (2.33), (2.35), (2.38) and (2.41), one can obtain individual energy contributions.

In general energy components are more sensitive to the choice of basis sets than the total interaction energy. Despite this, as is shown later, the energy decomposition analysis within a given basis set clearly reveals differences and similarities of the nature of interaction among various states.

D. BASIS SETS

In the following sections we will be calculating the interaction energy between relatively large molecules, which requires the size of basis sets to be limited. Since basis sets of accuracy of the minimal STO (Slater-type orbital) set have been succesfully used in the hydrogen-bond calculations, we will mainly use such a set. For comparison a larger basis set of near double zeta level, i.e. $4-31G$ set, will also be used. For excited of states of relatively high cnergy such as $\pi-\pi^*$ singlet state of H_2CO, a contribution of diffuse Rydberg type orbitals is considered essential (Buenker and Peyerimhoff, 1970; Whitten, 1972). So a set of diffuse p orbitals will be used to augment the basis functions mentioned above. The following are the list of basis sets actually used.

(a) STO minimal set. The exponents are optimized for individual molecules and was used before for H_2CO-H_2O in the ground state (Morokuma, 1971). The POLYCAL program (Stevens, 1970) is used for the atomic orbital integral calculation.

The following four sets use Gaussian-type orbitals (GTO) as primitive functions. The GAUSSIAN 70 program (Hehre *et al.*, 1973) is used for GTO integral and ground-state HF calculations.

(b) STO-3G set. Scale factors are those recommended by Hehre *et al.* (1969). This set gives the total energy higher than the STO minimal set, but valence properties calculated with this set are comparable to the STO set.

(c) STO-3$G+p$ set. The above set is augmented with a set of uncontracted GTO's on each first row atom, with the exponent of 0.106 for the oxygen and 0.06 for the carbon atom (Whitten, 1972).

(d) 4-31G set. This larger set produces results which are comparable to the double zeta STO set. The exponents and scale factors are those by Ditchfield *et al.* (1971).

(e) 4-31$G+p$ set. The 4$-$31G set is augmented with a set of diffuse GTO's used in (c).

3. Hydrogen Bonds between H$_2$O and Carbonyl Group
(Formaldehyde and Acrolein) in Ground and Excited States

In many organic and biological problems, the carbonyl group, $\diagdown$C=O, is the most $\diagup$

common base to form hydrogen bonds to proton donors such as the $-$OH and $\diagdown$NH $\diagup$

groups. Carbonyl compounds usually exhibit well defined n–π^* and π–π^* transitions. The fact that the former is shifted to a higher energy upon hydrogen bonding while the latter is shifted to a lower energy is commonly used as a method to assign new transitions experimentally to the n–π^* or π–π^* state. (Pimentel and McClellan, 1960; Herzberg, 1966)

The hydrogen bond energy of the ground state of H$_2$CO with H$_2$O was first studied by Morokuma using a minimal STO basis set (1971). More recently the H$_2$CO hydrogen-bonding with H$_2$O, CH$_3$OH, NH$_2$OH, HOOH and FOH has been studied with the STO-3G basis set (Del Bene, 1973).

Iwata and Morokuma (1973a) reported the first *ab initio* calculations of hydrogen-bonding between a ground state molecule and an excited molecule. Their calculations were for the ground state H$_2$O with n–π^* and other lower excited states of H$_2$CO with the minimal STO basis set and the EHP procedure described in Section 2. Their findings can be summarized as

(1) In the ground state the most favorable approach of H$_2$O to H$_2$CO is the one in which one of the HO bonds of H$_2$O is pointing toward the oxygen of H$_2$CO while keeping the H$_2$O molecule within the plane of the H$_2$CO molecule (PH model of Figure 4a). The largest stabilization energy, 3.5 kcal/mole, is obtained for this geometry with $R_{OH}=1.89$ Å and $<$C=O$\ldots$H$=\theta=64°$. This energy is about a half of that for the water dimer with the same basis set. In the n–π^* singlet and triplet excited states, this approach gives very little stabilization energy. The blue shift of the transi-

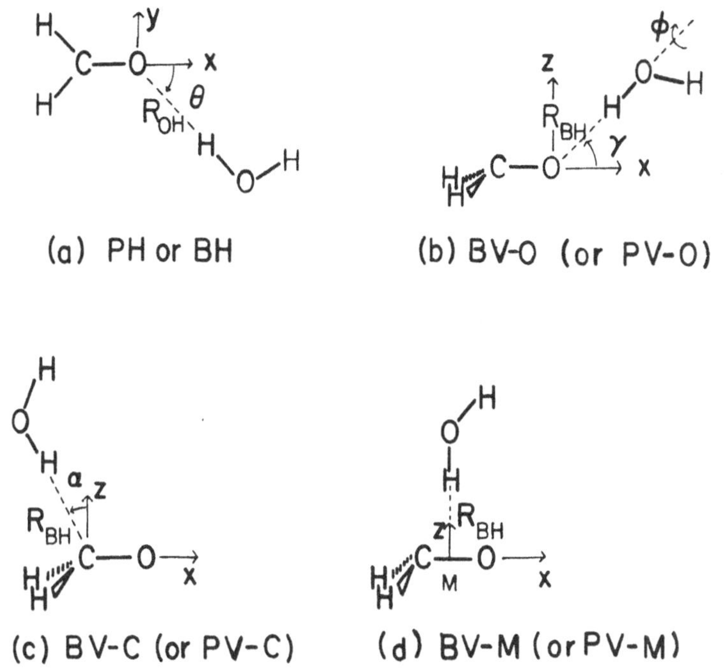

(a) PH or BH   (b) BV-O (or PV-O)

(c) BV-C (or PV-C)   (d) BV-M (or PV-M)

Fig. 4.   Models of hydrogen bond complex between H_2CO and H_2O.

tions in this complex model is calculated to be 0.14 eV (1100 cm^{-1}) and 0.18 eV (1400 cm^{-1}) for the singlet and triplet, respectively, qualitatively in agreement with experiments.

(2) For the same approach as mentioned above, the potential energy curves for the π–π^* singlet and triplet states show almost the same dependency on R_{OH} and θ. (Recognizing that the STO set is inadequate to describe the π–π^* singlet state of H_2CO, one should consider the calculated results for this state as a model calculation which may not correspond to the realistic state.) The calculated vertical transition energy shifts to red only by 50 cm^{-1} for the singlet π–π^* state and to blue by 400 cm^{-1} for the triplet π–π^* state. In the above calculations H_2CO has been assumed to retain its ground state geometry, corresponding to the vertical transition. Calculations have also been carried out for the excited complex by using the bent experimental H_2CO geometry of the $^1(n$–$\pi^*)$ state with a long C—O bond (1.323 Å) and an out-of-plane angle of 31° (Herzberg, 1966) (BH model in Figure 4a). Features mentioned in (1) and (2) above are insensitive to this geometry change.

(3) The π-hydrogen bond is the interaction in which one of the OH bonds of H_2O approaches the planar H_2CO in direction perpendicular to the molecular plane (PV models of Figure 4). Very little stabilization, if any, is obtained for the ground and excited states examined, regardless of where (the oxygen atom, the carbon atom or the midpoint of the two) the approach has taken place.

(4) For the bent H_2CO the above feature (3) is quite different. The n–π^* states are substantially stabilized when the carbon atom is coordinated by the HO bond, espe-

cially when the OH bond approaches the carbon atom to form a new tetrahedral conformation (PV-C model of Figure 4c).

(5) When a second H_2O molecule is coordinated to the H_2CO-H_2O complex (with the best geometry determined in (1)) at a position symmetric to the first (with respect to the $C=O$ axis), the increase in stabilization energy is approximately three-fourths of the initial stabilization.

(6) The $C-H...O$ hydrogen bond is extremely weak. From the above results it has been suggested that upon excitation of H_2CO to the $n-\pi^*$ states, two H_2O molecules originally bonded to the carbonyl oxygen will be removed and another or one of the two H_2O molecules will be attracted to the carbon atom.

In the present section (Iwata and Morokuma, 1973d) we want to (a) examine the basis set dependency of the hydrogen-bond energy and its decomposition analyses for the ground and $n-\pi^*$ and $\pi-\pi^*$ excited states of H_2CO-H_2O complex. (b) analyze energy components for various modes of approach of H_2O to planar or bent H_2CO in search of the origin of the hydrogen bonding, (c) carry out calculations for excited states of the complex $H_2CO...2(H_2O)$ and compare them with experimental red shifts for the $n-\pi^*$ state, and (d) perform calculations for the ground and $n-\pi^*$, $\pi-\pi^*$ excited states of the hydrogen-bonded complex between acrolein ($CH_2=CH-CHO$) and H_2O in order to account for red shifts of $\pi-\pi^*$ transitions experimentally observed.

A. BASIS SET DEPENDENCY OF THE HYDROGEN BOND ENERGY AND ITS COMPONENTS IN $H_2CO...H_2O$

When one decomposes the hydrogen bond energy into components following the scheme presented in Section 2C, one would expect they would be more sensitive to the choice of basis functions. An example which suggests such sensitivity is seen in energy components for $(H_2O)_2$ calculated with different basis functions (Morokuma, 1971; Kollman and Allen, 1970), although no direct test of basis function dependency has previously been made. Here we have carried out HF-EHP calculations for the ground, $n-\pi^*$ triplet, $n-\pi^*$ singlet, $\pi-\pi^*$ triplet and $\pi-\pi^*$ singlet states of H_2CO-H_2O with various basis functions discussed in Section 3D: STO-3G (abbreviated as $3G$), STO-3G augmented with a set of diffuse p orbitals (called $3G+p$), 4-31G (called 431) and 4-31G augmented with a set of diffuse p orbitals ($431+p$). The geometry used is the one found to be most stable in the ground state with the minimal STO set, i.e., the PH model of Figure 3a with $R_{OH}=1.89$ Å and $\theta=63.9°$.

Calculated hydrogen bond energies and their components are shown in Table II and Figure 5. Table II also shows calculated dipole moments for H_2CO and H_2O. For both molecules the ground state dipole moment is underestimated by the $3G$ set, whereas it is overestimated by the other sets used.

A glance at Figure 5 reveals that despite recognizable basis set dependencies of the hydrogen bond energy and its components, the gross characteristics of the ground and excited states are rather independent of basis set. The hydrogen bond energy is largest in the ground state, smaller in $\pi-\pi^*$ states and the smallest or even negative in

TABLE II

Hydrogen bond energy, its components (kcal/mole) and dipole moment μ (Debye)[a] for ground and excited states with various basis sets

State	Basis set	μ_{H_2CO}	E_H	E_{es}	E_{ex}	E_{pr}	E_{ct}	ΔQ
Ground	$3G$	1.53	3.4	4.6	-7.3	0.1	5.9	0.027
	$3G+p$	3.08	5.6	11.4	-10.9	1.2	3.9	0.047
	431	3.01	6.3	9.7	-6.9	0.8	2.8	0.031
	$431+p$	3.18	6.6	10.8	-7.9	1.4	2.3	0.010
$^3(n-\pi^*)$	$3G$	0.48	-1.0	0.3	-5.9	1.2	3.4	0.015
	$3G+p$	0.68	-0.6	2.7	-7.9	1.5	3.0	0.042
	431	0.84	0.1	2.0	-5.4	2.0	1.4	0.019
	$431+p$	0.87	0.1	2.2	-5.9	2.0	1.8	0.003
$^1(n-\pi^*)$	$3G$	0.38	0.1	0.4	-6.0	1.5	4.1	0.014
	$3G+p$	0.68	0.1	2.7	-8.3	2.5	3.2	0.041
	431	0.91	1.0	2.1	-5.5	2.6	1.9	0.018
	$431+p$	0.89	0.9	2.2	-6.1	2.8	2.0	0.002
$^3(\pi-\pi^*)$	$3G$	0.46	2.1	3.9	-7.3	0.0	5.5	0.027
	$3G+p$	1.14	1.4	8.5	-10.6	-0.1	3.7	0.047
	431	1.10	1.5	6.5	-6.9	0.1	1.8	0.030
	$431+p$	1.13	1.7	7.2	-7.7	0.0	2.1	0.010
$^1(\pi-\pi^*)$	$3G$	0.80	3.6	3.9	-7.3	0.8	6.2	0.027
	$3G+p$	1.22	-0.4	3.6	-9.9	2.6	3.3	0.037
	431	0.33	5.5	6.0	-6.7	4.0	2.2	0.028
	$431+p$	0.47	2.5	3.6	-7.3	4.4	1.8	0.001

[a] Dipole moment of the ground state H_2O: $3G$, $1.71D$; $3G+p$, $2.35D$; 431, $2.67D$; $431+p$, $2.79D$.

$n-\pi^*$ states. This overall trend is mainly controlled by the electrostatic energy E_{es}, which also decreases in the same order. The order of calculated dipole moments of H_2CO: $\mu_{ground} > \mu_{\pi-\pi^*} > \mu_{n-\pi^*}$ apparently determines this term. The exchange repulsion is only slightly state dependent. The $n-\pi^*$ state has a smaller repulsion, because in these states a lone pair electron which can overlap strongly with the H_2O electron cloud is transferred into the nonoverlapping π^* orbital. For the same reason the charge transfer energy E_{ct} is smaller for the $n-\pi^*$ states. The singlet state has a larger polarization-resonance energy E_{pr} than the corresponding triplet state; as discussed in Section 3C, the difference is very roughly a measure of the resonance contribution. Both $n-\pi^*$ states have E_{pr} larger then the ground state, probably due to the ease of mixing with other excited states or, in other words, an increased polarizability in the former states.

A detailed examination of Table II and Figure 5 shows some interesting basis set dependencies. The $3G$ set underestimates the dipole moment of the ground state molecules whereas the other sets used overestimate it. As a result, the electrostatic energy E_{es}, whose leading term in the expansion in the intermolecular distance r is $\mu_{H_2CO}\mu_{H_2O}/r^3$, is smaller in the $3G$ set than in other sets. The exchange repulsion E_{ex} is insensitive to the basis set. From the results, one sees that $E_{es}+E_{ex}$ is negative for the $3G$ set but is positive for other sets. This is consistent with similar discrepancies

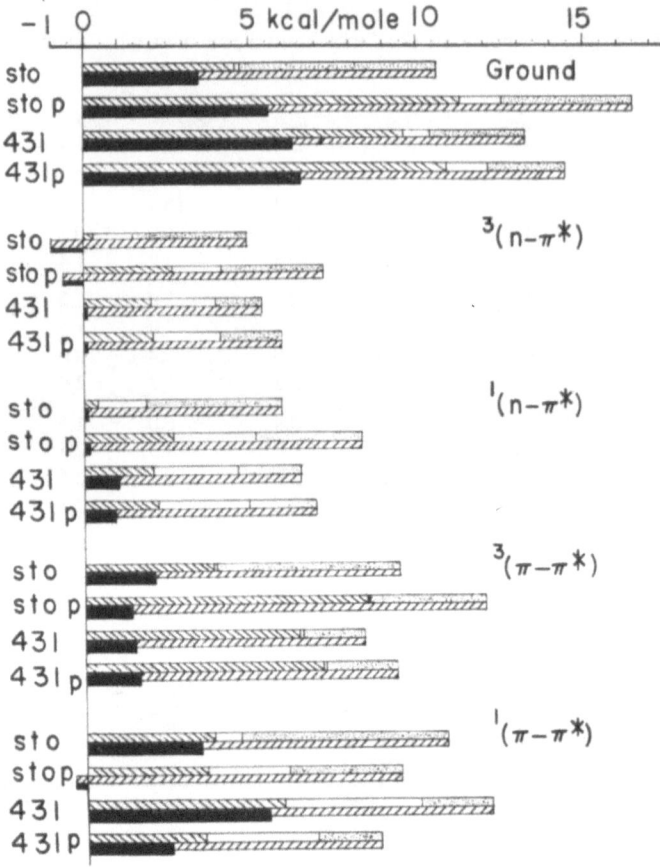

Fig. 5.   Hydrogen bond energy and its components with various basis sets for the PH model ($R_{OH} =$ = 1.89 Å and $\theta = 63.9°$) of $H_2CO - H_2O$. ■ E_H; ▨ E_{es}; ▢ E_{pr}; ▨ E_{ex}, and ▨ E_{ct}.

found for $(H_2O)_2$ (Morokuma, 1971; Kollman and Allen, 1970). The $3G$ set yields the largest charge transfer energy E_{ct}, which appears to be compensating the smaller electrostatic energy. The increases in E_{ct} in the $3G$ set compared to the other sets is not as large as the decrease in E_{es}, so that the $3G$ set gives the smallest hydrogen bond energy.

For the singlet state the difference of the basis set should make a large difference in the nature of the π^* orbital. Despite all the changes of the π^* orbital, no basis set predicts the red shift of the $\pi–\pi^*$ transition commonly observed experimentally for ketones and aldehydes. Additional calculations with $3G+p$ and $431+p$ sets using very diffuse p orbitals (exponent 0.02 for the carbon and 0.05 for the oxygen) led to even a larger blue shift. For further discussion on the red shift, additional calculations will be given in Section 3D.

B. ENERGY DECOMPOSITION ANALYSIS FOR $H_2CO...H_2O$ IN VARIOUS STATES

Encouraged by the insensitivity of the qualitative conclusions on the choice of basis functions, as was just shown above, we carried out an energy decomposition analysis for lowlying states of $H_2CO...H_2O$ for various models by using the minimal STO basis set which was used previously (Morokuma, 1971; Iwata and Morokuma, 1973a).

Models used are shown in Figure 4. For the labels of models P stands for the planar H_2CO (optimized for this basis set; Morokuma, 1971) and B for the bent H_2CO (experimental geometry of $^1(n-\pi^*)$ excited state; Herzberg, 1966). The planar molecule lies in the xy plane, while the bent molecule has the C—O bond on the x axis and the two protons below the xy plane, H is for the horizonal approach of H_2O, i.e., within the xy plane. V is for nonhorizontal approaches, subdivided into V-O: the OH is approaching the oxygen atom, V-C: the carbon atom and V-M: the midpoint of the C—O bond. Geometrical parameters are defined clearly in Figure 4.

Figures 6 and 7 show the hydrogen-bond energy and its components for PH models as functions of R_{OH} at $\theta = 30°$ and of θ at $R_{OH} = 1.797$ Å, respectively. One can clearly see that the difference between states is mainly determined by the electrostatic energy E_{es}. The larger stabilization energy of the singlet $n-\pi^*$ state compared to the triplet state can be attributed, as discussed in Section 2C, to the resonance energy. Results for BH models are very similar to the PH results.

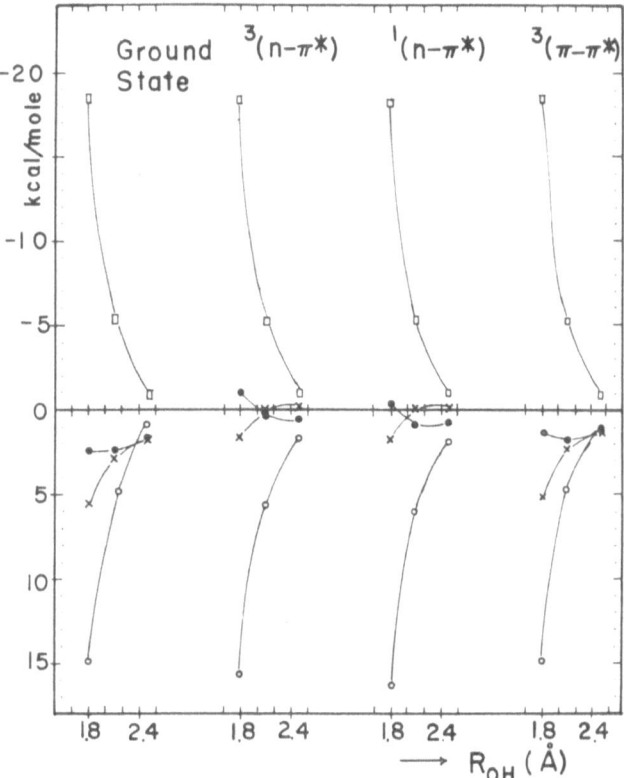

Fig. 6.   Hydrogen bond energy and its components as functions of the hydrogen bond distance R_{OH} for PH models at $\theta = 30°$ with the minimal STO basis set. ● E_H; × E_{es}; ⊓ E_{ex}; ○ $E_{pr} + E_{ct}$.

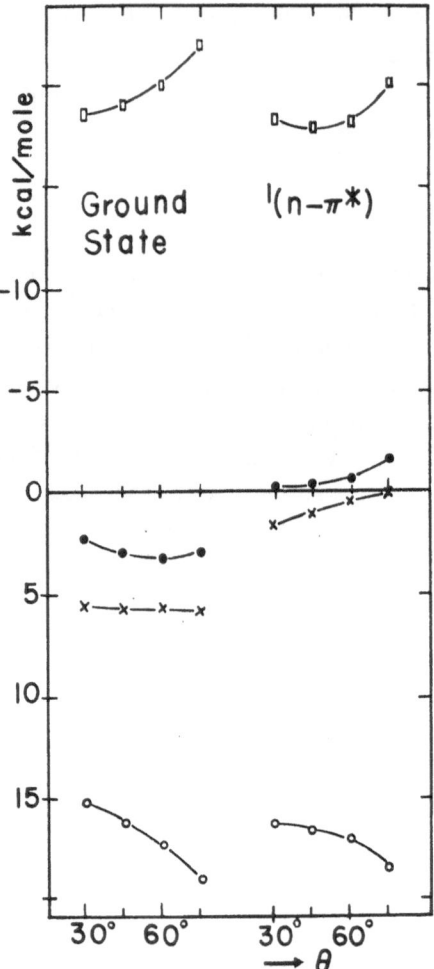

Fig. 7. Hydrogen bond energy and its components as functions of the angle θ for PH models at $R_{OH} = 1.797$ Å with the minimal STO set. For symbols, see Figure 6. The $^3(n-\pi^*)$ curve is not shown because it is very similar to the $^1(n-\pi^*)$ curve.

Table III shows a comparison of energy decomposition between some respresentative models. Without displaying the analysis for all 50 models we examined (Iwata and Morokuma, 1973b), we will summarize findings for V-type models.

All the states examined are unstable in the PV-O and BV-O models at the distance $R_{BH} \sim 1.9$ Å where the ground state in the PH model has an energy minimum. The energy decomposition indicates that a large exchange repulsion and a small electrostatic stabilization is responsible for the instability. For the PV-M and BV-M models one finds that E_{es} is very small and the balance of large E_{ex} and $E_{pr} + E_{ct}$ determines whether the net result is bonding or not. When H_2O approaches the carbon atom (V-C models) the $n-\pi^*$ singlet and triplet and $\pi-\pi^*$ singlet states gain the hydrogen bond energy. This is more enhanced for the bent H_2CO (BV-C models) than for the planar H_2CO (PV-C models). The most stable geometry of interaction for these excited states is found around the bond distance $R_{BH} \sim 2.3$ Å and the angle $\alpha \sim 15°$. This

TABLE III

Hydrogen bond energy, its components (in kcal/mole) and the charge transfer ΔQ

Model no. (geometry)	State	E_H	E_{es}	E_{ex}	E_{pr}	E_{ct}	$E_{pr}+E_{ct}$	ΔQ
PH 1	Ground	3.3	5.8	− 19.8	0.2	17.1	17.3	0.0302
($\theta = 60°$	$^3(n-\pi^*)$	− 1.9	0.4	− 18.2	1.6	14.4	6.0	0.0174
$R_{OH} = 1.797$ Å	$^1(n-\pi^*)$	− 0.6	0.5	− 18.1	2.2	14.9	17.1	0.0163
	$^3(\pi-\pi^*)$	1.8	5.0	− 19.9	0.1	16.7	16.7	0.0302
	$^1(\sigma-\pi^*)$	0.6	1.1	− 18.6	1.4	16.9	18.3	0.0108
	$^1(\pi-\pi^*)$	3.6	5.0	− 19.9	1.1	17.4	18.5	0.0300
PV-03	Ground	0.3	0.7	− 4.0	0.1	3.5	3.6	0.0069
($R_{BH} = 2.30$ Å,	$^3(n-\pi^*)$	− 0.1	− 0.1	− 4.1	0.7	3.4	4.0	0.0069
$\gamma = 90°$)	$^1(n-\pi^*)$	0.3	0.0	− 4.1	0.9	3.4	4.3	0.0069
	$^3(\pi-\pi^*)$	− 0.7	− 0.1	− 3.0	0.0	2.3	2.4	0.0022
	$^1(\sigma-\pi^*)$	− 0.0	− 0.1	− 3.9	0.7	3.2	4.0	0.0066
BV-02	Ground	0.9	0.9	− 2.8			2.8	0.0081
($R_{BH} = 2.30$ Å,	$^3(n-\pi^*)$	− 0.1	− 0.4	− 3.8			4.2	0.0082
$\gamma = 90°$)	$^1(n-\pi^*)$	0.4	− 0.4	− 3.8			4.6	
	$^3(\pi-\pi^*)$	− 0.4	0.3	− 2.7			2.0	0.0031
	$^1(\sigma-\pi^*)$	0.1	0.1	− 3.7			3.0	0.0075
BV-06	Ground	1.8	1.7	− 2.2			2.3	0.0044
($R_{BH} = 2.30$ Å,	$^3(n-\pi^*)$	0.2	− 0.8	− 2.3			3.3	0.0046
$\gamma = 30°$)	$^1(n-\pi^*)$	0.6	− 0.8	− 2.3			3.6	0.0047
	$^3(\pi-\pi^*)$	1.3	1.4	− 2.1			2.0	0.0041
	$^1(\sigma-\pi^*)$	− 0.2	− 1.1	− 1.8			2.7	0.0008
PV-M1	Ground	− 0.8	0.2	− 5.6			4.7	0.0081
($R_{BH} = 2.30$ Å)	$^3(n-\pi^*)$	− 0.1	0.6	− 5.9			5.2	0.0105
	$^1(n-\pi^*)$	0.0	0.7	− 5.9			5.2	0.0103
	$^3(\pi-\pi^*)$	− 1.6	− 0.6	− 4.4			3.4	0.0030
	$^1(\sigma-\pi^*)$	− 0.2	0.5	− 5.9			5.1	0.0098
PV-C1	Ground	− 1.4	− 0.2	− 6.0			4.7	0.0059
($R_{BH} = 2.3$ Å,	$^3(n-\pi^*)$	0.5	1.6	− 6.2			5.5	0.0123
$\alpha = 0°$)	$^1(n-\pi^*)$	0.4	1.6	− 6.6			5.4	0.0124
	$^3(\pi-\pi^*)$	− 1.5	− 0.5	− 5.2			4.2	0.0051
	$^1(\sigma-\pi^*)$	0.1	1.3	− 6.6			5.4	0.0118
BV-C2	Ground	− 0.5	− 0.3	− 2.3	0.1	2.0	2.1	0.0031
($R_{BH} = 2.5$ Å,	$^3(n-\pi^*)$	1.6	2.0	− 2.8	− 0.5	2.9	2.4	0.0076
$\alpha = 15°$)	$^1(n-\pi^*)$	1.2	1.9	− 2.7	− 0.8	2.8	2.0	0.0078
	$^3(\pi-\pi^*)$	− 0.6	− 0.2	− 2.1	− 0.1	1.8	1.8	0.0037
	$^3(\sigma-\pi^*)$	1.3	1.6	− 2.7	− 0.3	2.6	2.3	0.0073

is similar to results of calculations on the protonation of excited formaldehyde (Morokuma, 1972). In V-C models again the positive E_{es} is well correlated with the positive E_H; the ground and $\pi-\pi^*$ triplet state which have a negative E_{es} are actually repulsive.

C. $H_2CO...2H_2O$ AND $n-\pi^*$ BLUE SHIFT

A previous calculation (Morokuma, 1971) suggests that a carbonyl compound in the ground state is hydrated by two water molecules in aqueous solution. Then this would be a better model for calculating the hydrogen-bond shift in electronic transitions. We use a model which is the same as the PH model of Figure 4a with $R_{OH} = 1.9463$ Å and $\theta = 60°$ for the $H_2CO \ldots H_2O$ complex except that another H_2O is added to H_2CO symmetrically with the first H_2O. Table IV shows the comparison of 1:1 and 1:2 complex results in the minimal STO basis set. A hydrogen bond energy of 2.50, 1.06 and 2.64 kcal/mole is added by the interaction of the second H_2O molecule for the ground, $\pi–\pi^*$ triplet and $\pi–\pi^*$ singlet states respectively. These are as large as 73, 48 and 75% of the first hydrogen-bond energy for each state. The $n–\pi^*$ states of H_2CO-2H_2O are destabilized relative to the corresponding states of isolated molecules, and more destabilized than those of H_2CO-H_2O.

The calculated blue shift of the $n–\pi^*$ singlet transition, 0.264 eV (2130 cm^{-1}) for the 1:2 complex is in good agreement with the experimental value of 1900 cm^{-1} (Pimentel and McClellan, 1960) for the blue shift in water of the acetone $(CH_3)_2CO$ $n–\pi^*$ singlet transition.

In a 1:2 complex the hydrogen bond energy $E_H(WFW)$ can be divided into the sum of pair interactions and the nonadditive three-body interaction term $V(WFW)$ (Hankins et al., 1970):

$$E_H(WFW) = 2E_H(FW) + V(\text{W-W}) + V(WFW),$$

where $E_H(FW)$ is the 1:1 complex hydrogen bond energy at this geometry when the second H_2O molecule is removed, and $V(\text{W-W})$, -0.85 kcal/mole in this model, is the interaction energy of two H_2O molecules at this geometry in the absence of H_2CO. The calculated nonadditive energies for all the states in Table IV are smaller than that for the water trimer with a large basis set (Hankins et al., 1970).

D. $H_2C = CH - CHO \ldots H_2O$ AND $\pi–\pi^*$ RED SHIFT

As discussed in Section 3A, all the basis sets examined led to a blue hydrogen bond shift of the $\pi–\pi^*$ singlet transition, contrary to the commonly accepted red shift. We recognize the fact that experimental shift values are for conjugated ketones and aldehydes rather than simple ketones and aldehydes (Pimentel and McClellan, 1960). The conjugation would drastically change molecular orbitals of the carbonyl group. Therefore we have carried out HF-EHP calculations for acrolein $H_2C = CH - CHO$ interacting with H_2O. For $H_2C = CH - CHO$ standard distances and angles are assumed: $R_{C=C} = 1.36$ Å, $R_{C-C} = 1.46$ Å, $R_{CO} = 1.22$ Å, $R_{CH} = 1.09$ Å and all the bond angles $= 120°$; the model is the PH model with $R_{OH} = 1.89$ Å and $\theta = 63.9°$ which was used in Section 3A. Calculated hydrogen bond energies and shifts of vertical transition energies for various states with $3G$ and $3G + p$ basis sets are compared with corresponding $H_2CO \ldots H_2O$ results in Table V. It is encouraging to recognize that the $\pi–\pi^*$ singlet state of acrolein exhibits a substantial stabilization over the ground state, in clear contrast with formaldehyde. Experimentally the $\pi–\pi^*$ transition of crotonaldehyde $CH_3CH = CH - CHO$ ($v = 47400$ cm^{-1}) makes a red shift of 2280 cm^{-1} in water (Pimentel and McClellan. 1960). Though neither $3G$ nor $3G + p$ shift is as large

TABLE IV

Comparison between $H_2CO\ldots H_2O$[a] and $H_2CO\ldots 2H_2O$[b]

State	Hydrogen bond energy (kcal/mole)			Charge transfer ΔQ		Vertical transition energy and shift in parentheses (eV)	
	$H_2CO\ldots H_2O$ E_H(FW)	$H_2CO\ldots 2H_2O$ E_H(WFW)	V(WFW)	$H_2CO\ldots H_2O$	$H_2CO\ldots 2H_2O$	$H_2CO\ldots H_2O$	$H_2CO\ldots 2H_2O$
Ground	3.39	5.88	−0.05	0.0197	0.0386		
$^3(n-\pi^*)$	−0.68	−2.11	0.10	0.0113	0.0212	3.28 (0.17)	3.45 (0.34)
$^1(n-\pi^*)$	0.23	−0.29	0.10	0.0108	0.0202	4.35 (0.13)	4.48 (0.26)
$^3(\pi-\pi^*)$	2.22	3.28	−0.31	0.0196	0.0386	4.10 (0.05)	4.16 (0.11)
$^1(\sigma-\pi^*)$	0.89	1.15	0.22	0.0093	0.0167	9.45 (0.10)	9.55 (0.20)
$^1(\pi-\pi^*)$	3.50	6.14	−0.01	0.0196	0.0384	15.11 (−0.02)	15.10 (−0.03)

[a] Geometry of $H_2CO\ldots H_2O$: PH model ($\theta = 60°$, $R_{OH} = 1.9463$ Å).

[b] Geometry of $H_2CO\ldots 2H_2O$: Add to the above $H_2CO\ldots H_2O$ another water at the symmetric position.

TABLE V

Comparison of hydrogen bond energies (kcal/mole) and shifts of the vertical transition energies in parentheses (cm^{-1})

State	H_2CO-H_2O		$H_2CHCHCO-H_2O$		$(HCOOH)_2$
	3G	3G + p	3G	3G + p	3G + p
Ground	3.4	5.6	4.1	7.4	17.5
$^3(n-\pi^*)$	−1.0(+1530)	−0.6(+2180)	−1.4(+1940)	−0.6(+2780)	4.4(+4570)
$^1(n-\pi^*)$	0.1(+1160)	0.1(+1940)	−0.2(+1490)	0.5(+2420)	6.3(+3920)
$^3(\pi-\pi^*)$	2.1(+440)	1.4(+1450)	4.3(−60)	7.7(−100)	10.9(+1820)
$^1(\pi-\pi^*)$	3.6(−100)	−0.4(+2100)	7.2(−1090)	9.0(−570)	

as experiment, it is clear from the table that conjugation plays an essential role in the red shift of the π–π* transition. Knowing from Figure 5 that the 3G +p set tends to give weaker π–π* stabilization, and consequently smaller red shifts, then the larger 431 +p set, it would be very desirable to carry out a 431 +p calculation. It would be also instructive to have hydrogen bonding calculations for the ground and π–π* states of an aromatic system such as acetophenone and pyrazine.

4. Ground and Excited States of Formic Acid Dimer (HCOOH)$_2$

Excited states of a dimer provides an interesting problem in electronic spectroscopy. Because of the resonance interaction between the excitation of one molecule with the excitation of the other molecule as discussed in Section 2B, every excited state of the monomer is split into two states in the dimer. This splitting is observed in the electronic spectra and is related to the lifetime of excited states (Mataga and Kubota, 1970).

Excited states of hydrogen-bridged dimers have some biological implications. Using a semiempirical SCF method, Rein and Harris (1964–65) calculated the potential energy curve for the proton transfer in the guanine–cytocine base pair and suggested that in ionized and excited states the proton is more easily transferred from guanine to cytocine than in the ground state.

Recent very accurate SCF calculations by Clementi et al. (1971) for the ground state of the guanine–cytocine base pair and the formic acid dimer are impressive. They found that the transfer of a single proton from one molecule to another does give a potential energy curve which is monotonically increasing as the proton is transferred. The potential energy curve with a double minimum is obtained only when two bridging protons are transferred simultaneously.

An interesting question is what would happen in excited states: is there a double minimum and what is the height and position of the barrier? We would like to study this by using the TCEHP method discussed in Section 2B, which is suited to handle excited states of a dimer of two identical molecules. Since there exists a very accurate SCF calculation by Clementi et al., we use a rather small basis set to at first compare our ground state results with theirs in order to examine the basis set's applicability, and then use the same set to carry out TCEHP calculations for excited states.

The starting geometry of the dimer is that used by Clementi *et al.*, i.e., the experimental one. As shown in Figure 8, the protons H_2 and H_4 are kept collinear with O_2 and O_3 and with O_4 and O_1, respectively and are moved simultaneously (maintaining the inversion symmetry) from the experimental position, $R_{(O_2H_2)} = R_{(O_4H_4)} = 1.852$ Bohr and $R_{(O_3H_2)} = R_{(O_1H_4)} = 3.307$ Bohr, toward the opposing oxygen atoms, O_3 and O_1, respectively. The basis set used is the STO-3G set augmented with a set of diffuse p orbitals on the carbon and oxygen atoms. This $(3G+p)$ set was shown in Section 3 to improve excited state energies over the straight STO-3G set and to give reasonable hydrogen bond energies for the ground and excited states of the H_2CO-H_2O complex. At the experimental geometry the total energy of $(HCOOH)_2$ by Clementi's best set $(9/5+polarization)$, is -377.31018 Hartree and by the $(9/5)$ set, -377.14839 Hartree. The $(3G+p)$ set gives the total energy of -372.88571 Hartree.

The ground state potential energy curves for the symmetric transfer of the protons

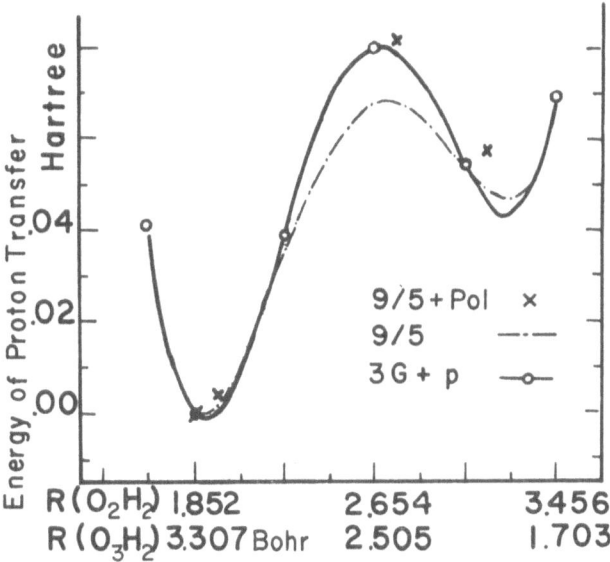

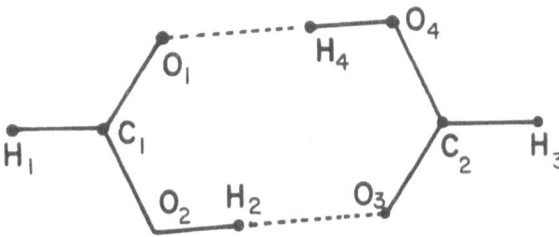

Fig. 8. Geometry of $(HCOOH)_2$ and the ground state potential energy curves as functions of $R_{O_2H_2}$ for three basis sets.

described above are shown in Figure 8. In the figure all the curves are shown relative to the energy at the experimental geometry. The solid $(3G+p)$ curve was determined by the fifth-order polynomial fit of six points we actually calculated. One finds that the agreement of the $(3G+p)$ result with the best $(9/5+\text{polarization})$ is excellent. This is not surprising since the STO-3G basis set, and the minimal STO set as well, have been successful in predicting the hydrogen bond energy rather well for the water dimer (Morokuma and Winick, 1970; Del Bene and Pople, 1970) and for H_2CO-H_2O (Morokuma, 1971). The first three columns of Table VI compare, for the $(9/5)$ and

TABLE VI

Positions (Bohr) of maximum and minima and barrier heights (kcal/mole) for the symmetric proton transfer in $(HCOOH)_2$

Basis Set	9/5	STO-$3G+p$			
State	Ground	Ground	$^3(n\text{–}\pi^*)$	$^1(n\text{–}\pi^*)$	$^3(\pi\text{–}\pi^*)$
First minimum, $R_{O_2H_2}$	1.873	1.896	1.889	1.893	1.868
$V_1 = E(\text{max}) - E(\text{1st min})$	43.0	51.0	76.8	73.6	66.2
Maximum, $R_{O_2H_2}$	2.714	2.682	2.693	2.690	2.642
$V_2 = E(\text{max}) - E(\text{2nd min})$	13.5	23.6	55.9	54.6	41.5
Second minimum, $R_{O_2H_2}$	3.241	3.251	3.304	3.300	3.323
$\qquad\qquad\qquad R_{O_3H_2}$	1.918	1.908	1.855	1.859	1.836
$V_2 - V_2$	29.5	27.4	20.9	19.0	24.7

$(3G+p)$ sets, the position of the first minimum, the position of the maximum, the barrier energy V_1 at the point relative to the first minimum, the barrier energy V_2 at the point relative to the second minimum, the position of the second minimum and the energy difference $V_1 - V_2$ between the first and second minima (positive when the first is lower than the second). The $(3G+p)$ values are from the fifth-order fit and the $(9/5)$ values are from seventh order polynomial fit of Clementi's eight calculated points.

Now we turn to excited states. Figure 9 shows the excitation energies for lower excited states for the monomer and the dimer in the vicinity of the most stable geometry. The $^3A'$, $^3A''$ and $^1A''$ states of the monomer (symmetry C_s) correspond in the conventional notation to $^3(\pi\text{–}\pi^*)$, $^3(n\text{–}\pi^*)$ and $^1(n\text{–}\pi^*)$ states, respectively. Each excited state splits into two states in the dimer (symmetry C_{2h}). In all the three cases the g state has a lower energy than the u state. Since the transition from the ground state is allowed only to u states, the excitation always brings the molecule to the upper of each pair of states, from which it can fall to the lower state. This lower state, because the transition from the ground state is forbidden, would be of long life to be able to show emission (Mataga and Kubota, 1970). The calculated oscillator strength for the dimer $^1A_g \rightarrow {}^1A_u$ transition is $f=0.0031$ (transition dipole 0.0742 Å) is not twice as large as the monomer $^1A' \rightarrow {}^1A''$ value $f=0.0019$ (0.0613 Å).

In Figure 10 we show the potential energy curves for the symmetric proton transfer for three excited states, $^3(\pi\text{–}\pi^*)$ $[^3A_g]$, $^3(n\text{–}\pi^*)$ $[^3B_g]$ and $^1(n\text{–}\pi^*)$ $[^1B_g]$, as well as

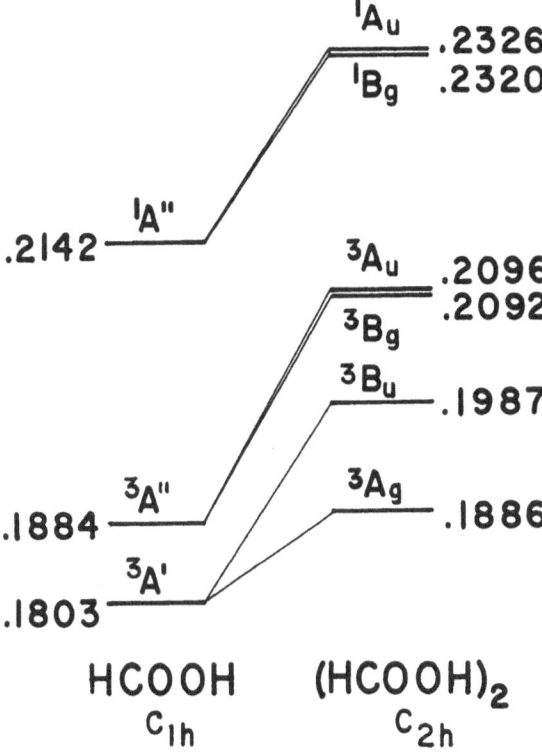

Fig. 9.   EHP excitation energies (Hartree) of the formic acid monomer HCOOH, and TCEHP excitation energies of the dimer $(HCOOH)_2$, at the geometry with symmetric hydrogen bonding.

the ground state for reference. Table VI also includes the positions of the maximum
and two minima, and barrier heights for the simultaneous, symmetric proton transfer
in excited states. These numbers are again determined by the fifth order polynomial
fit of six points. One can observe in Figure 10 and Table VI that in all the states
examined the positions of the maximum and the two minima are almost the same as
in the ground state. In general in excited states the barrier height V_1 of the first minimum increases rather significantly compared with the ground state, and this is more
marked for $n-\pi^*$ states. The second minimum is stabilized more in the excited states,
especially for the $n-\pi^*$ states.

In the last two columns of Table V we show the hydrogen bond energies E_H and
shifts of vertical transition energies for $(HCOOH)_2$ at the experimental geometry. For
the ground state E_H is larger than twice the hydrogen bond energy in H_2CO or acrolein. Even the $n-\pi^*$ states gain substantial hydrogen bond energy. Because of the
extreme stabilization of the ground state, $\pi-\pi^*$ as well as $n-\pi^*$ transitions exhibit blue
shifts, as is experimentally observed in dimers (Mataga and Kubota, 1970).

When one stretches an O—H bond, the OH σ orbital energy should increase while
the energy of the OH σ^* orbital should decrease. Therefore there should be states
corresponding to excitations from the OH σ orbital or to the σ^* orbital that may have
their minima near the maximum of the states shown in Figure 10. Such states may

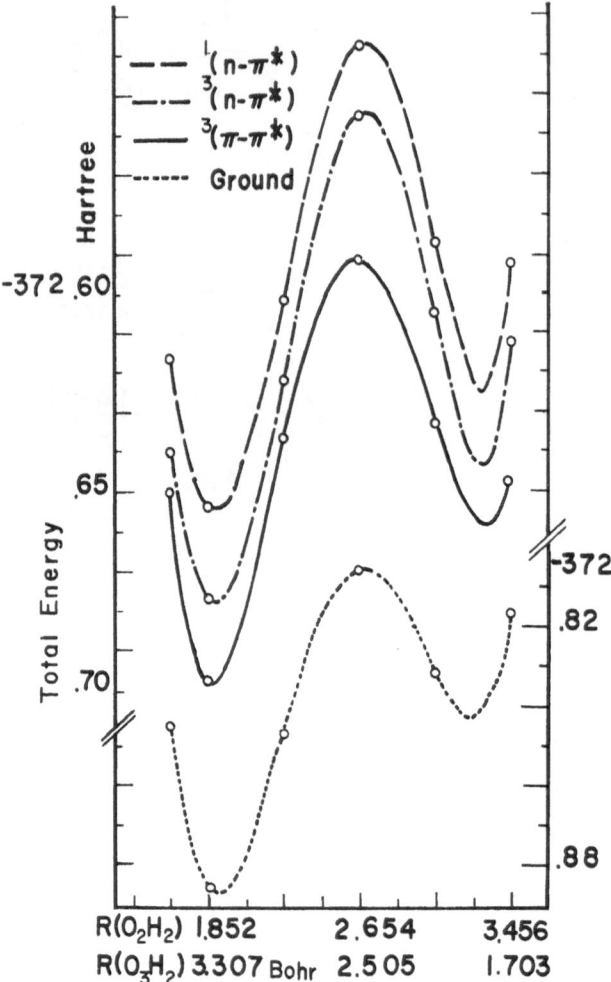

Fig. 10. Potential energy curves as functions of $R_{O_2H_2}$ for the ground and lower excited states of (HCOOH)$_2$ with the $3G + p$ basis set.

have some effects on the behavior of excited states. Calculations are in progress for such states.

5. Symmetric Hydrogen Bond in Hydrogen Maleate Ion HOOC–CH=CH–COO–

Hydrogen maleate ion is a unique system: its intramolecular hydrogen bond is symmetric. The structure of potassium hydrogen maleate determined by Darlow and Cochran (1961) is shown in Figure 11. The hydrogen atom is located on the center between two oxygen atoms and the O...O distance (2.437 Å) is unusually short for a hydrogen bonded system.

A similar symmetric hydrogen bond has been found for bifluoride ion (FHF)⁻ in KHF$_2$ (Pimentel and McClellan, 1960). The F...F distance is also very short (2.26 Å). This system has been studied in the HF method with a large STO basis set in a pio-

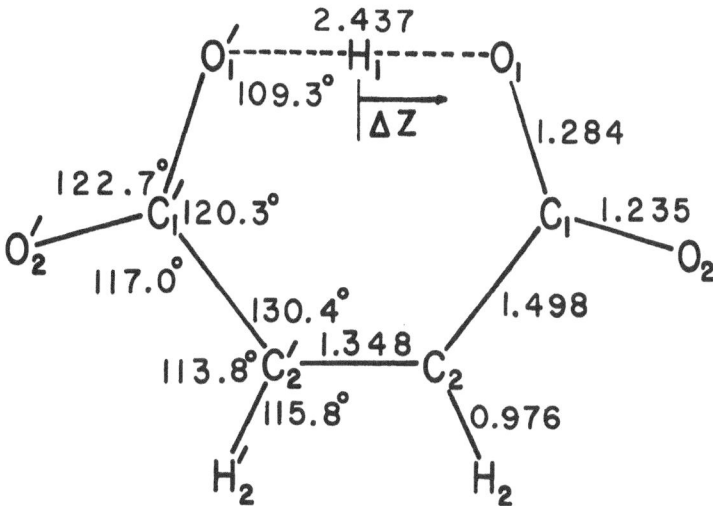

Fig. 11. Experimental geometry of hydrogen maleate and the definition of Δz, the deviation of proton from the symmetric position.

neering work on the theory of hydrogen bonding by McLean and Yoshimine (1967). By scanning the two bond distances, they actually found that the symmetric geometry with $R_{F...F} \sim 4.2$ to 4.3 Bohr ~ 2.22 to 2.28 Å is the most stable.

In addition to the geometrical interest, hydrogen maleate has spectroscopic interest in connection with the charge transfer mechanism of strong hydrogen bonds. Nagakura (1964) observed experimentally a strong absorption band at 211 m$\mu \sim 5.87$ eV and further found that the direction of the transition moment of this band is almost parallel to the O...H...O line. It has been suggested that this band is due to the transition between the ground and excited energy levels formed by the resonance

$$O-H\text{---}O^- \leftrightarrow O^-\text{---}H-O.$$

Here we will present some preliminary results for calculations on this system with the STO-3G basis set (Iwata *et al.*, 1973). The experimental geometry (Figure 11) is used except for the position of the hydrogen bonding proton H_1. In the first series of calculations H is moved on the $O_1 H_1 O_1'$ line from the center toward O_1. The relative total energy of the ground state is plotted in Figure 12 as a function of the shift Δz from the center. The $O_1 H_1$ distance is also shown on the abscissa. The curve has only one minimum at $\Delta z = 0$, in agreement with experiment. Since the curvature is rather small at the minimum, a rather large amplitude of proton vibration is expected.

To obtain information on the hydrogen-bond energy in hydrogen maleate, we have calculated the energy of the nonhydrogen bonded species by rotating the $O_1 H_1$ bond by 180° around the $O_1 C_1$ axis, so that the proton H_1 is not in the neighborhood of any proton acceptor group. With the $O_1 H_1$ distance optimized for this geometry to be 0.989 Å, the energy difference between this nonbonded species and the most stable hydrogen-bonded geometry ($\Delta z = 0$, i.e. symmetric) is 52.1 kcal/mole. Since the energy of the hydrogen-bonded species with $R_{O_1 H_1} = 0.989$ Å ($\Delta z = 0.229_5$ Å) is about

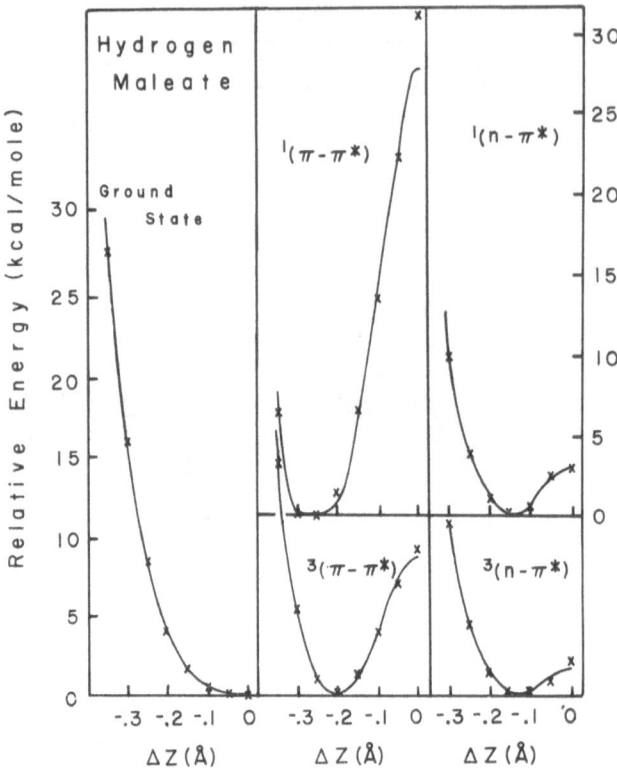

Fig. 12.   The relative total energy of hydrogen maleate in the ground and excited states as functions
of Δz. MO results are shown by ×'s and the curves are polynomial fits to these values.

5.9 kcal/mole higher than the most stable one, the energy difference due to the rotation of the O_1H_1 bond around the O_1C_1 axis is about 46.2 kcal/mole, which should
include the *cis-trans* isomerization energy and the hydrogen-bond energy at the regular
OH bond length. The above 5.9 kcal/mole is accomplished as a net gain of the
strengthening of the hydrogen bond against the weakening of the O_1H_1 σ bond. The
total stabilization energy 52.1 kcal/mole is extremely large for a 'molecular interaction'.

Contrary to the ground state, all four lower excited states (singlet and triplet A'',
n–π^* states and singlet and triplet A', π–π^* states) calculated by the TCEHP method
have an asymmetric hydrogen bond. The relative energies of these states as function
of Δz are shown also in Figure 12. Our qualitative interpretation of this drastic change
is as follows. Two essential conditions for a symmetric hydrogen bond seem to be (1)
a short distance between the proton acceptor and the proton donor atoms and (2) a
rich electron density in the hydrogen bonding region, so that a large interaction energy
can be gained against the weakening of the donor-proton σ bond. Both $(FHF)^-$ and
the ground state of hydrogen maleate satisfy these conditions. In all the excited states
discussed above, an excitation takes place from an MO (lone pair type or π type) mainly localized in the interacting CO^- to a π MO in the noninteracting $C{=}C{-}CO$. As

the result the interaction region becomes more or less neutral and can accommodate only a weak, nonsymmetric hydrogen bond. The geometry of excited states of hydrogen maleate is not known experimentally. Further experimental studies as well as theoretical studies are urged on this subject.

6. Ground and Charge-Transfer States of Molecular Complex Between Cyanocarbonyl $CO(CN)_2$ and ROR

Since Mulliken's famous paper (1952) molecular complexes have attracted much attention from chemists in varieties of fields (recent reviews: Mulliken and Person, 1969; Andrew and Keefer, 1964; Foster, 1969; Mataga and Kubota, 1970). In the molecular complex mixing of the charge transferred structure $(D^+ A^-)$ gives stability to the ground state, and the excitation from the ground state to the charge-transferred state gives rise to the characteristic charge transfer absorption band.

The hydrogen bonding discussed in Sections 3 and 4 can be considered a kind of molecular complex formation. *Ab initio* molecular orbital methods have contributed significantly to its understanding. Apart from this, a large number of experiments have been performed for what are sometimes called π complexes, which involve a π electron system as the electron acceptor or donor. Aromatic hydrocarbons-halogens and aromatic hydrocarbons-tetracyanoethylene are two such examples. Because relatively large molecules are involved in the complex, *ab initio* calculations for such complexes are rather expensive to carry out.

In this section we wish to report preliminary results of calculations for the ground and charge transfer excited states of the π complex between cyanocarbonyl, $CO(CN)_2$, and H_2O (Lathan and Morokuma, 1973). The choice of this system is based on the fact that the cyanocarbonyl complexes, cyanocarbonyl-aromatic hydrocarbons and cyanocarbonyl-ethyl ether, have had their charge-transfer spectra measured in the gas phase (Prochorow and Tramer, 1966) as well as in solution (Prochorow *et al.*, 1966). In the calculation we used H_2O as the model instead of ethyl ether. With H_2O, the system contains only eight first row atoms, making it one of the smallest π complexes for *ab initio* calculation.

We used the STO-3G basis set in this preliminary study. The geometry of $CO(CN)_2$ is not known. So we combined the geometry of tetracyanoethylene (planar, $R(N\equiv C) = 1.162$ Å, $R(C-C) = 1.435$ Å, $<NCC = 180°$, $<CCC = 117.8°$; Hope, 1968) and that of formaldehyde ($R(C=O) = 1.203$ Å). The bisector of the water HOH angle was kept perpendicular to the plane of $CO(CN)_2$ with the water protons directed away from the acceptor. This seems to be the most reasonable mode of approach when each proton of H_2O is replaced by an ethyl group. The distance between the $CO(CN)_2$ plane and the H_2O oxygen atom is defined as R, the distance of interaction, as shown in the inset of Figure 13. Figure 13 also indicates by arrows the positions above which we have actually placed the H_2O molecule.

Figure 14 shows the stabilization energy (kcal/mole) of the ground state of the complex as the function of the distance of interaction R and the position R_A (mea-

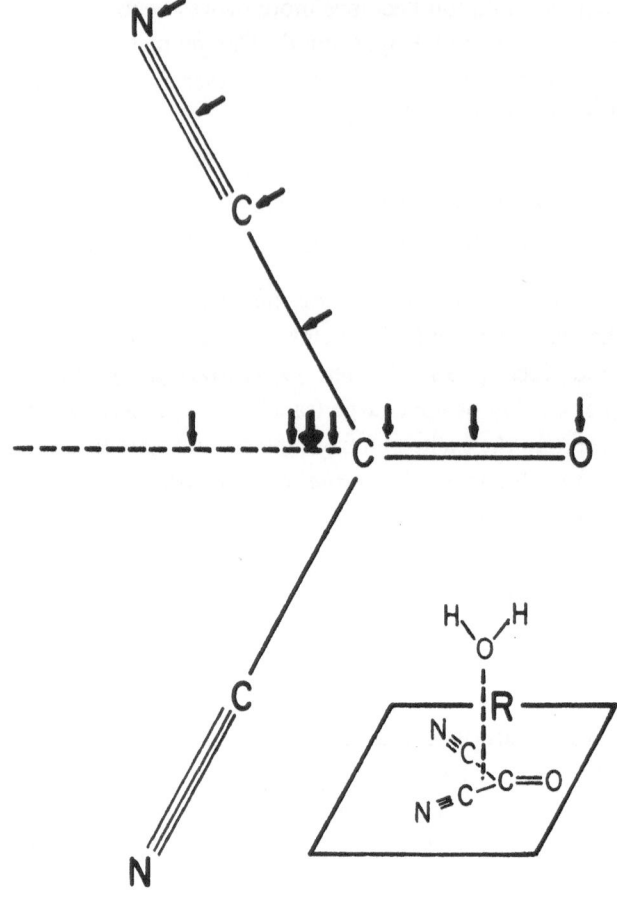

Fig. 13. Model geometries of cyanocarbonyl CO(CN)₂–ROR molecular complex. The bisector of ROR, with the oxygen atom facing CO(CN)₂, is kept perpendicular to the CO(CN)₂ molecular plane. The ROR plane is kept parallel to the bond axis. The arrows indicate the positions to which ROR has been approached in actual calculations. The heavy arrow shows the least energy approach for the ground state.

sured from C in Å) of approach on the C=O bond and its extension. In these calculations the plane of H_2O is assumed to be parallel to the C=O bond.(A calculation at $R=2.7$ Å and $R_A=-0.3$ Å with the plane of H_2O perpendicular to the C=O axis gives a slightly larger stabilization energy, 3.57 kcal/mole than the best one in Figure 14 but this orientation of H_2O was not pursued in detail.) Figure 15 is a similar stabilization energy chart as the function of R and the position R_B (measured from the carbonyl C in Å) of approach on the C—C≡N bonds. Here again the H_2O plane is assumed to be parallel to the C—C≡N axis.

From Figures 14 and 15 it becomes clear that a substantial stabilization is achieved only when the H_2O approaches near the carbonyl carbon atom. The largest stabilization energy from the figures is 3.53 kcal/mole at $R=2.7$ Å and on the extension of C=O away from the carbon end by 0.3 Å (i.e. $R_A=-0.3$ Å and is shown in Figure 13 by the largest arrow). The energy decomposition analysis for various values of R at

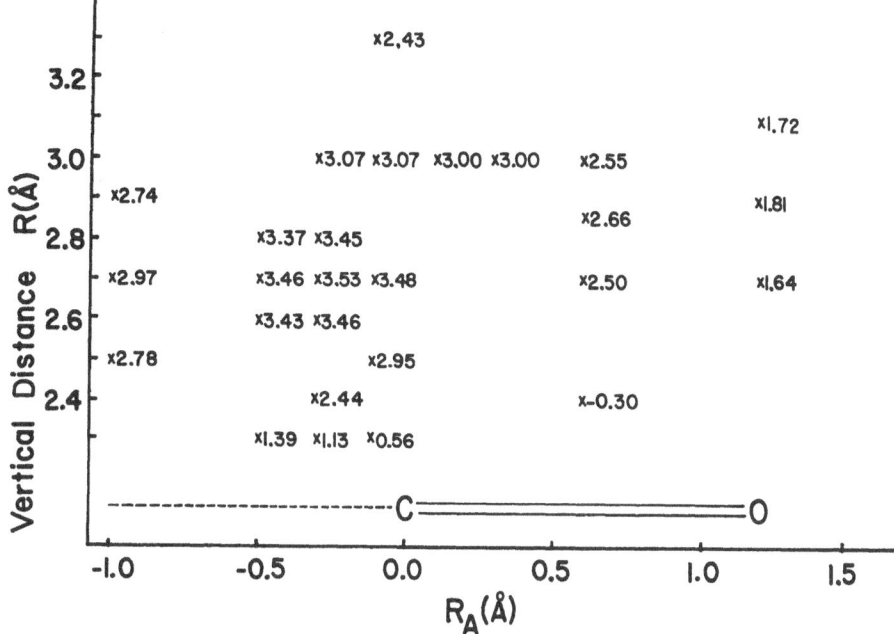

Fig. 14. The energy (kcal/mole) of interaction as functions of the distance R between ROR and the $CO(CN)_2$ plane and of the position of approach on (the extension of) the $C=O$ axis.

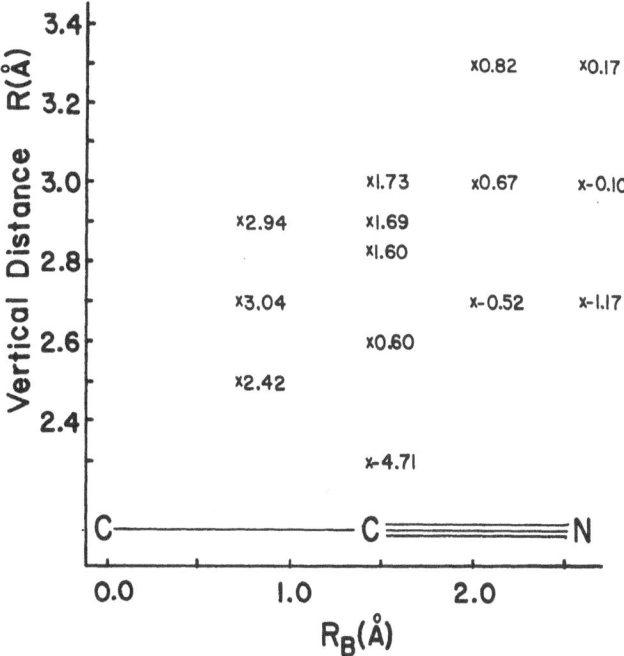

Fig. 15. The energy (kcal/mole) of interaction as functions of the distance R between ROR and the $CO(CN)_2$ plane of the position of approach on the $C-C=N$ axis.

$R_A = -0.3$ Å is shown in Table VII and for the near optimum R for various other approaches in Table VIII. The comparison of Tables VII and VIII for similar values of R suggests that the stability of the best approach ($R_A = -0.3$ Å) is mainly due to the electrostatic energy. This is well understood qualitatively in terms of the electron population of isolated molecules, shown in Figure 16. The three carbon atoms at the center are all positively charged, and therefore the negatively charged oxygen atom of H_2O would be attracted to the vicinity of the center of the CCC triangle. The above suggestion that the stabilization of this complex is mainly due to the electrostatic energy is supported by the decomposition scheme at the most stable geometry ($R = 2.7$ Å

TABLE VII

Energy decomposition analysis for least energy approach
($R_A = -0.3$ Å) in kcal/mole

R(Å)	2.0	2.4	2.6	2.7	2.8	3.0
Ground state						
ΔE	-9.57	2.44	3.46	3.53	3.45	3.07
E_{es}	17.2	6.6	4.8	4.2	3.8	3.1
E_{ex}	-42.2	-8.0	-3.2	-2.0	-1.3	-0.5
E_{pr}	0.9	0.4	0.3	0.2	0.2	0.1
E_{ct}	14.5	3.4	1.6	1.1	0.7	0.3
Singlet CT state						
ΔE	68.32	87.80	88.08	87.22	86.00	82.95
E_{es}	110.5	93.0	88.2	86.1	84.2	80.7
E_{ex}	-52.4	-10.3	-4.2	-2.7	-1.7	-0.6
E_{pr}	6.8	4.4	3.7	3.4	3.1	2.7
E_{ct}	3.4	0.6	0.4	0.4	0.3	0.2

TABLE VIII

Energy decomposition analysis for various modes of approach (kcal/mole)

Mode of approach	N of CN	Midpoint of CN	C of CN	Midpoint of CC	Midpoint of CO
R(Å)	3.3	3.3	3.0	2.7	2.86
Ground state					
ΔE	0.17	0.82	1.73	3.04	2.66
E_{es}	0.2	0.8	1.9	3.9	3.0
E_{ex}	-0.1	-0.1	-0.5	-2.0	-0.7
E_{pr}	0.1	0.1	0.1	0.2	0.1
E_{ct}	0.0	0.0	0.2	1.0	0.3
Singlet CT state					
ΔE	69.0	71.94	78.88	85.99	86.7
E_{es}	68.7	71.3	77.8	85.4	84.1
E_{ex}	-0.1	-0.1	-0.5	-2.5	-0.8
E_{pr}	0.2	0.7	1.6	3.2	3.5
E_{ct}	0.1	0.0	0.0	-0.1	0.0

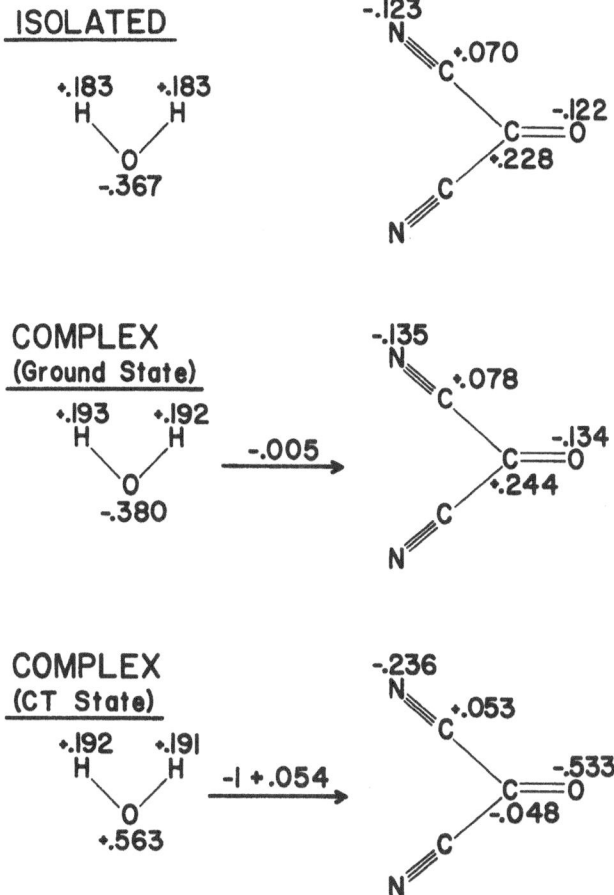

Fig. 16. Electron population analysis for isolated molecules and for the ground and CT states near the equilibrium geometry of the ground state complex.

in Table VII). The electrostatic energy $E_{es}=4.25$ kcal/mole is the most important contribution to the stabilization energy and is a few times larger than the charge transfer energy $E_{ct}=1.09$ kcal/mole. The small E_{ct} is reflected by the small amount of the charge transfer $\Delta Q=0.005$ (Figure 16). (Compare with H_2CO-H_2O of Section 3.) Considering the general trend of the STO-3G basis set as discussed in Section 3, underestimating dipole moments and, consequently, the electrostatic interaction while somewhat overestimating the charge transfer, one would be inclined to suggest that the importance of E_{es} might be real for this complex. In this connection one should call attention to Hanna's proposal (1968; Hanna and Williams, 1968) that the electrostatic force is the major contributor to complex formation between benzene and halogens.

Experimentally the geometry of the $CO(CN)_2$–ROR complex is not known. The dissociation energy in the ground state D_g is estimated to be about 5.7 kcal/mole (Prochorow and Tramer, 1966). The agreement of theoretical value 3.6 kcal/mole is quite reasonable.

KEIJI MOROKUMA ET AL.

Now we would like to turn our attention to the charge transfer (CT) state. This is not the lowest excited state of the symmetry; the intramolecular $n \to \pi^*$ type excited state has lower energy. Nevertheless by starting from an appropriate initial excitation, we were able to obtain the EHP convergence to the desired CT state without difficulty except for small R. Table VII lists the stabilization energy and its decomposition analysis of the singlet CT state as a function of the separation R at the approach most favored for the ground state $(R_A = -0.3 \text{ Å})$. As is naturally anticipated, the electrostatic stabilization between the donor cation and the acceptor anion accounts for almost all the stabilization energy, though the amount of charge transfer back from the acceptor anion to the donor cation is substantially larger than in the ground state.

An examination of Figure 16 reveals that in the CT state H_2O^+ has a large positive charge on the oxygen atom, which would prefer approaching the most negatively charged atom, the oxygen of $CO(CN)_2$ anion. This appears indeed the case, as is shown by the stabilization energy and energy decomposition analysis listed as of function of R for this mode of approach in Table IX. (See also Table VIII for analogous information on other modes of approach.) Figure 17 gives the potential energy curves for the ground and CT states to the same scale. The CT energy curve for the approach favored for the ground state $(R_A = -0.3 \text{ Å})$ has a minimum at $R \sim 2.5 \text{ Å}$ (0.2 Å shorter than in the ground state) with a stabilization energy of about 3 eV. However, the CT state curves for approach at the carbonyl oxygen (the curve through the filled-in circles) shows a stabilization energy of about 4 eV at $R \sim 2.3 \text{ Å}$. These values are in reasonable agreement with experimental results of Prochorow and Tramer (1966) who estimate the dissociation energy of the excited complex to be about 5 eV and the separation distance R for the CT complex to be 0.4 Å shorter than for the ground state. Furthermore, Prochorow and Tramer suggest that the CT state's R_{min} corre-

TABLE IX

Energy decomposition analysis for approach on oxygen atom (kcal/mole)

$R(\text{Å})$	1.9	2.1	2.3	2.5	2.7	2.9	3.1
Ground state							
ΔE	−24.02	−9.01	−2.07	0.73	1.64	1.81	1.72
E_{es}	10.9	5.5	3.5	2.7	2.2	1.9	1.7
E_{ex}	−43.9	−17.9	−7.0	−2.6	−0.9	−0.3	−0.1
E_{pr}	0.8	0.5	0.3	0.2	0.2	0.1	0.1
E_{ct}	8.2	2.8	1.1	0.4	0.2	0.1	0.0
Singet CT state							
ΔE	75?	85?	93.83	92.26	88.84	84.95	81.77
E_{es}			95.8	90.6	86.2	82.2	78.7
E_{ex}			−7.0	−2.6	−0.9	−0.3	−0.1
E_{pr}			5.0	4.2	3.6	3.0	2.6
E_{ct}			0.0	0.1	0.1	0.0	0.0

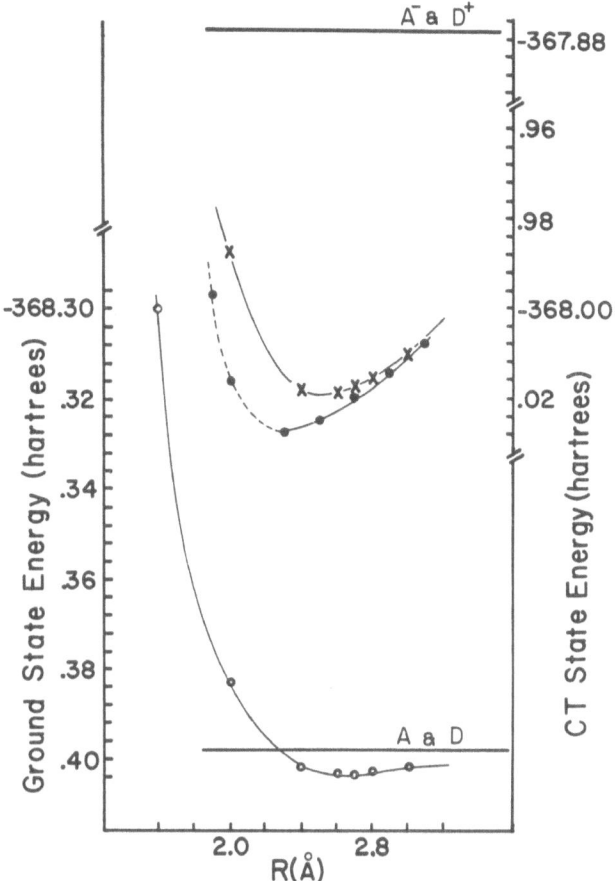

Fig. 17. Potential energy curves as function of the separation R of the ground and CT states of $CO(CN)_2$–ROR complex. ○ ground state, × CT state at $R_A = -0.3$ Å, ● CT state on the carbonyl oxygen. The broken line part is not very reliable because of existence of a nearby state of the same symmetry.

sponds to a repulsive part of the ground state curve. This is borne out by the calculations as seen in Figure 17.

7. Conclusions and Perspective

In the preceding sections we have summarized results of our recent calculations on lower excited states as well as the ground state of hydrogen-bond complexes and a charge-transfer molecular complex.

Though the hydrogen bond energy and its components are dependent on the choice of basis sets, qualitative conclusions one can draw from the energy decomposition analysis within a given set is rather insensitive to the basis set. Very roughly speaking, the difference of the electrostatic energy E_{es} among different states and different relative orientations of interacting molecules tends to determine the most preferred orientations of the interaction. This electrostatic energy control of the preferred orientation is reflected to the fact that the classical dipole–dipole picture often successfully predicts the relative orientation (Pimentel and McClellan, 1960). This does not mean that the hydrogen bond is electrostatic. At the geometry where a substantial hydrogen bond energy is obtained, three components–electrostatic (attraction), exchange repulsion and charge transfer or delocalization (attraction) – are often roughly of the same order of magnitude and cancel two of the three. In this sense the hydrogen bond is the balance of three different contributions.

It was a little bit of a surprise to find that a 'charge-transfer' complex between cyanocarbonyl and ROR turned out to be rather electrostatic. Not only the position of the approach in the ground state seems to be dictated by the electrostatic force, but the energy at the calculated equilibrium geometry has only a small charge transfer contribution. Since the experimental geometry of this complex is not known, the validity of the prediction is yet to be seen. Also more molecular orbital studies for the ground state of weak molecular complexes are badly needed.

For excited states of hydrogen-bond complexes and charge-transfer complexes, the work has just started. The blue shift of the $n-\pi^*$ transition appears to be under control in *ab initio* calculations. The red shift of the $\pi-\pi^*$ transition is more complicated because of the diffuse, Rydberg nature of the excited state in the isolated carbonyl group. Though the conjugation appears to be playing an important role in the red shift, more molecular orbital studies are needed to understand this phenomenon. The present calculation of the formic acid dimer predicts that the barrier of the symmetric proton transfer in the lowest $n-\pi^*$ singlet and triplet and $\pi-\pi^*$ triplet states is higher than in the ground state. There would be some higher states involving $O-H\sigma$ or σ^* orbital that would give a lower barrier or even a single minimum potential curve. The existence and the energy of such states are yet to be studied. Considering the amount of experimental work done in the field and the importance in spectroscopy and photochemistry, we would like to call theorists' attention to molecular interactions in excited states.

At the end of the paper we would like to mention here two interesting approaches to problems of molecular interactions. Molecular orbital studies of molecular interactions have not been limited to the complex of two or three molecules. Complexes of several molecules have been studied (Del Bene and Pople, 1970; Haukins *et al.*, 1970; also see Kollman and Allen, 1972). Yet it is practically impossible to include *all* the molecules in the solution. The model recently used by Newton (1973) for calculation of the hydrated electron is a blend of the molecular approach, such as we have been discussing, and the polarizable continuum approach. A certain number of molecules (here four water molecules and an electron) are enclosed in a spherical cavity which is surrounded by a continuum characterized by a dielectric constant. Molecular orbitals are solved interatively for explicitly considered electrons under the influence of the polarizable media. This model seems to have an interesting applicability to the solvation of ions, and we are pursuing such calculations.

Another is related to the electronic transition during the collision process. To study this semiclassically, collision theorists are interested in the analytical continuation of potential energy surfaces to complex values of nuclear coordinates. Two adiabatic surfaces of the same symmetry, which because of the noncrossing rule cannot intersect each other in the real nuclear coordinate, can intersect for complex values of the coordinate. A conventional way of analytical continuation is by inserting the desired values of the complex nuclear coordinate R into the surface $V(R)$ derived for only real values of R. But the most rigorous way of analytical continuation is the actual diagonalization of the electronic Hamiltonian for complex values of R. We have actually carried out such calculations for the two adiabatic curves, $E_{3\sigma}$ and $E_{4\sigma}$, of HeH^{++}, and are able to find the complex intersection point (Morokuma and George, 1973). We are now applying the method to the H_3^+ in which two adiabatic surfaces corresponding to $H_2^+ + H$ and $H_2 + H^+$ exhibit a similar avoided crossing.

Acknowledgements

Acknowledgement is made to Mr. A. D. Isaacson for his assistance in maleate calculations. One of the authors (S.I.) is on leave from the Institute for Physical and Chemical Research, Wako, Saitama, Japan.

This research is supported in part by the National Foundation Grant (GP-33998X) and by the Center for Naval Analyses of the University of Rochester. National Science Foundation Travel Grant GP-39369 and A. P. Sloan Foundation Fellowship in part supported the travel expenses. We also acknowledge The University of Rochester Computing Center for the use of computing facilities.

References

Andrews, L. J. and Keffer, R. M.: 1964, *Molecular Complexes in Organic Chemistry*, Holden-Day, San Francisco, Calif., U.S.A.
Bender, C. F., Dunning, T. H., Schaefer, H. F., Goddard, W. A., and Hunt, W. J.: 1972, *Chem. Phys. Letters* **15**, 171.
Buenker, R. J. and Peyerimhoff, D. S.: 1970, *J. Chem. Phys.* **53**, 1368.
Clementi, E.: 1967, *J. Chem. Phys.* **46**, 3851, **47**, 2323.
Clementi, E., Mehl, J., and Von Niessen, W.: 1971, *J. Chem. Phys.* **54**, 508.
Darlow, S. F.: 1961, *Acta Crystallog.* **14**, 1257.
Darlow, S. F. and Cochran, W.: 1961, *Acta Crystallog.* **14**, 1250.
Das, G. and Wahl, A. C.: 1972, *J. Chem. Phys.* **56**, 1769, and references therein.
Davidson, E. R.: 1973, *Chem. Phys. Letters* **21**, 565.
Del Bene, J.: 1973, *J. Chem. Phys.* **58**, 3139.
Del Bene, J. and Pople, J. A.: 1970, *Chem. Phys. Letters* **4**, 426; *J. Chem. Phys.* **52**, 4858.
Dierksen, G.: 1970, *Chem. Phys. Letters* **4**, 373.
Ditchfield, R., Hehre, W. J., and Pople, J. A.: 1971, *J. Chem. Phys.* **54**, 724.
Dyke, T. R. and Muenter, J. S.: 1972, *J. Chem. Phys.* **57**, 5011.
Dyke, T. R. and Muenter, J. S.: 1973, Presented at 28th Symposium on Molecular Structure and Spectroscopy, Columbus, Ohio.
Foster, R.: 1969, *Organic Charge-Transfer Complexes*, Academic Press, London.
Hankins, D., Moskowitz, J. W., and Stillinger, F. H.: 1970, *Chem. Phys. Letters* **4**, 581; *J. Chem. Phys.* **53**, 4544.
Hanna, M. W.: 1968, *J. Am. Chem. Soc.* **90**, 285.
Hanna, M. W. and Williams, D. E.: 1968, *J. Am. Chem. Soc.* **90**, 5358.
Hehre, W. J., Stewart, R. F., and Pople, J. A.: 1969, *J. Chem. Phys.* **51**, 2657.
Hehre, W. J., Lathan, W. A., Ditchfield, R., Newton, M. D., and Pople, J. A.: 1973, Quantum Chemistry Program Exchange, Indiana University.

Herzberg, G.: 1966, *Electronic Spectra of Polyatomic Molecules*, Van Nostrand, Princeton.
Hirschfelder, J. O.: 1967, *Adv. Chem. Phys.* **12**.
Hirschfelder, J. O., Curtis, C. F., and Bird, R. B.: 1964, *Theory of Gases and Liquids*, Wiley, New York.
Hope, H.: 1968, *Acta Chim. Scand.* **22**, 1057.
Iwata, S. and Morokuma, K.: 1973a, *Chem. Phys. Letters* **19**, 94.
Iwata, S. and Morokuma, K.: 1973b, *J. Am. Chem. Soc.* **95**, 1563.
Iwata, S. and Morokuma, K.: 1973c, 'On Restricted Hartree–Fock Methods for General Open-Shell Molecules', unpublished.
Iwata, S. and Morokuma, K.: 1973d, 'Two-Configurational Electron-Hole Potential Method for Excited States', *Theor. Chim. Acta*, in press.
Iwata, S. and Morokuma, K.: 1973e, 'Molecular Orbital Studies of Hydrogen Bonds. VI. Formic Acid Dimer in the Ground and Excited States' (to be submitted).
Iwata, S., Isaacson, A. D., and Morokuma, K.: 1973, 'Molecular Orbital Studies of Hydrogen Bonds. VII. Symmetric and Asymmetric Hydrogen Bonds in Ground and Excited States (to be submitted).
Kollman, P. A. and Allen, L. C.: 1969, *J. Chem. Phys.* **51**, 3286.
Kollman, P. A. and Allen, L. C.: 1970, *Theor. Chim. Acta* **18**, 399.
Kollman, P. A. and Allen, L. C.: 1972, *Chem. Rev.* **72**, 283.
Lathan, W. A. and Morokuma, K.: 1973, 'Molecular Orbital Studies of Charge Transfer Complex' (to be submitted).
Löwdin, P. O.: 1963, *Rev. Mod. Phys.* **35**, 724.
Margenau, H.: 1939, *Rev. Mod. Phys.* **11**, 1.
Margenau, H. and Kestner, N. R.: 1971, *Theory of Intermolecular Forces*, 2nd ed., Pergamon Press, New York.
Mataga, N. and Kubota, T.: 1970, *Molecular Interactions and Electronic Spectra*, Marcel Dekker, New York.
McLean, A. D. and Yoshimine, M.: 1968, 'Tables of Linear Molecular Wave function', supplement to *IBM J. Res. Dev.* **12**, 206.
Morokuma, K.: 1971, *J. Chem. Phys.* **55**, 1236.
Morokuma, K.: 1972, presented at 27th Symposium on Molecular Structure and Spectroscopy, Columbus, Ohio.
Morokuma, K. and George, T. F.: 1973, *J. Chem. Phys.* **59**, 1959.
Morokuma, K. and Iwata, S.: 1972, *Chem. Phys. Letters* **16**, 192.
Morokuma, K. and Konishi, H.: 1971, *J. Chem. Phys.* **55**, 402.
Morokuma, K. and Pedersen, L.: 1968, *J. Chem. Phys.* **48**, 3275.
Morokuma, K. and Winick, J.: 1970, *J. Chem. Phys.* **52**, 1301.
Morokuma, K., Iwata, S., and Isaacson, A. D.: 1973, 'Hydrogen Bonding in the Ground and Excited States of Hydrogen Maleate Anion' (to be submitted)
Mulliken, R. S.: 1952, *J. Am. Chem. Soc.* **74**, 811.
Mulliken, R. S. and Person, W. B.: 1969, *Molecular Complexes*, Wiley-Interscience, New York.
Nagakura, S.: 1964, *J. Chim. Phys.* **60**, 15.
Newton, M. D.: 1973, private communication.
Pimentel, G. C. and McClellan, A. L.: 1960, *The Hydrogen Bond*, W. H. Freeman, San Francisco, Calif., U.S.A.
Prochorow, J. and Tramer, A.: 1966, *J. Chem. Phys.* **44**, 4545.
Prochorow, J., Tramer, A., and Wierzchowski, K. L.: 1966, *J. Mol. Spectrosc.* **19**, 45.
Rein, R. and Harris, F. E.: 1964, *J. Chem. Phys.* **41**, 3393.
Rein, R. and Harris, F. E.: 1965, *J. Chem. Phys.* **43**, 4415.
Roothaan, C. C. J.: 1951, *Rev. Mod. Phys.* **23**, 69.
Roothaan, C. C. J. and Bagus, P. S.: 1963, *Methods Computational Phys.* **2**, 47.
Stevens, R. M.: 1970, *J. Chem. Phys.* **52**, 1397.
Tusi, A. and Nixon, E.: 1970, *J. Chem. Phys.* **52**, 1521.
Van Thiel, M., Becker, E. D., and Pimentel, G.: 1957, *J. Chem. Phys.* **27**, 486.
Whitten, J. L.: 1972, *J. Chem. Phys.* **56**, 5458.